VARIORUM COLLECTED STUDIES SERIES

Studies on Eighteenth-Century Geology

Rhoda Rappaport
1935–2009
(Photograph courtesy of Blossom and Irwin Primer)

Rhoda Rappaport

Studies on Eighteenth-Century Geology

Edited by Kenneth L. Taylor
and Martin J.S. Rudwick

Routledge
Taylor & Francis Group
LONDON AND NEW YORK

First published 2011 by Ashgate Publishing

2 Park Square, Milton Park, Abingdon, Oxfordshire OX14 4RN
711 Third Avenue, New York, NY 10017

Routledge is an imprint of the Taylor & Francis Group, an informa business

First issued in paperback 2018

British Library Cataloguing in Publication Data

Rappaport, Rhoda.
    Studies on eighteenth-century geology.
    – (Variorum collected studies series)
    1. Geology – Europe – History – 18th century.
    I. Title II. Series III. Taylor, Kenneth L. IV. Rudwick, M.J.S.
    551'.094'09033–dc22

    ISBN 978-1-4094-2959-3 (hbk)
    ISBN 978-1-138-38261-9 (pbk)

Library of Congress Control Number: 2011930084

VARIORUM COLLECTED STUDIES SERIES CS986

CONTENTS

UNDERSTANDING THE EARTH AND ITS HISTORY

THE LANGUAGE OF EARTH SCIENCE

SCIENTIFIC PURSUITS IN EARLY MODERN EUROPE

This volume contains xxiv + 340 pages

PUBLISHER'S NOTE

The articles in this volume, as in all others in the Variorum Collected Studies Series, have not been given a new, continuous pagination. In order to avoid confusion, and to facilitate their use where these same studies have been referred to elsewhere, the original pagination has been maintained wherever possible.

Each article has been given a Roman number in order of appearance, as listed in the Contents. This number is repeated on each page and is quoted in the index entries.

A Note on Article V

The three figures in Prof. Rappaport's article V ("The geological atlas of Guettard, Lavoisier, and Monnet: conflicting views of the nature of geology", as this appeared in the 1969 volume *Toward a History of Geology*) are images of maps at a scale making it possible to discern no more than their most grossly general features. For the present volume, in which all the articles are reproduced as originally published, the publisher and editors have undertaken to supplement the original article with new and larger reproductions of each of the three maps. These are presented as separate plates at the end of article V.

Because the original maps which these illustrate are quite sizable, the reproductions printed here in larger format still do not allow scrutiny of the finest details; however they do afford the reader a fuller appreciation of each map. Lacking specification of the particular sources from which Prof. Rappaport drew her original reproductions, we have tried to secure images of the maps from copies that are as nearly identical as possible to those represented in the original article.

The redone forms of figures 1 and 2 come from the Ewell Sale Stewart Library of the Academy of Natural Sciences of Philadelphia – a set carrying the title *Premières feuilles de l'atlas minéralogique: ouvrage commencé en 1766, par MM. Guettard et Lavoisier, tous deux de l'Académie des Sciences*. The new image for figure 3, from Monnet's *Atlas et description minéralogiques de la France*, is provided by the Bibliothèque, École nationale des Mines de Paris. For this last image, the original map is bound in a volume making a flat reproduction impossible, but this does not hinder a view of the entire map.

INTRODUCTION

In a scholarly career spanning five decades, Rhoda Rappaport published some two dozen articles and essays in the history of science. This reckoning does not count the substantial contributions she made to reference works, particularly the *Dictionary of Scientific Biography*. Nor does it count her important book, *When Geologists Were Historians, 1665–1750*, where she examined the ways thinkers of the seventeenth and early eighteenth centuries grappled with the problem of establishing reliable knowledge of the past – a province where historical and antiquarian scholars were then making significant headway – and how this entered deeply into the formative processes by which geological science took its beginnings. The majority of Professor Rappaport's articles addressed questions about geology's early development, frequently with emphasis on the scientific scene in eighteenth-century France. Her articles consistently broke new ground, bringing clarity to seemingly confusing issues, and offering perceptions and new documentation that have enduring value for readers and researchers. Sixteen of these papers are selected for reproduction in this volume.

Six months before her death in late October 2009, Dr Rappaport was invited by the editors of Ashgate's Variorum Collected Studies Series to prepare a volume of this kind. She declined, saying she could not muster the enthusiasm necessary for the task of arranging and providing commentary on her past scholarly productions. With the clarity of hindsight, we see that the poor state of her health must have been the main reason. We are therefore gratified that following news of Rhoda's passing – we will take the liberty of familiarity hereafter in referring to Rhoda by her first name, as she was a cherished colleague and friend to both of us – the Ashgate editors agreed to a renewal of the plan to produce this collection of her papers. We trust this decision will be welcomed by others, students and scholars now and well into the future, who take an interest in the topics to which she applied her considerable talents as historical investigator and analyst.[1]

---

[1] For views on her life and work as a teacher and scholar, see the éloges, "Rhoda Rappaport, 1935–2009," by Kenneth L. Taylor and Alice Stroup, *Isis* 101, 2010: 833–7; and "Rhoda Rappaport, historian of geology, 1935–2009," by Jill S. Schneiderman, *Earth Sciences History* 29, 2010: 171–3. The present volume's Bibliography includes a complete listing of her published writings (except reviews) other than those reproduced here. As the reader will readily discern,

The President of the Geological Society (London) stated in 2003, on the occasion of conferring upon Rhoda the Society's Sue Tyler Friedman Medal, that her papers are of a quality and scholarly influence "out of all proportion to their bulk."[2] The studies in this volume exhibit the intelligence and craftsmanship underlying the lasting impact of Rhoda's work. They display her consistent dedication to rigorous historical standards of argument and documentation, her good judgment in formulating historical questions it is useful to ask, and the disciplined vision and imagination with which she saw possible routes toward instructive answers.

In common with many other historians of science of her generation, Rhoda approached her chosen research topics with a decidedly intellectualist focus. That is, she considered the core of her subject to lie in the scientific ideas advanced and debated by the historical figures she studied. Prioritizing scientific conceptualization did not mean, however, that she saw fit to undervalue (let alone ignore) the placement of scientific endeavors within their cultural framework, or the practical deeds and social interactions through which scientific work has always been conducted. Nor did she permit herself the error sometimes imputed to historians of ideas, of treating concepts as disembodied entities capable of acting independently of their flesh-and-blood advocates and adversaries. This is evident in the close attention her research gave to varying interests and purposes standing behind scientific activities, to the investigative practices used in their pursuit, to the institutional structures within which they have been organized, and to the modes of communication by which they have been transmitted and evaluated.

Nonetheless, Rhoda was an avowed defender of the history of ideas. And she did often express concern – it rose on occasion to the level of distress – about some of the consequences of the "social turn" in the history of science that was so conspicuous a development during much of her career. She was by no means opposed to "contextualist" history, indeed she practiced a sophisticated form of it herself. However, she was dismayed by the more radical contentions for the "social construction" of scientific knowledge, which she viewed as an unsatisfactory doctrine of social determinism. And she was disturbed about the possible ascendancy of what some during the early 1980s were calling a

some of the details in this introduction are based on the authors' personal knowledge, some of it recorded in surviving correspondence.

  [2] The presentation remarks of the Geological Society President, Sir Mark Moody-Stuart, as well as Rhoda Rappaport's reply (made *in absentia* since she was unable to attend the Society's Annual Meeting), were at the time of this writing (June 2011) accessible at the Society's web site: http://www.geolsoc.org.uk/gsl/site/GSL/lang/en/page2984.html

  A more detailed acceptance speech, which she made closer to home a few months later, can be read in *Northeastern Geology & Environmental Sciences* 26, 2004: 107–9.

"history of science without science." She felt there was a discernibly polemical and hegemonic element to this socially-focused reorientation within scholarly work in the history of science, which had the effect of threatening interest in, or worse yet even the perceived legitimacy of, direct historical engagement with the substance of historical characters' ideas. She often thought of her own work as part of an effort to maintain a prominent place for intellectual history within the abiding traditions of historians of science. The ways people thought scientifically in the past, then, their confrontation with natural phenomena and their efforts to arrive at satisfactory explanations of them – these were usually at the root of what she sought to understand.[3]

Rhoda's notion of what qualified as "scientific" was suitably expansive. It mattered little to her whether or not the sorts of things her subjects were doing and talking about would be judged unequivocally scientific by modern observers. What did matter was whether such actions and statements reflected genuine aspirations to comprehend the world according to the standards of the historical moment. If an author three centuries ago believed, for example, that knowledge of the remotest periods of nature's past required a record of human testimony (a common enough conviction as her research helped make clear), then the exercise of this attitude, as well as its historical undoing, must be the object of careful historical study, not dismissal. Similarly, in her research on invocation of the Flood as a plausible agent of geological change, and the manner in which that idea was undermined, Rhoda demonstrated the importance of historical approaches to "diluvialism" animated by sensibilities more reflective than regret, embarrassment, or condemnation.

In Rhoda's sustained interest in the history of diluvial thinking, incidentally, it is possible to glimpse an aspect of her historical motivation. Her initial determination to pursue research on this topic grew out of her dissatisfaction with the oft-encountered claim that geology's "progress" during the eighteenth century was blocked by religious orthodoxy. She became convinced that this was an unfounded proposition, sanctified through a historical literature tinged with a kind of geological ancestor-veneration, passed down from tendentious writings of influential nineteenth-century geological authors (notably Cuvier and Lyell) trying to support their own views through rhetorical strategies that included stigmatizing competitors and setting up straw men. She succeeded

---

[3] Discussions during the 1980s of concern about a "history of science without science" are usually traced back to Charles C. Gillispie's (unpublished) Sarton Lecture at the 1980 annual meeting of AAAS, as reported in *Science* 207, 1980: 389. Commentators in the aftermath included Nathan Reingold, "Science, Scientists, and Historians of Science," *History of Science* 19, 1981: 274–83; Robert Siegfried, "Historiography of Science" [book review], *Annals of Science* 39, 1982: 201–3; and Charles Rosenberg, "Science in American Society: A Generation of Historical Debate," *Isis* 74, 1983: 356–67.

in showing that early geologists' supposed subservience to religious authority has been greatly exaggerated; along the way she demolished the notion that there existed a coherently univocal religious orthodoxy exercising the alleged scientific constraint. A good part of Rhoda's work should in fact be understood as her way of rectifying historical misconceptions about geology that were essentially the historical residue of past scientific disputations, recitations of inherited accounts of controversy written, inevitably, by the victors. Throughout her career, she felt a compulsion to help undo the anti-historical influence she thought whiggish formulas and attitudes had exercised unnecessarily long in the history of geology, more so than in most other areas of science's history. Instances of explicit exhortation on these historiographical matters may be seen now and then in her articles, but on the whole rather than rant against whiggism, Rhoda generally resolved to do her part by striving to publish research that might stand as a salutary corrective example. In this respect as in others she emulated her mentor Henry Guerlac, who evidently believed it was better to demonstrate one's historiography in scholarly action than to talk about it.

***

We have chosen to organize this volume's sixteen papers under five section headings. These reflect our sense of some significant patterns in their subject matter and treatment. Readers will easily recognize that the ten studies in the three middle sections are those addressed most directly to topics in the history of geology. But it should not escape notice that matters relevant to geology's early development arise in most of the other six articles as well, those found in the first and last sections. In the following brief discussion of the articles in this volume, we will try to provide some perspective on how Rhoda came to the main subjects of her research, how she viewed the connections among them, and how they have enlarged our historical understanding.

The order of the six articles presented here in the first two sections ("Chemistry," and "The *Mineralogical Atlas*") accords generally with the chronological sequence of Rhoda's early research. When she decided to pursue graduate-level study in the history of science, she had no specific plan to focus on geological science's history. That she came in due course to dedicate her research efforts especially to eighteenth-century geology was, in a very real sense, accidental. As an undergraduate student at Goucher College Rhoda had majored in mathematics and physics. Her first encounter with the history of science came in a history course taught by Dorothy Stimson, who was thus mainly responsible, as Rhoda liked telling, for her seduction (the word was hers) to this field. Entering the history of science graduate program at Cornell University, she became a student of Henry Guerlac, the distinguished interpreter of science in the French Enlightenment, and particularly of the work of the great chemist Antoine-Laurent Lavoisier. Guerlac suggested that for her Master's thesis she investigate one of the eighteenth century's most prominent teachers of chemistry, Guillaume-François Rouelle, who counted Lavoisier and many other luminaries of the period among his students. Completing that thesis by the end of her second year in graduate school, Rhoda presented the results of her precocious research in two substantial articles, published in successive annual volumes of *Chymia* (1960, 1961). These (studies I and II) were her first publications, if we leave aside a short article in the Cornell alumni magazine on a forgotten eighteenth-century French pioneer in the physical study of air (1958). She showed, among other things, how features of Rouelle's chemical system represented adaptations of the chemical doctrines of Georg Ernst Stahl, and how in Rouelle's work one sees steps toward a deepened and subtler chemical understanding in France well before Lavoisier himself set to work.

Rhoda saw that Rouelle's chemistry was in effect an attempt at systematic comprehension of the entire mineral realm. In this light it was not surprising that Rouelle had a great deal of influence on the succeeding generation of French mineralogists and physical geographers, not least through his codification of three types of rocks ("ancient," "intermediate," and "new") distributed over the earth's surface. It was apparently because of the geological dimension of Rouelle's chemical teachings, and also because she knew Lavoisier began his scientific career with a distinctly geological orientation, that Rhoda was drawn for her doctoral research to study the project of a French mineralogical atlas undertaken by Jean-Étienne Guettard, Lavoisier, and (in due course) Antoine-Grimoald Monnet. Her first opportunity for extensive research work in France came during the academic year 1960–1961, which she spent in Paris with the support of a fellowship from the American Association of University Women. (Her Masters research on Rouelle had been based mainly on microfilms of notes taken by Rouelle's students. Rhoda made a first brief visit to Paris in 1959, attending a three-day conference on eighteenth-century chemistry [Kent 1959].) Building upon the work she did in Paris during that doctoral research year, Rhoda completed her Ph.D. thesis in 1964. It analyzed the *Atlas* both as a cartographic venture and as an example of state involvement in scientific projects, and it situated Guettard, Lavoisier, and Monnet within a framework of contemporary geological thinking.

This thesis was the main foundation for Rhoda's four articles (III through VI) gathered here in the second section, "The *Mineralogical Atlas*." They are presented here not in chronological order of publication, but rather in an order reflecting the centrality of her interest in Lavoisier's role in the mapping project. Lavoisier is the only one of the three principals of the *Atlas* project with a major profile in all four of these essays, and the articles provide in succession a thorough accounting of Lavoisier's work and ideas in geology.

They show how Lavoisier's geological work was begun using ideas borrowed from older contemporaries (particularly Rouelle, Buffon, and Guettard), and how he rapidly developed a distinct perspective of his own, with a flair for field research and for quantification, particularly for purposes of charting the positions of strata, the spatial relationships of which he liked to depict in sections. He contemplated the integration of local data within a regional scheme, which he envisioned leading ultimately to a comprehensively global one, to show how strata have been produced and subsequently altered. Lavoisier came to realize that a satisfactory accounting of the strata of northern France would not be achieved through so simple a differentiation of past depositions as Rouelle's tripartite scheme. A more complex series of repeated marine incursions and retreats would be required. And he judged these fluctuations to have cycled over periods of hundreds of thousands of years. With these studies detailing Lavoisier's substantial and sustained interest in geological questions (he returned to these researches in the last years of his life), Rhoda helped to redress an imbalance in Lavoisier scholarship, which had for some time fixed its attention, unsurprisingly, mainly on his role as chief architect of the "chemical revolution."

Knowing details such as these about Rhoda's early scholarly trajectory helps us understand how she considered herself, at the outset, at least as much a researcher on Lavoisier as on the geology of Lavoisier's time. It also helps us see how, when she did decide to continue her concentration on early geology, her approach differed from that shown in much of geology's existing historical literature. Unlike the back-to-front or "tunnel history" sort of examination of modern geology's ancestry typified in classic books by Geikie (1905) and Adams (1938), Rhoda's approach to the subject was, historically speaking, sideways. Using terms probably not familiar to her at the time, her academic training and also perhaps her own intellectual dispositions inclined her toward "synchronic" studies rather than "diachronic" ones. As applied to her early researches, this meant that for an understanding of Lavoisier's work as a geologist, the most important connections included contemporaneous chemical and mineralogical teaching and royally-sponsored scientific projects, more than later consequences of Lavoisier's geological work for nineteenth-century stratigraphers.

As Rhoda completed her doctoral thesis she realized what rich possibilities existed for historical exploration of the nascent earth sciences in eighteenth-century France, and in European culture more generally. She soon produced a valuable survey on problems and sources in the history of geology from the time of Buffon to that of Cuvier (1964). In 1967 she participated in the landmark international conference on the history of geology in New Hampshire, organized by Cecil J. Schneer (article V), which brought together a sizable fraction of a small but growing community of researchers in this rather neglected field. For a time, around 1970, Rhoda was considering preparation of a book on geological cartography during the eighteenth century. She remarked on how difficult it might be to find a suitable publisher for such a project, which in addition to specimens of pertinent maps would include contemporary essays on problems of making such maps. Perhaps for this reason she redirected a good part of her effort toward related issues, including questions about how scientific work was encouraged and funded in this period, particularly by the French royal government. Among the results of that interest are study XV in this volume, and some of Rhoda's articles for the *Dictionary of Scientific Biography*, notably those on Malesherbes and the Turgot brothers.

Soon after embarking in 1961 on her teaching career at Vassar College, Rhoda formed what became a fixed habit of spending about one month each year working in Paris libraries and archives, a custom to which she adhered with rare exceptions for over four decades, until a few years beyond her 2000 retirement. From time to time she would include London on her itinerary. She made regular annual trips to the Cornell University Library, as well. When academic leave time permitted, she extended her Paris research sojourns to several months. In one notable departure from this pattern, during one of her early sabbatical leaves Rhoda remained at Vassar to take an intensive course in Italian, an investment whose eventual dividends are most obviously seen in her study of Vallisneri (article XIII) and in her book, *When Geologists Were Historians* (1997).

Rhoda's range of vision in history of geology widened steadily from the 1970s forward, soon attaining a highly inclusive cosmopolitan European perspective. At the same time, the topical scope of her studies also grew beyond early modern natural history and physical science, encompassing the period's historical and philosophical scholarship, as well as theology and religious apologetics. Increasingly she tried to formulate research problems crossing both national and disciplinary boundaries, and engaging more widely varying sorts of thinkers. Her interests – and particularly her ongoing study of thinkers wrestling with the difficulties of reconstructing the past – also tended to expand the chronological scope of her research some way back in time, into the second half of the seventeenth century.

In what is no doubt a gross oversimplification, it can be said that the work she pursued for a good many years as a mature scholar falls largely within three broad (and admittedly overlapping) categories: One of these centered on the evolving understanding of petrifactions, or fossil objects. Rhoda investigated how acceptance of the organic derivation of many kinds of fossils was incorporated in early earth science, especially as this drew naturalists into offering explanations for the preservation and observed distribution of fossils, and how these considerations reflected on alterations the earth must have undergone. Such problems connected directly with a second focus of Rhoda's study, on struggles during the seventeenth and eighteenth centuries with the problem of establishing reliable knowledge of the earth's past: What sorts of methods and evidence did the period's researchers think might bear on such problems, and how much confidence did thinkers have in them? These two closely-related kinds of researches are represented especially in the middle section of this volume, "Understanding the Earth and its History."

A third and comparatively small category, but an important one, within which some of Rhoda's work may be classed centers upon language. She was persistently attentive to questions about the specific sorts of language through which early modern writers addressed geological issues, and the terminology historians have employed in interpreting those efforts. While this special linguistic alertness is perceptible in a number of her essays, there are two articles for which this is the main organizing principle. These two studies constitute the fourth section, "The Language of Earth Science."

Central to Rhoda's inquiries composing the section on "Understanding the Earth and its History" was what she called "the essential insecurity of naturalists when dealing with the remote past."[4] She examined how Hooke, Fontenelle, and Leibniz addressed the problem of establishing knowledge of the earth's past condition and operations (VII, VIII, IX). This entailed their respective ways of treating the relevance of historical evidence (human testimony) alongside the testimony of nature, as well as their ideas about the degree of confidence one can reasonably have in any reconstruction of past events. In the case of Hooke, as Rhoda showed, the emphasis was squarely on achieving an understanding of processes relevant to past and ongoing changes in the earth, rather than on developing a knowledge of sequences of past events. The latter sorts of question were, on the other hand, more clearly the preoccupations of Leibniz and especially Fontenelle. Her studies involving these three figures also feature a "context-of-presentation" component. Rhoda did detective work to uncover a previously obscure historical order to Hooke's lectures before the Royal Society, thus clarifying his objectives and tactics in trying to persuade his listeners regarding the nature of fossil bodies and the earthquake processes – by which he meant crustal dislocations – relevant to understanding their modern locations. Fontenelle, the Paris Academy's perpetual secretary, seized the opportunities afforded by this office as he fulfilled his duty to summarize and interpret the academicians' and correspondents' work in its annual *Histoire*. In Rhoda's analysis Fontenelle emerges as an important earth-science thinker essentially through his commentary on and adaptation of the work of others. Among those others, one outstanding instance was Leibniz, whose "lost" 1706

---

[4] "Fontenelle interprets the earth's history" (VIII), p. 293.

communication to the Academy, upon which Fontenelle built an interpretive discussion, Rhoda found in the institution's *Procès Verbaux*.

The last article within "Understanding the Earth and its History" is her study on the Flood in eighteenth-century thought (X). This is perhaps the most often cited of all her essays (and it was reprinted in a volume on *The Flood Myth* in 1988). This seminal essay encompasses many of the issues taken up later in her book, *When Geologists were Historians*: the relative weights given to physical evidence and human testimony in constituting knowledge about the past, and the tug of war between traditional reliance on authority and independent-minded pursuit of secular explanation within the communities of scholars in coming to grips with information about fossils and the strata enclosing them. Also typifying Rhoda's way of conducting historical research, it exhibits her critical scrutiny of common but vague historical formulas, subjecting these to close examination. As was already mentioned above, one cliché whose emptiness she here exposes is the notion that those trying to attain natural knowledge of the earth were greatly constrained by religious "orthodoxy."

The Flood was a source of recurring fascination for Rhoda. More than any other single factor, it may have been the exegetical aspects of the project to put the Flood into proper historical perspective that motivated her to write directly about particular features of the language used by her geological subjects and their historical interpreters. "Borrowed words" (XI) is the first of two essays of this kind, presented here under the rubric "The Language of Earth Science." This article may be, along with "Geology and orthodoxy," one of the most memorable and suggestive of Rhoda's essays. In her study of eighteenth-century thinking about the Flood she had already given some consideration to the special uses of terms like "monuments" and "revolutions" in the discourse of geological naturalists. Now she elaborated and extended the discussion, in particular with analysis of the notion of geological "accidents."

Rhoda's second lengthy essay on historical vocabulary, "Dangerous words" (XII), differs from "Borrowed words" in its critical appraisal of interpretive labels employed mainly by historians, even though these were terms used in varying degrees by some of the relevant historical characters as well. The words she examined – "diluvialism," "neptunism," and "catastrophism" – have been historically more troublesome than helpful, she showed, not only on account of the confusing elusiveness of their definitions, but also because they have so often provided historians with tempting cover for unsupported simplifications, and because of their susceptibility for use as weapons of condemnation rather than comprehension. This discussion is vintage Rhoda Rappaport: It expresses her characteristically discriminating sensitivity to what authors write, her alertness to problems created in transfer of terms from one context to another, and her appreciation of semantic dangers in disregard of such problems – all set within a richly developed body of examples and references.

Of the five sections of this book, the last – "Scientific Pursuits in Early Modern Europe" – may strike some readers as possessing the least transparent topical or thematic coherence. It consists of four arguably rather disparate studies produced during different phases of Rhoda's career. These four studies do share common ground, however, in what may be thought of as their cultural focus. Without neglecting the conceptual issues that pervade her research, for these articles Rhoda oriented herself more directly toward understanding problems of community and cultural identity on the part of those engaged in science. Two of the articles focus on notable scientific characters (Antonio Vallisneri, and the Baron d'Holbach) with some emphasis on what may be called their programmatic interests. The other two derive from Rhoda's concern to understand how eighteenth-century scientific work was encouraged and supported, but also channeled and disciplined through its institutions and forms of patronage.

Rhoda took satisfaction in the notion that her 1991 study of Vallisneri, "Italy and Europe" (XIII), helped to elevate the profile of a major early modern Continental scientific figure who, while far from forgotten in Italian and French scholarship, for a long time has had only a nominal presence in Anglophone historical literature. It pleased her also to call attention to what she considered the remarkable vitality of current Italian research in cultural history. She saw the eighteenth-century Italian scene as full of opportunities for research: a good part of this article amounts to a review of archival resources currently being made available and of historical questions their study may help clarify. Characteristically interested in evidence bearing on trans-national intellectual commerce in early modern Europe, Rhoda discussed Vallisneri as an enthusiast for Italian cultural nationalism yet also a thinker well connected with foreign groups and individuals. And she examined his geological views, which had relevance to her ongoing studies on varieties of diluvial thinking. Her attention to the journalistic activities of this Paduan physician and naturalist must surely have been linked with the writing she may already have had under way, on review journals as crucial elements of a pan-European "Republic of Letters" during Vallisneri's lifetime; this topic was to become integral to the first section of her book *When Geologists were Historians.*

If research on Vallisneri had a connection to her later book's opening chapter, Rhoda's article on d'Holbach (XIV) was linked with the way the book closes, in the chapter "Buffon and the rejection of history." In an effort to identify others besides Buffon with whom to associate a mid-eighteenth-century re-direction of geological thinking – from descriptive historical sequences of strata toward some more physically-grounded way of incorporating strata within geological explanation – Rhoda decided to read all 800 of the *Encyclopédie* articles written by d'Holbach. (Many of the articles are brief, amounting to dictionary definitions, and she was already familiar with many of the long ones.) For d'Holbach the key to geological understanding lay in chemistry, a field in which German and Swedish researchers were acknowledged leaders. In an effort to promote a kind of merger of this knowledge with French science d'Holbach produced, during the same fifteen years when he was churning out the *Encyclopédie* articles, annotated French translations of eight German and Swedish chemical, mineralogical, and metallurgical works. Rhoda argued that this monumental effort on d'Holbach's part should be understood as a campaign for reform and renewal of French science, including geological science in particular, through assimilation of the advanced chemical and technological knowledge extant in the German world, as distinct from interpreting it as arising out of a taste for popularization, or as a prelude to d'Holbach's formulation of his materialist philosophy. (Presenting an offprint of this article to her sister and brother-in-law, Rhoda wrote: "This is one I really enjoyed, both for the research and writing.")

Alone among the sixteen papers assembled here, "Government patronage of science in eighteenth-century France" (XV) is a review. It treats the massive three-volume study by André J. Bourde, *Agronomie et Agronomes en France au XVIIIe Siècle* (1967). Rhoda's interest in Bourde's examination of scientific farming lay mainly in what it has to say about the cultural reputation of science during the eighteenth century, about science's place in the period's projects for reform, and thus about government patronage of science perceived as economically and socially useful. These were subjects to which she had been led in her doctoral research on the French *Mineralogical Atlas*. She realized how much might be learned about eighteenth-century science by attending to how leading figures in the administrations under Louis XV followed policies seeking to stimulate and utilize scientific knowledge, harnessing it for the advantage of industrial and economic interests. About three-quarters of this review is in fact a discussion of paths to be followed in further research on these issues.

"The liberties of the Academy of Sciences," the volume's final essay (XVI), was Rhoda's contribution to a *Festschrift* in honor of her mentor, Henry Guerlac. Her chapter opens a window onto life within the Academy. It does so chiefly through a study of elections and promotions in the Academy, across a lengthy span of seven of the institution's most illustrious decades, a time marked by stability in its regulations. The central consideration, which Rhoda pursued applying results of her extensive research in the Academy's archives, was the functioning of a conspicuous academic ideal, namely the freedom of the Academy's members to exercise their best judgment on what we would call personnel issues. Government ministers and academicians alike affirmed this ideal as essential to achieving and maintaining the Academy's high quality. Rhoda's examination of the extent of this meritocratic principle's actual operation, amidst the realities of personal ambitions and rivalries, of lobbying and intrigues, and of time-honored habits of institutional practice, reveals much about the Academy's complex inner workings and its relationship to its royal patron.

***

The foregoing remarks provide, we hope, some sense of the paths by which Rhoda determined what to study and write about, some indications of the sorts of interpretive themes she chose in addressing her subjects, and some signs of her convictions as a historian. In sounding the studies themselves, of course, the reader will apprehend more fully what she tried to do and how she did it.

If others' experiences in encounters with Rhoda's work are at all like ours, they will include the satisfaction of being instructed by a scholar who had both a refined curiosity about her subjects and a high level of skill in presenting the results of her research. Rhoda liked historical detective work, but one sees that the puzzles to which she tracked down answers were linked with larger questions she always kept in view. She sought to reduce the complexities inevitably disclosed by historical research to levels of useful simplicity, but she refused to resort to facile generalizations. Composing her articles with great care, she strived to express her ideas economically, often succeeding in condensing complications down to few words. The clarity of her writing is such that solidly documented assertions are never confused with the clearly-marked probabilities or conjectural ideas that helped animate her analyses.

The fundamental seriousness of Rhoda's work by no means entirely conceals the enjoyment she found in research. Scattered here and there in her articles are understated signals of what she found amusing. One of our favorites is her formulation about a now little-known historical writer, the abbé Vertot, who was popular during the eighteenth century (article XI, p. 32): "In an age when history is not easily divorced from scholarship, the modern reader can still understand why Vertot's works were so successful: they are unfailingly lively, informative, and intelligent, and Vertot's gift for narrative is not marred in any important way by demands upon the reader's mind."

Rhoda Rappaport's articles *do* make demands upon the reader's mind. That is as she intended.

KENNETH L. TAYLOR and MARTIN J.S. RUDWICK

Norman, Oklahoma; Cambridge, England
June 2011

BIBLIOGRAPHY

Adams, Frank Dawson. 1938. *The Birth and Development of the Geological Sciences*. Baltimore: William & Wilkins. Reprint: New York, Dover Publications, 1954.

Bourde, André J. 1967. *Agronomie et agronomes en France au XVIIIe siècle*. 3 vols. Paris: S.E.V.P.E.N.

Geikie, Archibald. 1905. *The Founders of Geology*. 2nd edition. London & New York: Macmillan. Reprint: New York, Dover Publications, 1962. [Originally published by Macmillan in 1897.]

Guettard, Jean-Étienne, and Antoine-Laurent Lavoisier. 1766. *Premières feuilles de l'atlas minéralogique: ouvrage commencé en 1766, par MM. Guettard et Lavoisier*. [Paris]: De Prony. [Volume with hand-written title page, in Ewell Sale Stewart Library, Academy of Natural Sciences, Philadelphia.]

Guettard, Jean-Étienne, and Antoine-Grimoald Monnet. 1780. *Atlas et description minéralogiques de la France, entrepris par ordre du Roi*. Paris: Didot l'aîné; Desnos; A. Jombert jeune. [Some copies include maps dating from later than 1780.]

Hamm, Ernst P. 2001. "Of 'histories written by the hand of nature itself'." *Annals of Science* 58, pp. 311–17. [A particularly insightful essay review of Rhoda Rappaport, *When Geologists were Historians*.]

Kent, Andrew. 1959. "Chemistry in the Eighteenth Century." *Nature* 184, p. 1452. [A report on an international meeting held in Paris, 11–13 September 1959, and attended by Rhoda Rappaport (two photographs of participants at this meeting are deposited in the History of Science Collections, University of Oklahoma Libraries).]

Rappaport, Rhoda. 1958. "An Early Study of Air," *Cornell Alumni News* 60, no. 18, p. 630. [On *La Manière de rendre l'air visible et assez sensible ...*, by P. Moitrel d'Élément (Paris, 1719).]

—. 1958. G.-F. Rouelle: His *Cours de Chimie* and Their Significance for Eighteenth Century Chemistry. M.A. thesis, Cornell University.

—. 1963. "Antoine-Laurent Lavoisier – A Biographical Sketch," *Journal of Nutrition* 79, pp. 3–8.

—. 1964. Guettard, Lavoisier, and Monnet: Geologists in the Service of the French Monarchy. Ph.D. thesis, Cornell University.

—. 1964. "Problems and Sources in the History of Geology, 1749–1810," *History of Science* 3, pp. 60–77.

—. 1965. Appendix, in *Supplement to a Bibliography of the Works of Antoine Laurent Lavoisier, 1743–1794,* by Denis I. Duveen. London: Dawsons of Pall Mall, pp. 129–32. [Notes regarding the *Atlas et description minéralogiques de la France.*]

—. 1971. "Commentary on the Paper of M.J.S. Rudwick," in *Perspectives in the History of Science and Technology*, edited by Duane H.D. Roller (Norman, Oklahoma: University of Oklahoma Press), pp. 228–31. [Remarks on Martin J.S. Rudwick's "Uniformity and Progression: Reflections on the Structure of Geological Theory in the Age of Lyell," pp. 209–27.]

—. 1972. "Guettard," in *Dictionary of Scientific Biography*, vol. 5, pp. 577–9.

—. 1974. "Malesherbes," in *Dictionary of Scientific Biography*, vol. 9, pp. 53–5.

—. 1974. "Monnet," in *Dictionary of Scientific Biography*, vol. 9, pp. 478–9.

—. 1975. "Rouelle, Guillaume-François," in *Dictionary of Scientific Biography*, vol. 11, pp. 562–4.

—. 1975. "Rouelle, Hilaire-Marin," in *Dictionary of Scientific Biography*, vol. 11, p. 564.

—. 1975. "Soulavie," in *Dictionary of Scientific Biography*, vol. 12, pp. 549–50.

—. 1976. "Turgot, Anne-Robert-Jacques," in *Dictionary of Scientific Biography*, vol. 13, pp. 494–7.

—. 1976. "Turgot, Étienne-François," in *Dictionary of Scientific Biography*, vol. 13, p. 497.

—. 1985. "Geology in the Eighteenth Century." Audiotape of a lecture given at Cornell University. 73 minutes. Cornell University Library. [Available for consultation only at the Cornell Library.]

—. 1988. Geology and Orthodoxy: The case of Noah's Flood in eighteenth-century thought," reprinted in *The Flood Myth*, edited by Alan Dundes (Berkeley, Los Angeles, London: University of California Press), pp. 383–403.

—. 1997. *When Geologists were Historians, 1665–1750.* New York & London: Cornell University Press.

—. 1997. "Questions of Evidence: An Anonymous Tract attributed to John Toland," *Journal of the History of Ideas* 58, pp. 339–48.

—. 2000. "Fontenelle," in *Encyclopedia of the Scientific Revolution: From Copernicus to Newton*, edited by Wilbur Applebaum. New York & London: Garland Publishing, pp. 235–6.

—. 2000. "Vallisneri," in *Encyclopedia of the Scientific Revolution: From Copernicus to Newton*, edited by Wilbur Applebaum. New York & London: Garland Publishing, p. 663.

—. 2003. "The Earth Sciences," in *The Cambridge History of Science*, vol. 4: *Eighteenth-century Science*, edited by Roy Porter. Cambridge: Cambridge University Press, pp. 417–35.

—. 2004. "Acceptance Speech of Sue Tyler Friedman Medal of the Geological Society of London," *Northeastern Geology & Environmental Sciences* 26, pp. 107–9.

ACKNOWLEDGEMENTS

Grateful acknowledgement is made to the following persons, journals, institutions and publishers for their kind permission to reproduce the articles included in this volume: The University of California Press, Berkeley, California, on behalf of *Chymia* (for essays I and II); The University of Chicago Press, Chicago, Illinois, on behalf of *Isis* (III); The Cambridge University Press and the British Society for the History of Science, on behalf of *The British Journal for the History of Science* (IV, VI, VII, X, and XI); The MIT Press, Cambridge, Massachusetts (V); Michel Blay, Editor, *Revue d'Histoire des Sciences*, Paris (VIII); Jürgen Herbst, Editor, *Studia Leibnitiana*, Hannover (IX); Paolo Galluzzi, Director, Istituto e Museo di Storia della Scienza, Florence (XII); Michael Hoskin, Director, Science History Publications Ltd, Cambridge, on behalf of *History of Science* (XIII and XV); The Voltaire Foundation, Oxford University (XIV); The Cornell University Press, Ithaca, New York (XVI).

The editors wish to thank John Smedley, Claire Jarvis, and Lindsay Farthing, of Ashgate Publishing, for all they have done to bring about the realization of this volume.

For assistance in securing images of the three maps reproduced anew with article V, we are grateful to Eileen C. Mathias, The Academy of Natural Sciences, Philadelphia, Pennsylvania; Marie-Noëlle Maisonneuve, Ecole Nationale des Mines, Paris; and David W. Corson, Cornell University Library, Ithaca, New York. Thanks also to Kerry V. Magruder, University of Oklahoma Libraries, who generously provided technical help relating to the texts and the maps.

We wish to express our appreciation to Rhoda Rappaport's nephew, Jeremy Primer, the executor of her estate; and to the estate's legatees, the Cornell University Library and The New York Public Library, Astor, Lenox and Tilden Foundation. Finally, we thank Rhoda's sister and brother-in-law, Blossom and Irwin Primer, for providing valuable information as well as the frontispiece photograph and many of the offprints reproduced here.

I

G.-F. ROUELLE: AN EIGHTEENTH-CENTURY CHEMIST AND TEACHER

I T IS frequently true in the history of science that the figures to excite the most interest are those in whom distinct signs of "modernity" are apparent. Until a short time ago this generalization could be applied in all its force to the field of the history of chemistry and the results are well known: the precursors of Lavoisier, the "founder" of modern chemistry, were left unstudied and were relegated to an unenlightened age in the history of a relatively new science. But it is to ignore the continuity of history to believe that chemistry was born *de novo* at the end of the eighteenth century. Studies of Lavoisier's predecessors, such men as Mayow, Hales, Black, and Bayen, have only recently contributed to a better understanding of the origins of the Chemical Revolution.

One of the outstanding figures in the so-called "pre-modern" era of chemistry was Guillaume-François Rouelle, teacher of chemistry at the Jardin du roi in Paris from 1742 until 1768. Once considered a truly eminent scientist, Rouelle is remembered today principally as the teacher of Lavoisier. That this man, who was in fact the teacher of an entire generation of French chemists, is at present relatively unknown and unstudied seems a strange oversight on the part of modern historians of science. It is hoped that this article, which is intended to serve as an introduction to further study of Rouelle's place in the history of chemistry, will supplement and bring up to date past research into the life and work of an important eighteenth-century scientist.[1]

The teaching tradition in which Rouelle was working had its origins in the early years of the seventeenth century. Almost from its start and as it evolved in the course of the century, this tradition had three major components. The first of these was the "private" course in chemistry, which was initiated in Paris by Jean Béguin and which consisted of lectures open to a fee-paying audience. Such courses were generally held at the homes or laboratories of the instructors. Second to develop was the "public" course offered without charge to anyone who might care to attend. In Paris, where both types of instruction originated and flourished, the first public courses were conducted by William Davisson at the Jardin du roi. The writing and publication of textbooks of chemistry—

---

[1] A bibliography of the works and portions of works devoted to Rouelle is given below, note 7.

the third element in this tradition—was a common practice among many of the instructors and dated from the earliest of the private courses.

The teachers of these early courses were trained as apothecaries or physicians, since chemistry was at that time an adjunct to the fields of pharmacy and medicine. Similarly the textbooks, which were in part concerned with demonstrating the uses of the "art of chymistry," laid considerable stress upon pharmaceutical preparations and their curative properties. In short, students entering the fields of pharmacy and medicine found chemistry a valuable ancillary discipline and these chemical lectures a convenient source of instruction.

The popularity of chemical lectures rose perceptibly during the career of Nicolas Lémery (1645–1715), whose private courses and extraordinarily successful textbook remained enviable models until the middle of the eighteenth century. Lémery, who stayed within the medico-pharmaceutical tradition of his age, was particularly noted for the clarity of his style and the variety and soundness of his experiments. His reputation, gained through both his lectures and his textbook, survived the eclipse of his chemical theory. When the corpuscular philosophy of Robert Boyle—the theory espoused by Lémery—gradually came into disrepute, Lémery's *Cours de chymie* (1675) still remained one of the most popular of laboratory manuals.[1a]

Although after Lémery's death there was no break in the continuity of the teaching tradition, no outstanding teaching personality appeared in Paris until the 1740s. By 1742 a controversial figure was attracting the attention of the Parisian public. Guillaume-François Rouelle (1703–1770), by his colorful personality and explosive temperament, charmed and excited his growing audiences, while at the same time, by the quality of his lectures and demonstrations, he gradually established a reputation as the foremost teacher of chemistry in France.

Rouelle's lectures were frequented by members of Parisian society as well as scientists from all parts of France. While he published no textbook, he achieved fame through his courses and became the master of an entire generation of French chemists. Rouelle's gifts as an inspiring teacher were reported in glowing terms by Louis-Sébastien Mercier, the chronicler of Paris who was also an apothecary and a Rouelle pupil:

When Rouelle spoke, he inspired, he overwhelmed; he made me love an art about which I had not the least notion; Rouelle enlightened me, converted me; it is he who made me a supporter of that science [of chemistry] which should regenerate all the arts, one after the other . . . ; without Rouelle, I would not have known how to look above the mortar of the apothecary.[2]

[1a] For an excellent discussion of the seventeenth-century teachers of chemistry, see John Read, "Humour and Humanism in Chemistry," London, 1947, Chapter 6.

[2] Louis-Sébastien Mercier, "Tableau de Paris," 12 Vols., Amsterdam, 1782–88, Vol. 11, pp. 178–179.

Yet praise such as Mercier's, based as it was upon personal contact with Rouelle, was insufficient to keep alive the latter's reputation as a teacher after his retirement from public life. The oral tradition and the written descriptions of Rouelle's lectures by his pupils were forgotten soon after his death in 1770.

As I have already suggested, Rouelle's reputation among his contemporaries and pupils did not stem wholly from sheer teaching ability, but also from the content and quality of his lectures. His few scientific publications also contributed to the esteem in which he was held and helped to secure his status as an eminent scientist. However, close examination of the contemporary writings about Rouelle—and, what is often more significant, those works which should mention Rouelle and do not—reveals that among his few articulate disciples there was some difference of opinion as to just how original Rouelle's chemistry actually was. There seems to have been no dispute about the originality of his published papers, but his more general chemical theories, contained in his lectures, provoked reactions of two very different kinds. For some pupils Rouelle was "the creator" and "the founder" of French chemistry and a theorist whose insights started a new chemical school.[3] A second group, consisting principally of professional chemists, considered Rouelle's theories to be identical with the work of his scientific predecessors; his rôle was thought to be that of a mere popularizer of already formulated ideas. It was unhappily this latter view of Rouelle's work that was perpetuated in the textbooks and treatises of his pupils, where Rouelle is usually omitted from accounts of the history of chemical thought. So effective was this complete and probably dispassionate silence that by 1800 the chemist Fourcroy, surveying the history of his field, saw no reason to devote to Rouelle more than a few lines describing a small portion of his work in organic chemistry.[4] Modern histories of chemistry are likewise silent

[3] Friedrich M. Grimm, "Correspondence littéraire, philosophique et critique par Grimm, Diderot, Raynal, Meister, Etc.," ed. M. Tourneux, 16 Vols., Paris, 1877–82, Vol. 9, p. 106, and Denis Diderot, "Œuvres complètes de Diderot," ed. J. Assézat and M. Tourneux, 20 Vols., Paris, 1875–77, Vol. 6, p. 407.

[4] Antoine Fourcroy, "Système des connaissances chimiques, et de leurs applications aux phénomènes de la nature et de l'art," 10 Vols., Paris, [1800], Vol. 7, p. 39; Vol. 8, p. 83; Vol. 9, pp. 37–38, 50.

These general remarks on Rouelle's reputation cannot be fully documented here. In general Diderot can be considered representative of the first group and Macquer of the second. Diderot's judgment is found in "Cours de chymie de M. Rouelle, rédigé par M. Diderot et éclairci par plusieurs notes," Bibliothèque de Bordeaux, MS 564, pp. 33–34: Rouelle "a ajouté beaucoup d'idées neuves et utiles à la Doctrine de ses maîtres Stahl et Becher et il occupe le premier rang parmi les chimistes modernes." Macquer's neglect of Rouelle is apparent in his "Dictionnaire de chimie," 2nd ed., 2 Vols. (in-4°), Paris, 1778, esp. arts. "Feu," Vol. 1, pp. 481–500, and "Principes," Vol. 2, pp. 293–297; and in P.-J. Macquer and Antoine Baumé, "Plan d'un cours de chymie expérimentale et raisonnée, avec un discours historique sur la chymie," Paris,

71   G.-F. ROUELLE: 18TH-CENTURY CHEMIST AND TEACHER

about Rouelle, mentioning him, if at all, only in connection with his published work and in his capacity as the teacher of Lavoisier.[5]

The reasons for Rouelle's descent from relative prominence to obscurity are not difficult to determine. That his originality was not fully recognized by many of his fellow chemists was certainly a decisive factor; since these included the only significant writers about chemistry during this period, their neglect of Rouelle effectively damaged his reputation. A second determining condition was Rouelle's failure to publish anything more than a few memoirs on subjects of limited importance and interest; he never realized his plan to issue a textbook of chemistry.[6] After his death a group of his disciples undertook this publication, but the project was never completed. With the acceptance of the new chemistry of Lavoisier, Rouelle's theories rapidly became obsolete and hardly worth publishing. There still existed numerous manuscript copies of his once-popular lectures, but their usefulness had disappeared and they remained shelved and forgotten.

It is the purpose of this study to show that the prevailing judgment of Rouelle's contemporaries was not an entirely adequate evaluation of his achievements and significance. Because Rouelle is now a relatively unfamiliar figure even to historians of chemistry, this first of two articles will be limited to a discussion of his biography and the general character of his chemical lectures. Once his position as an influential teacher and a scientist of some originality has been established, the second article will be concerned with the details of Rouelle's chemical theories and their place in the context of eighteenth-century chemistry.

If the sources for Rouelle's biography are rather slim and at times unsatisfactory, it is nevertheless possible on the basis of the available material to sketch the broad outlines of Rouelle's life.[7] Guillaume-François Rouelle was born in the Norman village of Mathieu in 1703.

1757, pp. lv–lvii, lxii–lxiii. (Unless otherwise specified, all references to Macquer's "Dictionnaire" are to the *quarto* 1778 edition.)

[5] *Cf.* Raoul Jagnaux, "Histoire de la chimie," 2 Vols., Paris, 1891, and Henry M. Leicester, "The Historical Background of Chemistry," New York, [1956].

[6] For Rouelle's plans to publish a textbook, see below, pp. 89.

[7] For contemporary biographies of Rouelle, see Diderot, "Œuvres," Vol. 6, pp. 405–510; Grimm, Vol. 9, pp. 106–109; and Grandjean de Fouchy, "Eloge de M. Rouelle," *Histoire de l'Académie royale des sciences*, pp. 137–149 (1770). Diderot and Grimm are the sources of anecdotes reported in modern works. A useful biography is Paul-Antoine Cap's "Rouelle," *Journal de pharmacie et de chimie*, Paris, Sept. 1842; more detailed and well documented accounts are those of Jean-Paul Contant, "L'Enseignement de la chimie au Jardin royal des plantes de Paris," Cahors, 1952, pp. 103–113, and Paul Dorveaux, "Apothicaires membres de l'Académie Royale des Sciences—IX. Guillaume-François Rouelle," *Revue d'histoire de la pharmacie*, **4,** 169–186 (Dec., 1933). The best accounts in general histories of chemistry are those of Ferdinand Hoefer, "Histoire de la chimie," 2 Vols., Paris, 1842–43, Vol. 2, pp. 386–391, and Eric J. Holmyard, "Makers of Chemistry," Oxford, [1946], pp. 189–196. Unfortunately, the latter work is

According to Diderot, he soon showed an inclination for scientific observation and at the age of fourteen rented a coppersmith's forge and proceeded to conduct experiments.

He attended the Collège Dubois of the University of Caen, where he began to study medicine, soon giving it up because he found chemistry and pharmacy better suited "à la sensibilité de son cœur" than medicine and surgery.[8] In 1725 he went to Paris to continue his studies and there became apprenticed to J.-G. Spitzley, a German pharmacist who succeeded to the laboratory of Nicolas Lémery. The dates of his association with Spitzley are uncertain, but it is generally agreed that his apprenticeship lasted seven years.[9] During his early years in Paris he became acquainted with the Jussieu brothers: Antoine who was professor of botany at the Jardin du roi, and Bernard, the demonstrator in botany; Rouelle may even have studied with them.[10]

After 1737 Rouelle was giving lectures in both pharmacy and chemistry in the place Maubert, not far from the Jardin du roi.[11] These lectures soon became very popular and, as Diderot observed, "the quarter [of the city] with the meanest populace became the meeting place for all classes, not excluding the children of nobles who wanted to learn."[12] Eventually

---

not well documented. A recent article on Rouelle is Douglas McKie's "Guillaume-François Rouelle (1703–1770)," *Endeavour*, **12**, 130–133 (1953). Although he makes use of the Rouelle manuscript at Clifton College, McKie adds nothing of importance to Holmyard, who has used the same manuscript. Two recent articles devoted to Rouelle's life and work are Claude Secrétan's "Coup d'œil sur la chimie prélavoisienne," *Bulletin de la Société vaudoise des Sciences naturelles*, Lausanne, **61**, 329–354 (1941), and the same author's "Un aspect de la chimie prélavoisienne (Le cours de G.-F. Rouelle)," *Mémoires de la Société vaudoise des Sciences naturelles*, Lausanne, **7**, 219–444 (1943). A portrait of Rouelle is reproduced by Contant, p. 105, and Edgar F. Smith, "Forgotten Chemists," *J. Chem. Education*, **3**, 29–40 (1926), portrait on p. 30; this portrait differs from that in McKie, p. 131. The portrait labeled G.-F. Rouelle in René Sordes, "Histoire de l'Enseignement de la Chimie en France," Paris, 1928, p. 63, is actually that of his younger brother, Hilaire-Marin Rouelle (1718–1779).

[8] Diderot, "Oeuvres," Vol. 6, p. 406. Grandjean de Fouchy, p. 138, states that Rouelle originally chose the medical profession principally because of his love for chemistry. Pierre Lemay, in "Les cours de Guillaume-François Rouelle, *Revue d'Histoire de la pharmacie*, No. 123, pp. 434–442 (mars 1949), asserts (p. 438) that Rouelle obtained the degree maître-ès-arts; the statement is undocumented.

[9] I am indebted to Dr. Glenn Sonnedecker of the University of Wisconsin for his aid in identifying the elusive Spitzley. Diderot, "Oeuvres," Vol. 6, p. 406, suggests Rouelle's apprenticeship dated from 1725 until about 1732. Dorveaux, pp. 171 f., prefers 1730–37, as does Contant, p. 103. Macquer's remark that Rouelle opened his courses soon after 1737 tends to confirm the 1730–37 dating; see Macquer, "Dictionnaire de chimie," art. "Phosphore d'Angleterre ou de Kunckel," Vol. 2, p. 214.

[10] Contant, p. 103; he cites no authority.

[11] Cf. Macquer, cited above, note 9. That Rouelle was teaching by 1740 is recorded by Jean Darcet, "Mémoire sur la calcination de la Pierre calcaire," *Observations sur la physique, sur l'histoire naturelle et sur les arts*, **22**, 19–34 (jan. 1783); see p. 33.

[12] Diderot, "Oeuvres," Vol. 6, p. 406.

73   G.-F. ROUELLE: 18TH-CENTURY CHEMIST AND TEACHER

Rouelle's lectures attracted the attention of people other than aspiring young students, with the result that both his doctrines and his personality became controversial subjects.

The new chemical doctrine which he sought to introduce brought him zealous supporters, and raised against him some adversaries whom he did not always treat with enough respect; which retarded the progress of that doctrine and the success of his courses. Nevertheless he succeeded in obtaining the esteem which was due to him, and his reputation secured for him not only the position of demonstrator at the Jardin [du roi] in 1742, but also entrance into the Academy of Sciences in 1744.[13]

Rouelle's reputation rapidly became such that, at the death of G.-F. Boulduc on January 17, 1742, he was appointed Boulduc's successor at the Jardin du roi.[14] Particularly noteworthy is the title of Rouelle's position: "Démonstrateur en Chimie au Jardin des Plantes, sous le titre de Professeur en Chimie."[15] The demonstrator, as has been frequently pointed out, was supposed to assist the professor of chemistry—at that time, Louis-Claude Bourdelin (1696–1777)—by carrying out experiments to illustrate the professor's lecture on chemical theory. Thus, the demonstratorship was a position subordinate to the professorship; but with Rouelle's appointment "the [demonstrator's] chair becomes more important in fact, if not by right, than chair number one."[16] Rouelle actually performed both functions simultaneously, taking it upon himself to conduct experiments and lecture on theoretical matters.

A much repeated anecdote reported by Grimm is concerned with Rouelle's relation to Bourdelin. Rarely in agreement with the views of Bourdelin, Rouelle would begin his own lectures with the announcement: "Gentlemen, all that *monsieur le professeur* has just told you is absurd and

---

[13] A.-L. de Jussieu, "Notice historique sur le Muséum d'histoire naturelle: V. Depuis 1739 jusqu'en 1760," *Annales du Muséum National d'Histoire Naturelle*, **6,** 1–20 (1805); see p. 5, n. 2. The "new chemical doctrine" referred to in this passage probably means Stahlian chemical theory which was introduced into France in modified form by Rouelle. To show that this doctrine was in fact new—i.e., different in some degree from Stahl's and different from the chemistry being taught at that time—will be the burden of a second article on Rouelle.

[14] The royal letter of appointment, dated June 9, 1743, is reproduced by Dorveaux, p. 172. This letter was a confirmation of the appointment already made by the intendant of the Jardin, Buffon, and presumably could have been issued at any time after Rouelle took over his duties at the Jardin. For Rouelle's presence at the Jardin as early as March, 1743, see Jean-Jacques Rousseau, "Les Institutions chymiques de Jean-Jacques Rousseau," *Annales de la Société Jean-Jacques Rousseau*, Vols. **12–13** (1918–21), **13,** 134, n. 1.

[15] Cf. Contant, pp. 42–45, for discussions of the two chairs in chemistry at the Jardin. After Rouelle's retirement in 1768, the title *Professeur en Chimie* was never again applied to a demonstrator; Hilaire-Marin Rouelle, who succeeded his brother, only received the title of demonstrator.

[16] *Ibid.*

false, as I will prove to you."[17] Too often interpreted as evidence of Rouelle's ill-temper and lack of respect for his colleagues and superiors, this anecdote is perhaps better understood by reference to Rouelle's appointment as "Professeur" as well as "Démonstrateur." As a professor of chemistry, Rouelle had the authority—and, incidentally, the knowledge and inclination—to oppose his views to those of Bourdelin.[18]

During his long tenure at the Jardin du Roi—he retired in 1768—Rouelle continued to give private instruction in chemistry and pharmacy in the place Maubert, moving his laboratory and courses to the rue Jacob in the Saint-Germain quarter in 1746. As A.-L. de Jussieu observed some years later, Rouelle's private courses had earned for him the reputation that secured his appointment to the Jardin. Thereafter, both the private and public courses, which continued to attract growing audiences, served as Rouelle's major vehicle for the dissemination of his knowledge and theories. Since these lectures contain the bulk of Rouelle's scientific work, it will be of value to discuss here in some detail the circumstances surrounding the lectures: when, where, and how frequently they met, and the nature of his audience.

A few details about Rouelle's private courses in chemistry (*cours particuliers*) have been supplied by Rouelle himself in an announcement made at the opening of the academic year, and by Rouelle *le cadet*, who supplanted his brother at the rue Jacob as well as at the Jardin du roi. The elder Rouelle's advertisement simply states that his course will begin "le lundi 26 novembre 1764, à trois heures après midi, en sa maison, rue Jacob, au coin de la rue des Deux-Anges, fauxbourg S. Germain."[19] Rouelle *le cadet* offers somewhat greater detail in an announcement which appeared in July, 1772, to herald the opening of his own

. . . private course which he will begin in the month of December. He asks the people who want to take it to enroll with him. He will give these lessons from eleven o'clock in the morning until one o'clock in the afternoon, on Monday, Wednesday, Friday and Saturday, at least if the people who want to take it do not prefer that he choose another hour and other days.[20]

[17] Grimm, Vol. 9, p. 109.

[18] Bourdelin has been described as a rather mild-mannered individual who realized that he was out of touch with the new doctrines in chemistry and, therefore, made no objection to being overruled. See Contant, pp. 62–63, and Condorcet's "Eloge de M. Bourdelin," *Histoire de l'Académie royale des sciences* (1777), p. 120. During the latter half of Rouelle's tenure at the Jardin, Bourdelin often allowed P.-J. Malouin to lecture for him. Lemay ("Les cours de Guillaume-François Rouelle," p. 434) reports that Bourdelin was replaced by both Malouin and Macquer. For Rouelle's attacks against Malouin and Macquer, see Grimm, Vol. 9, p. 106, and below, p. 90.

[19] *Journal de médecine, chirurgie, pharmacie* (1764), pp. 478–479. See also, Henry Guerlac, "Lavoisier and his Biographers," *Isis*, **45**, 51–62 (1954); esp. p. 59, n. 28.

[20] Contant, p. 117.

The announcement states that enrollment in the private courses was required, a procedure probably instituted to insure the collection of fees by the professor. This practice suggests that among the as yet undiscovered Rouelle documents there may survive a record of all or some of the persons privately instructed by the brothers.

Rouelle *le cadet* has also described the teaching techniques enployed by the elder Rouelle in his laboratory.

From the first private courses that my late brother gave in Paris, he felt the need of presenting to his hearers a short but adequate description of all the operations which formed the links of his demonstrations, and served as the foundation of his doctrine.

He believed that there was nothing more appropriate to fulfill this object, than to exhibit in glass bottles and jars the different and separated products of each operation, while the sign which he glued on the vessels contained very simple, very brief, but very clear information about each of his experiments, and the results obtained from them; and he gave them the name of Processes of the Course in Chemistry. Since these Processes followed step by step the order and progression of his lessons, he put them, as they increased in number, on shelves in his laboratory, to remain there during the entire Course. Experience made him aware of how useful this exhibition was for his hearers. They found there a very short, but adequate repetition of what they had already seen; so that the operation was repeated again before them, so to speak, a second, a third time, in a word, as often as they wished.[21]

Not only did Rouelle employ this useful pedagogic device, but he also posted in large letters in a conspicuous place in the laboratory the Peripatetic motto adopted as his own: *Nihil est in intellectu quod non prius fuerit in sensu.*[22] This motto, by which most of his work can be characterized, evidently was intended to impress upon his students that the only sound way to learn chemistry was through a strictly empirical approach.

In contrast to the private courses, all lectures at the Jardin du roi were free and open to the public.[23] Unfortunately for the historian, there was no need to enroll in these courses; prospective students did not even have to obtain permission to attend the lectures, but could come and go as they pleased. There is, therefore, no record of those who attended Rouelle's public lectures.

Throughout most of the eighteenth century, the Jardin had one amphitheatre which held about 600 students. In the earlier years the

[21] Quoted from H.-M. Rouelle's "Tableau de l'Analyse chimique" by Contant, pp. 114–115.

[22] A.-L. Lavoisier, "Oeuvres de Lavoisier," 6 Vols., Paris, 1864–93, Vol. 1, p. 246. Maurice Daumas, "Lavoisier," n.p., [1941], p. 19, says the motto was displayed in the Jardin's amphitheatre; this is unlikely since Rouelle was not the only lecturer to use the amphitheatre.

[23] Contant, pp. 22, 25–27.

amphitheatre was reserved for the teaching of anatomy and surgery in the winter, botany and chemistry in the summer. Rouelle thus began his course in June and probably completed the series of lectures and demonstrations before opening his private course in November or December.[24] No fixed length was prescribed for courses taught at the Jardin, the number of lectures varying according to the demands of the subject and the wishes of the professor.[25]

The audiences at the Jardin du roi must have presented a heterogeneous appearance. The amphitheatre was probably filled by aspiring young chemists and students in closely related fields, as well as by the curious Parisians who were attracted by Rouelle's colorful personality, his penchant for anecdotes, and his habitual disrespect for many prominent men of science and letters. There is no way of learning the identities of most of Rouelle's auditors, but among those who have left some record of their attendance at the public and private courses are many of the famous names in eighteenth-century science, philosophy, and letters. Besides Lavoisier, Jean Darcet, and Rouelle *le cadet*, the chemists and physicians taught by Rouelle include Théodore Baron, Pierre Bayen, A.-L. Brongniart, J.-B. Bucquet, the marquis de Courtanvaux, J.-F. DeMachy, the comte de Lauraguais, the Belgian Limbourg, P.-J. Macquer, the apothecary Sébastien Mercier, Jacques Montet, Augustin Roux, B.-G. Sage, the Swiss P.-F. Tingry, G.-F. Venel, and the Englishman Peter Woulfe. Other known disciples and visitors to Rouelle's lectures include the botanist A.-L. de Jussieu, the mineralogists Nicolas Gobet and Antoine Monnet, the geologists Nicholas Desmarest and H.-B. de Saussure, and three outstanding nonscientists: Diderot, Rousseau, and Turgot.[26]

The relations between Rouelle and many of his pupils do not permit or deserve detailed treatment; indeed, about some of his pupils little more can be said than that they learned from him the rudiments of chemistry. Among the less well-known disciples, Limbourg, for example, is known to have constructed a table of chemical affinities similar to

[24] The private course was a more elaborate affair lasting seven or eight months, or from mid-November until about the end of June. See Diderot, "Œuvres," Vol. 6, p. 409. The "Avant-Coureur," 7 July 1760, pp. 391–392, reports the opening of Rouelle's course at the Jardin on June 30; I am indebted to Mr. William Smeaton, University College, London, for bringing this reference to my attention.

[25] Contant, p. 27, cites a regulation in effect at the Jardin after 1731 which required the professor of chemistry to give approximately sixty-five lectures and to terminate his course by November.

[26] The reader is referred to the appropriate national biographies. Parmentier in his eulogy for Bayen implies other famous names among Rouelle's pupils when he notes: "[Bayen] revoyoit le laboratoire de *Rouelle*, dont il ne pouvoit s'approcher sans un souvenir attendrissant pour son illustre maître, et sans se rappeler ces conférences instructives avec les *Jussieu*, les *Malesherbes*, les *Turgot*, les *Dolbach* [sic] . . ." See Pierre Bayen, "Opuscules chimiques de Pierre Bayen," 2 Vols., Paris, [1798], Vol. 1, p. xlvii.

Rouelle's; Mercier apparently remained an apothecary and an intelligent reader of chemical literature; Sage opened his own course in chemistry, with the crystallographer Romé de l'Isle among his pupils, and remained a die-hard chemist of the old school years after Lavoisier's theories had gained wide acceptance.[27]

At least three of the chemists taught by Rouelle deserve some discussion: Lavoisier, Macquer, and Venel.

Lavoisier attended Rouelle's private course in the rue Jacob during the year 1762–63 or 1763–64. Although he produces no evidence in support of this statement, Lavoisier's biographer, Edouard Grimaux, claims that Lavoisier possessed a copy of Rouelle's lectures, used the manuscript in conjunction with Rouelle's private lectures, and filled the margins of his copy with annotations.[28] If such a manuscript exists, its discovery might well show that Lavoisier's interest in chemistry—which only becomes evident in his work of 1770—was inspired by Rouelle. At present, however, as Henry Guerlac has recently shown, Lavoisier's activities in the late 1760's and the contexts of his references to Rouelle in his writings strongly suggest that Lavoisier sought and obtained from Rouelle's lectures instruction in the elements of geology and mineralogy, rather than chemistry.[29]

One of the earliest active disciples of Rouelle, Pierre-Joseph Macquer (1718–84) probably attended Rouelle's lectures before 1742.[30] The publication of Macquer's *Elemens de chymie théorique* (1749)—a popular textbook which went through several editions and translations—was in all likelihood the most important single event in the spread of Rouelle's chemical theories. While Rouelle was not acknowledged as the source of Macquer's theories, the debt to Rouelle is so marked that Macquer's textbook and his own private courses in chemistry can be considered two of the principal channels through which Rouelle's teachings reached a still wider audience.[31]

---

[27] Limbourg's treatise: "Dissertation de Jean Philippe de Limbourg sur les affinités chimiques, qui a remporté le prix de physique de l'an 1758 . . . de l'Académie de Rouen," Liège, 1761. Sage's textbook: "Analyse chimique et concordance des trois règnes," 3 Vols., Paris, 1786.

[28] Edouard Grimaux, "Lavoisier 1743–1794," Paris, 1888, p. 5. Denis I. Duveen, who possesses a manuscript copy of the catalogue of Lavoisier's library, informs me that no such item is listed, but that Lavoisier may have owned one nevertheless; the catalogue does record a copy of H.-M. Rouelle's "Tableau de l'Analyse chimique."

[29] Lavoisier's attendance at Rouelle's lectures is treated fully by Henry Guerlac, "A Note on Lavoisier's Scientific Education," *Isis*, **47**, 211–216 (1956). After completing his work with Rouelle, Lavoisier collaborated with Guettard in gathering material for the projected "Atlas minéralogique de la France."

[30] Macquer, "Dictionnaire de chimie," art. "Phosphore d'Angleterre ou de Kunckel," Vol. 2, p. 214.

[31] Although I have not consulted any Rouelle manuscripts dated pre-1749, at least one such does exist. (See Rousseau, *Annales*, Vol. 13, p. 134, n. 1.) The existence of an

Independently of Macquer, Gabriel-François Venel (1723–1775) worked to spread the chemical doctrines of Rouelle.[32] Also one of Rouelle's earliest pupils—he first heard Rouelle in 1746[33]—Venel had two means of disseminating his teacher's ideas: his private courses in chemistry at Montpellier and his numerous articles on chemistry for the *Encyclopédie* of Diderot and d'Alembert. Little is known about Venel's course in chemistry other than that his assistant and demonstrator was Jacques Montet, another Rouelle pupil.[34] For the *Encyclopédie*, Venel's more important writings include the articles "calcination," "chymie," "combustion," "eau," "feu," "menstrue," "mixte," and "principes."

Between them, Macquer and Venel represent two currents of chemical thought, both of them based on Rouelle's teachings. Macquer's work can appropriately be labeled *la saine chimie*; his textbook displays balanced judgment and great care in the use of experiment as a base for theory.[35] Venel, on the other hand, added a neo-Paracelsian flavor to the fundamentally unchanged doctrines of Rouelle. A recent summary of Venel's position accurately states that Venel

. . . invokes a new Paracelsus, who will make of chemistry the science that *understands* nature and displaces geometry from that pretension. He will be gifted, this Paracelsus, with the sheerly technical insight to penetrate beyond physics, but he will have a spirit and imagination like that of the pre-Newtonian philosophers.[36]

---

early manuscript and Jussieu's remarks about Rouelle's "new chemical doctrine" suggest that in all likelihood Rouelle was the innovator and Macquer the imitator, rather than the reverse. Priority has been claimed for Macquer by L. J. M. Coleby, "The Chemical Studies of P. J. Macquer," London, [1938], p. 14.

Douglas McKie's recent article, "Macquer, the first lexicographer of chemistry," *Endeavour*, **16**, 133–136 (1957), is a useful summary of Macquer's achievements; McKie discusses the popularity of Macquer's textbook and its use in Joseph Black's courses at the University of Edinburgh. Black's practice had been initiated by his own teacher, William Cullen.

[32] E.-H. de Ratte, "Eloge de Monsieur Venel, Prononcé à l'Académie de Montpellier," *Observations sur la physique, sur l'histoire naturelle et sur les arts*, **10**, 3–14 (1777).

[33] Venel, "Mémoire sur l'analyse des eaux de Selters ou de Seltz. Seconde Partie," *Mémoires de mathématiques et de physique, Présentés à l'Académie Royale des Sciences, par divers Sçavans, & lûs dans ses Assemblées*, **2**, 80–112 (1755); esp. p. 109.

[34] Eulogy by Poitevin in R.-N.-D. des Genettes (ed.), "Eloges des académiciens de Montpellier," Paris, 1811, pp. 242–249. For Venel's teachings at the University of Montpellier, see J.-A. Chaptal, "Mes souvenirs sur Napoléon," Paris, 1893, p. 15.

[35] See, for example, Macquer's cautious rejection of all theories explaining the increased weight of a metallic calx on the grounds that they have not been sufficiently well established and are unsatisfying; P.-J. Macquer, "Elemens de chymie-pratique," 2 Vols., Paris, 1751, Vol. 1, pp. 307–308.

[36] Charles C. Gillispie, "The *Encyclopédie* and the Jacobin Philosophy of Science: A Study in Ideas and Consequences," in *Critical Problems in the History of Science*, ed. Marshall Clagett, Madison, Wis., 1959, p. 258.

Fortunately, Venel's desire to reduce chemistry once more to a probing after the hidden secrets and qualities of nature did not bear fruit, and it was the Rouelle-Macquer tradition that became dominant.

Among the nonscientists at Rouelle's lectures were Diderot, Rousseau, and Turgot. The only one of the three whose chemical writings have been examined at all is Rousseau, who at one time in his career had scientific ambitions. In 1744 he attended Rouelle's lectures and in 1747 started what was to be his own chemical opus, *Les Institutions chymiques*.[37] The manuscript, never completed and published only in this century, contains nothing of any originality; most passages have in fact been traced to the writings of Boerhaave, to the Stahlian chemists whose works were available to Rousseau, and to a manuscript of Rouelle's lectures owned by Rousseau.[38]

Turgot's interest in chemistry has been ignored by political and economic historians and remains largely unknown to historians of science. In his correspondence with Condorcet, Turgot displays considerable chemical knowledge and insight. Although he probably had little experience in the laboratory, he suggested original experiments and was able to interpret data to formulate intelligent hypotheses.[39]

While Diderot's biological writings have been studied fairly carefully, no attempt has yet been made to collect and examine the references to chemistry which are scattered through many of his works. He wrote no chemical treatise, as Rousseau did, but he attended Rouelle's lectures for three consecutive years (1754–57), edited and reworked his notes, and used his knowledge of chemistry to confirm—if it did not help him to formulate —certain of his philosophical ideas.[40] His respect for Rouelle's chemistry is evident in his "Plan d'une Université pour le gouvernement de Russie" (1775–76), where he recommends the notebooks of Rouelle, "reviewed,

[37] For Rousseau's studies in chemistry, see Théophile Dufour, "Les Institutions chimiques de Jean-Jacques Rousseau," Geneva, 1905.

[38] Rousseau, *Annales*, **13**, p. 134, n. 1; passages from Rouelle manuscript, Vol. 13, pp. 136, 139.

[39] In Turgot's article "Expansibilité" in the *Encyclopédie* (volume published in 1756), he mentions Rouelle as if he is currently attending or has recently attended Rouelle's lectures. See A.-R.-J. Turgot, "Oeuvres de Turgot et documents le concernant," ed. Gustave Schelle, 5 Vols., Paris, 1913–23, Vol. 1, pp. 571n, 572, 573. Turgot's ability is evident in this article and in several letters to Condorcet; cf. M.-J.-A.-N.-C. Condorcet and A.-R.-J. Turgot, "Correspondance inédite de Condorcet et de Turgot, 1770–1779," ed. Charles Henry, Paris, 1883, pp. 59–63, 109–116.

[40] D. Fernand Paitre, "Diderot biologiste," thesis, Lyon, 1904. Diderot's use of chemistry cannot be fully documented here, but see "Oeuvres," Vol. 2, p. 64, for Naigeon's introduction to Diderot's work on matter and motion; also, see below, p. 87. Other sources of Diderot's materialism are treated by Aram Vartanian, "Trembley's Polyp, LaMettrie, and Eighteenth-century French Materialism," *Journal of the History of Ideas*, **11**, 259–286 (1950); and Marx Wartofsky, "Diderot and the Development of Materialist Monism," in *Diderot Studies II*, ed. Fellows and Torrey, Syracuse, [1952].

corrected and augmented by his brother and doctor Darcet," as the best possible textbook to use in teaching a course in chemistry.[41]

Outside the lecture hall, Diderot and Rouelle became friends, both attending the gatherings of the *salon d'Holbach*, where science and philosophy were popular topics of conversation.[42]

The contacts between Rouelle and d'Holbach introduce a very important area of Rouelle's activities and influence: his interest in having French translations made of German and Swedish treatises on chemistry and mineralogy. D'Holbach, who produced a considerable number of such translations in the course of about twenty years, was eminently qualified for work of this kind. He had received some scientific training at Leyden and was a native German who spent most of his adult life in France. That d'Holbach was acquainted with Rouelle's scientific work is clear from his frequent citations of Rouelle in his numerous articles for the *Encyclopédie*.[43] Despite his having studied at Leyden, the center of Boerhaave's influence, d'Holbach favored the rival Stahlian doctrines taught by Rouelle.[44]

This combination of Rouelle's interests and d'Holbach's capabilities produced one of the first of d'Holbach's translations, Wallerius' *Mineralogie* (1753), which contains an acknowledgment of Rouelle's aid in reading and commenting upon the manuscript of the translation.[45] The association of Rouelle and d'Holbach probably continued after 1753 and it is possible that Rouelle served as technical advisor for other such translations.

Additional examples of Rouelle's rôle in promoting editions and translations of scientific works are to be found in the activities of his pupils

[41] Diderot, "Œuvres," Vol. 3, pp. 463–464.

[42] André Morellet, "Mémoires inédits de l'abbé Morellet," 2nd ed., 2 Vols., Paris, 1822, Vol 1, pp. 118, 132–134. For this famous gathering of French intellectuals, see also Charles Avezac-Lavigne, "Diderot et la société du baron d'Holbach," Paris, 1875, and René Hubert, "D'Holbach et ses amis," Paris, [1928].

[43] D'Holbach's articles are listed by Pierre Naville, "Paul Thiry d'Holbach et la philosophie scientifique au XVIIIe siècle," 5th ed. [Paris], [1943], p. 407. For references to Rouelle, see esp. arts. cuivre, étain, fer, gypse, mer, mercure, métal, or, orpiment, plomb, régule d'antimoine, safre, sel, and soufre, in Denis Diderot and Jean le Rond d'Alembert, "Encyclopédie, ou Dictionnaire raisonné des sciences, des arts et des métiers," Paris, 1751–65.

[44] Naville, pp. 181–200, treats d'Holbach's relations with German science. Asked in 1761 which untranslated works he would suggest having translated into French, d'Holbach replied with works of Kunckel, Stahl, Becher, and Lehmann; the first three were in the same chemical tradition as Rouelle, and Lehmann's geological work was used by Rouelle. This list is reproduced by Max Pearson Cushing, "Baron d'Holbach: A Study of Eighteenth Century Radicalism in France," Columbia U. Diss., New York, 1914, p. 31.

[45] J. G. Wallerius, "Mineralogie, ou description générale des substances du regne mineral," trans. d'Holbach, 2 Vols., Paris, 1753, Vol 1, p. viii; the aid of both Rouelle and Bernard de Jussieu is acknowledged by d'Holbach. A list of d'Holbach's translations is given by Naville, p. 407ff.

I

and friends. The Abbé Nicolas Lenglet-Dufresnoy translated from the Spanish Alonzo Barba's *Métallurgie* (1751) at Rouelle's request. Admittedly inspired by the lectures of Rouelle, the pharmacist P.-F. Dreux decided to try to reconcile the theories of his teacher with those of the German chemist Johann Friedrich Meyer; Dreux produced a translation of Meyer's *Essais de Chymie* (1766), a work which soon attracted a school of supporters and aroused the admiration of Lavoisier. Two members of the *côterie holbachique* contributed to the wave of translations, the Abbé Morellet attempting to translate Stahl's *Zymotechnia* and Augustin Roux editing the newly translated works of Henckel. To combat the popularity of Boerhaave's *Elementa Chemiae*, which appeared in French in 1754, J.-F. DeMachy, one of Rouelle's pupils, translated Juncker's *Conspectus chemiae* (1730), a work of the rival Stahlian school preferred by Rouelle; this work appeared in 1757 under the title *Eléments de chimie suivant les principes de Beccher et de Stahl*. In 1756 Théodore Baron, another Rouelle pupil, produced a new edition of the long-popular *Cours de chymie* (1675) of Nicolas Lémery.[46] Of these seven works just listed, only three translations—those of d'Holbach, Lenglet, and Dreux—can be traced directly to Rouelle's advice or inspiration. The remaining four, however, were products of Rouelle's associates and show, furthermore, preference for the school of chemistry advocated by Rouelle; these circumstances suggest that Rouelle played some part in having works of this kind edited and translated.

As I have suggested, Rouelle's lectures were instrumental both in disseminating his ideas among a wide audience that included many capable students and in securing for him several academic distinctions. The first of these honors was his appointment to a chair at the Jardin du roi in 1742. Soon afterward, in December of 1743, Rouelle read a paper on neutral salts before the Paris Academy of Sciences. This paper was given a favorable report by two Academicians, Bourdelin and Hellot, and in the following year Rouelle was elected to the Academy as *adjoint* in the class of chemistry. His paper appeared in the *Mémoires* of the Academy in 1744 and was followed by four more memoirs, which form the major part of Rouelle's publications.[47]

[46] Baron's notes frequently refer to Rouelle's work on salts and to the discoveries of Stephen Hales (see below, p. 94); Nicolas Lémery, "Cours de chymie," ed. T. Baron, Paris, 1756, pp. 81n(a), 115n(a), and 17n(c). Lenglet's work and Rouelle's rôle in having the translation made are mentioned in the Bordeaux MS, p. 1178. For DeMachy, see Hélène Metzger, "Newton, Stahl, Boerhaave et la doctrine chimique," Paris, 1930, p. 192. For Morellet and Stahl, see Morellet, Vol. 1, p. 96. For Dreux and Rouelle, see J. F. Meyer, "Essais de Chymie," trans. P. F. Dreux, 2 Vols., Paris, 1766, Vol. 1, pp. xvii–xviii. For the rôle of Roux as an aide to d'Holbach, see Naville, p. 431, n. 16.

In *Chymia*, 5, 73–112 (1959), "Some French Antecedents of the Chemical Revolution," Henry Guerlac treats Rouelle and the translators.

[47] A list of all publications attributed to Rouelle is appended to this study.

Of Rouelle's five published memoirs, three deserve the particular attention of the historian of science since they deal with the classification of salts—long a complex issue in the history of chemistry. The earliest of these memoirs (1744) appeared at a time when there existed at least three schools of thought about the nature of the ill-defined and unclassified substances known as salts. Despite the attacks of Robert Boyle, the notion of salt as one of the three spagyric principles had not completely vanished from chemistry. A second theory, associated with Stahl and his disciples, held salt to be compounded of water and one or more of three kinds of earth principles. Still other chemists spoke of the shape and motion of constituent particles, which determined the relative acidity or alkalinity of salts. To complete the picture of confusion, the term "salt" was frequently used loosely in reference to a variety of chemical substances.[48]

In his paper of 1744, "Mémoire sur les Sels neutres," Rouelle set up a classification of neutral salts using crystal form and chemical composition as bases for his system. He first distinguished six "sections" in the family of neutral salts, each section defined by crystal form. Each such division was then further subdivided into "genera" according to the acid component of the salt, and into "species" according to the base involved.[49] This scheme was based upon careful study of the phenomena of crystallization, reported in his memoir of 1745 ("Sur le sel marin"), and was later extended by equally painstaking experiments with acids and alkalis, published in 1754 (second "Mémoire sur les Sels neutres").[50] In this latter study Rouelle stated the definition of neutral salts presupposed in 1744: "[a] neutral salt [is] a salt formed by the union of an acid with any substance, which serves it as a base and gives it a concrete or solid form."[51] He then proceeded to distinguish between neutral salts with an "excès d'acide" and those with "très-peu d'acide," testing these compounds with chemical indicators, and thereby completing his answer to the controversial question of the nature of acid, alkali, and salt. Although Rouelle's work was not at once universally accepted, his classification soon became firmly established in chemical literature.[52]

[48] Marie Boas, "Acid and Alkali in Seventeenth Century Chemistry," *Archives Internationales d'histoire des sciences*, No. 34, 13–28 (Jan.–Mar. 1956). Despite its title, this article includes some discussion of Rouelle's published memoirs.

[49] G.-F. Rouelle, "Mémoire sur les Sels neutres, dans lequel on propose une division méthodique de ces Sels, qui facilite les moyens pour parvenir à la théorie de leur crystallisation," *Mémoires de l'Académie royale des sciences*, pp. 353–364 (1744); esp. pp. 359–363 and plate.

[50] G.-F. Rouelle, "Sur le Sel marin (Première partie.) De la crystallisation du Sel marin," *Mémoires de l'Académie royale des sciences*, pp. 57–79 (1745), and "Mémoire sur les Sels neutres, dans lequel on fait connoître deux nouvelles classes de Sels neutres, & l'on développe le phénomène singulier de l'excès d'acide dans ces sels," *Mémoires de l'Académie royale des sciences*, pp. 572–588 (1754).

[51] *Ibid.*, pp. 573–574.

[52] Cf. Macquer's citations of Rouelle in his "Elemens de chymie-pratique," Vol. 2,

In 1750 Rouelle the teacher and Academician was elected to the *Compagnie des apothicaires de Paris,* which enabled him to add an apothecary shop to the laboratory in the rue Jacob.[53] In the same year he became a member of the academies of Stockholm and Erfurt, and in 1752 was elevated to the rank of *associé* in the Academy of Sciences.

During the 1740's and 1750's Rouelle's scientific talents were sought after by the French government. When the position of apothecary to the king became vacant upon the death of Boulduc in 1742, it was offered to Rouelle, who declined this lucrative appointment.[54] He later accepted the position of *inspecteur de la pharmacie* at the Hôtel-Dieu, because these duties did not interfere with his other activities.[55] When, in 1744, a cattle plague was spreading havoc in France, Rouelle was appointed to a committee of scientists asked to study the disease and find its cure.[56] In 1753 he was charged by the minister of war with investigating a new method of making and refining saltpeter, and in the following year he and his brother were asked by the minister of finance to examine the alloys of gold used in minting coins.[57]

Rouelle's activities during the last fifteen years of his life remain obscure. Although he continued to give his lectures, he had become ill while working with saltpeter and perhaps could do little more than teach. At the Jardin he was assisted in his demonstrations by his brother and their nephew, and in the last two years of his appointment was often replaced by his brother.[58] Because of poor health he refrained from

p. 31, and in "Elements of the Theory and Practice of Chymistry," tr. Andrew Reid, 2 Vols., London, 1758, Vol. 1, p. 22n; also Baron in Lémery, p. 81n(a). Baumé's criticisms of Rouelle's work on salts are detailed by Macquer, "Dictionnaire de chymie," 2nd ed., 4 Vols. (in-8°), Paris, 1778, art. "Sel," Vol. 3, pp. 442–444.

[53] His seven years with Spitzley had not entitled him to an apothecary's license, for which four years *apprentissage* and six years *compagnonnage* were required. See Maurice Bouvet, "Histoire de la pharmacie en France des origines à nos jours," Paris, [1937], p. 79.

[54] *Ibid.,* p. 327 and Dorveaux, p. 179. He reputedly declined because it would have meant giving up his teaching to live at Court. In view of the fact that Boulduc held this position while teaching at the Jardin, it seems more likely that Rouelle simply could not accept the appointment in 1742 because he only became a licensed apothecary in 1750.

[55] Dorveaux, p. 179, dates this appointment 1753; Contant, p. 104, uses 1755–56.

[56] Mauguin, "Etudes historiques sur l'administration de l'agriculture en France," 3 Vols., Paris, 1876, Vol. 1, p. 261.

[57] H.-M. Rouelle in Lavoisier, "Œuvres," Vol. 1, p. 542, n. 1, and Contant, p. 104. Dorveaux, p. 180, dates these appointments 1754 and 1755, respectively.

[58] Grimm, Vol. 9, p. 107, Dorveauz, p. 181, and Contant, p. 114. The nephew, Jean or John Rouelle, emigrated to America in 1788 to take his place as the only appointed professor for the projected, but never realized, Académie des sciences et beaux-arts des Etats-unis de l'Amérique, Richmond, Virginia. Cf. Henry E. Sigerist, "The Rise and Fall of the American Spa," *Ciba Symposia,* **8,** 313–326 (1946).
Rouelle is said to have owned a *glacerie* near Langres; see G.-F. Venel, "Instructions

applying to the Academy of Sciences for the post of *pensionnaire*, left vacant in 1766 by the death of Hellot. For the same reason, he retired from the Jardin in 1768 and, at his request, was succeeded by his brother, Rouelle *le cadet*. Moving from Paris to Passy for his health, he died there in 1770.

Little is known about Rouelle's personal life. He was married to Anne Mondon, who assisted Rouelle *le cadet* in maintaining the apothecary shop after her husband's death. In 1771 their daughter married Jean Darcet, a noted chemist and one of Rouelle's pupils.[59]

In contrast to the few facts known about Rouelle's life, a rather clear picture of his personality has been recorded by his pupils and other contemporaries. The picture is of a man with pronounced, even violent, opinions about everything concerned with science, but comparatively mild and even-tempered in all other respects. Diderot, who knew him from his lectures and at the gatherings of the *salon d'Holbach*,[60] observed that Rouelle could not bear to watch suffering and, therefore, abandoned medicine and surgery for chemistry and pharmacy. He was a would-be poet, musician, philosopher, politician, and theologian. His scientific work was occasionally invaded by his notable piety, as when Diderot remarks:

One thing in the Bible appeared to him to be difficult to believe, that is where Noah had gotten the bitumens with which he had supplied the Ark, since it had been demonstrated to him that the formation of bitumens post-dated the Flood.[61]

Although a dynamic lecturer, Rouelle was often extremely absent-minded and easily distracted. His lectures were inspiring, but disorganized and sometimes difficult to follow.

He began one subject, but he was soon distracted by a multitude of ideas which presented themselves. . . . He applied his experiments to a general system of the world; he encompassed the phenomena of nature and the works of the arts; he connected them by the acutest analogies; he got lost, one got lost with him, and one never returned to the particular object of the day's demonstration, without being astonished at the immense space that had been surveyed.[62]

Such digressions were nearly disastrous at times; an experiment which had to be watched carefully would be neglected, and a sudden explosion

---

sur l'usage de la houille," Avignon, 1775, p. 534. The period of his ownership is not specified, but this may have been one of the activities occupying the latter part of Rouelle's life.

[59] Contant, p. 114. The Rouelles had twelve children in all. Cf. A.-L. de Jussieu, "Sixième notice historique sur le Muséum: VI. Depuis 1760 jusqu'en 1788," *Annales du Muséum National d'Histoire Naturelle*, **11**, 1–41 (1808); esp. p. 6, n. 1.

[60] Morellet, Vol. 1, p. 132.

[61] Diderot, "Œuvres," Vol. 6, p. 408.

[62] *Ibid.*, p. 407.

85   G.-F. ROUELLE: 18TH-CENTURY CHEMIST AND TEACHER

would send lecturer and audience scurrying for shelter. Rouelle's habitual inattention doubtless did nothing to improve his lack of dexterity as a demonstrator; Diderot calls him a "manipulateur distrait et maladroit" whose younger brother excelled him by far in the demonstration of experiments.[63]

Rouelle's mildness in nonscientific matters stands in sharp contrast to the lack of patience and the violent temper of Rouelle the scientist. Grandjean de Fouchy observed in his *éloge* of Rouelle that "he was naturally affable, but it was not advisable to contradict him in chemistry; the least blunder of this kind, provoked him more than an insult."[64] He was obsessed with the idea that he was being plagiarized, and indiscriminately termed his contemporaries *plagiaires*; yet he always absentmindedly revealed his scientific *arcanes* to his students, protesting all the while that he would never disclose them to anyone.[65]

In his lectures Rouelle attacked his predecessors, contemporaries, and former students, criticizing Buffon for his *beau parlage*, the physicians for their ignorance and incompetence, and many of his predecessors for their errors or lack of originality.[66] Although his vitriolic and tactless outbursts must have antagonized many contemporaries, the very targets of his attacks could not help but concede Rouelle's considerable merit. There is little doubt that the chemist P.-J. Macquer was referring to Rouelle when he complained of

. . . the expressions [which] escaped a celebrated man in the heat of discourse, and by which he imperilled his reputation, at the same time that he was establishing it by the real services he otherwise rendered to chemistry.[67]

II

The chief sources of our knowledge of Rouelle's teachings are the manuscripts of lecture notes taken down by his pupils and preserved in numerous copies in European libraries and in private collections.[68]

---

[63] *Ibid.*, pp. 408, 410.

[64] Grandjean de Fouchy, p. 147.

[65] Grimm, Vol. 9, p. 107. Cf. Bordeaux MS, pp. 605, 725, 840, and 1000 for examples of Rouelle's *arcanes*.

[66] For his attacks on Buffon, see Grimm, Vol. 9, p. 108; for the physicians, Pierre Lemay, "Le Cours de Pharmacie de Rouelle," *Revue d'Histoire de la Pharmacie*, No. 152, 17–21 (mars 1957). A compliment from Rouelle was evidently highly prized; Condorcet remarks in his eulogy of the famous physician, Théodore Tronchin, "M. Rouelle a souvent répété qu'aucun médecin ne prescrivait de meilleurs formules, et un tel suffrage nous dispense de tout éloge." M.-J.-A.-N.-C. Condorcet, "Oeuvres," ed. A. C. O'Connor and F. Arago, 12 Vols., Paris, 1847–1849, Vol. 2, p. 506.

[67] Macquer, "Dictionnaire de chimie," Vol. 1, p. vii.

[68] A checklist of all the Rouelle manuscripts I have been able to locate may be consulted in my thesis, "G.-F. Rouelle: His *Cours de Chimie* and Their Significance for Eighteenth Century Chemistry," Cornell University, 1958, Appendix A.

While the existing lectures on pharmacy are evidently products of the private courses held at Rouelle's laboratory and apothecary shop, manuscripts of the chemical lectures may represent either the private teaching or the public courses he delivered at the Jardin du roi.

In my investigations of Rouelle's chemical lectures, I have consulted seven manuscripts that date from the period 1751–60/61 and therefore span only the middle portion of Rouelle's teaching career. That earlier manuscripts or others dating from his last years of lecturing would prove of considerable interest and value is suggested by Rouelle's contemporaries in their casual references to the content of his courses. A.-L. de Jussieu, for example, mentions "the new chemical doctrine" taught by Rouelle before 1742.[69] Jean Darcet, Rouelle's son-in-law, summarizes briefly the geological ideas held (and taught) by Rouelle as early as 1740; that these geological theories were radically revised in about 1763 is reported by Nicholas Desmarest.[70]

My investigation has been further circumscribed by the condition of the texts. Two of the seven manuscripts—including the earliest, which dates from 1751—are fragmentary and are thus of small value.[71] A third text, while complete, was available to me only in part.[72] The four remaining manuscripts all give evidence of having undergone later editing, in the course of which Rouelle's lectures have undoubtedly suffered some distortion. These four documents all belong to a single family of manuscripts and are derived from a lost "ancestral" manuscript dating from the period 1754–57.[73]

At least four of these Rouelle manuscripts probably had their origin in the notes taken by Denis Diderot during his attendance at Rouelle's lectures over a three-year period, 1754–57. These notes may no longer be extant—at least, Diderot scholars have been unable to determine their present whereabouts. Unfortunately, a manuscript bearing the

---

[69] Above, p. 73.

[70] Darcet, pp. 32–33. Nicholas Desmarest's article "Rouelle. Notice sur sa doctrine relative à plusieurs points importans de l'histoire naturelle de la terre," *Encyclopédie méthodique*, tome I of Géographie Physique, Paris, 1794, pp. 409–431.

[71] "Cours de Chimie, ou Leçons de Mr Rouelle, Recueillies pendant les années 1754, et 1755; Rédigées en 1756; revues et corrigées en 1757 et 1758," Bibliothèque nationale, n.a.fr. 4043–4044; and "Cours de Chymie par M. Rouelle. 1751," Bibliothèque nationale, n.a.fr. 4045.

[72] "Cours de chimie, ou leçons de M. Roüelle, démonstrateur au Jardin du Roy, récueillies en 1754 et 1755, . . . corrigées en 1757 et 1758," Bibliothèque nationale, fr. 12303–12304.

[73] Bordeaux, "Cours de chymie"; "Cours de Chimie, ou Leçons de Monsieur Rouelle, Recueillies pendant les années 1754, 1755 et rédigées en 1756. Revuës, et corrigées, en 1757, et 1758," Science Library, Clifton College, Bristol; "Cours de Chimie," Bibliothèque de Nancy, MS 265 (307); and "Traité de Chymie de Rouelle," New York City, private collection of Denis I. Duveen. A complete collation and detailed discussion of all seven manuscripts is included in my thesis, "G.-F. Rouelle," Appendices B and C.

notation "Copié sur l'original écrit de la propre main de Diderot" has also dropped from sight.[74]

Among the manuscripts I was able to consult, the Bordeaux "Cours de chymie" is clearly identified by its full title as a prominent example of the Diderot manuscript tradition. This entire manuscript underwent three general revisions: (1) the editing by Diderot during the period 1754–58; (2) the additions made between 1758 and 1769, when the manuscript was copied by François Latapie de Paule, whose name appears on the title page of tome I; and (3) the addition of interfoliated notes, the latest of which is dated 1779.[75] Diderot's editing appears in three places in the Bordeaux manuscript. The first of these is the historical introduction, which occupies the initial thirty-four pages of the manuscript and which was published by Charles Henry.[76] Principally because this passage is not found in most Rouelle manuscripts, Henry attributed the entire section to Diderot. Edouard Grimaux, on the other hand, disputed Henry's conclusion and claimed that while the introduction is in the style of Diderot, it is based upon Rouelle's ideas.[77] Following Henry's suggestion, I think it likely that the historical introduction is Diderot's version of part of Venel's article "Chymie" in the *Encyclopédie*. This article, published in 1753, and the manuscript's historical introduction contain so many similar phrases and sentences that Diderot's borrowing from Venel seems almost a certainty.

[74] For a discussion of this lost manuscript, see Charles Henry, "Introduction à la chymie. Manuscrit inédit de Diderot," *Revue scientifique*, pp. 97–108 (2e semestre 1884); esp. p. 98. Diderot's chemical papers, if they are extant, are not in Leningrad or in the Fonds Vandeul. See Maurice Tourneux, "Les Manuscrits de Diderot conservés en Russie," *Archives des Missions Scientifiques et Littéraires*, 3me série, 12, 439–474 (1885), esp. p. 463; and Herbert Dieckmann, "Inventaire du Fonds Vandeul et Inédits de Diderot," Geneva, 1951.
Diderot attended Rouelle's lectures for three consecutive years ("Oeuvres," Vol. 6, p. 407). Since he was still working in chemistry in 1758, it is possible that he spent that year editing his lecture notes; see Deleyre's letter to Rousseau, 28 February 1758, in Denis Diderot, "Correspondance," ed. Georges Roth, 3 Vols., [Paris], [1955–57], Vol. 2, p. 43.
[75] The microfilmed Bordeaux MS does not show the title page, but Henry notes the inscription *Latapie delineavit*, 1769; see Charles Henry, "Le cours de chimie de Rouelle avec des pages inédites de Diderot," *Revue scientifique*, p. 801 (1er semestre 1885). Latapie (1739–1823) was a botanist who taught for some years at the Jardin des Plantes de Bordeaux.
The date 1779 appears in the Bordeaux MS, p. 626a.
[76] Henry, "Introduction à la chymie," pp. 99–108.
[77] Edouard Grimaux, "Le cours de chymie de Rouelle," *Revue scientifique*, pp. 184–185 (2e semestre 1884). Grimaux' argument is based on a manuscript he owned which contains such an introduction, but in an abbreviated, outlined form. He printed parallel passages from his manuscript and the Bordeaux "Cours de chymie," and their close similarity seems to me to weaken the basis for his conclusion. There is no reason to think that the two manuscripts are independent and that two different students took these notes from Rouelle's lectures—which Grimaux would apparently have liked to believe.

The second area in which Diderot's editing appears is in the pages of the Bordeaux manuscript labeled by Charles Henry "L'Utilité de la chymie."[78] This section, too, Henry would attribute to Diderot. There may be an as yet undiscovered source for this passage, comparable to Venel's article, but in any case it is clear from manuscripts not in the Diderot tradition that Rouelle himself did lecture briefly on the utility of chemistry.[79] Rouelle's remarks on this subject were, however, simple statements concerning the debt of other sciences and various industries to chemistry, while the Bordeaux manuscript contains a good deal of detail about the ways in which chemistry has benefited mankind. It seems, therefore, that Diderot either wrote an expanded version of Rouelle's remarks or adapted the work of some unknown writer for this purpose.

Diderot's remaining contributions to the Bordeaux manuscript are difficult to identify with any certainty. Many explanatory phrases, added refinements of meaning, and little displays of erudition, which are not found in other manuscripts, may stem from Diderot's editing. Such additions generally occur in the introductory matter prefaced to separate topics, rather than in the experiments. In the following passage from Rouelle's discussion of the elements, the additions probably made by Diderot are enclosed in parentheses:

The first chemists recognized with the Dutchmen, Basil Valentine and Paracelsus three principles, mercury, sulfur and salt; but they only considered them secondary principles (and perhaps only admitted them in an allegorical sense). Van Helmont admitted (with Thales) only water as the principle of all substances. Other chemists (among whom is Willis) have added two principles to those of Paracelsus, so that they have admitted five of them: mercury or spirit, sulfur or oil, salt, phlegm or water and earth.[80]

What I have called the second general revision of the Bordeaux manuscript occurred between 1758 and 1769. Because the date 1762 appears in one of the many square-bracketed passages [* . . .] of this manuscript, these marked sections can be considered a second class of revisions.[81] Although square-bracketed passages deal with the same subjects as the texts in which they occur, there is sufficient discontinuity

[78] Bordeaux MS, pp. 36–42; Henry, "Le cours de chimie de Rouelle," pp. 802–804.

[79] Bib. nat. MS 4045, pp. 2–4.

[80] Bordeaux MS, pp. 42–43. Phrases added to the Bordeaux MS have been identified by comparison with the same passage in Bib. nat. MS 4043 (p. 5v), Clifton College MS (p. 8), and the Lausanne MS used by Claude Secrétan, "Un aspect de la chimie prélavoisienne," p. 296.

[81] The date 1762 appears in Bordeaux MS, p. 792. Many of the bracketed passages, but not all, are also found in the Nancy MS.

of thought to indicate that these are later insertions—either comments on the text or additions based on Rouelle's post-1758 lectures.[82]

The authorship of these interpolated passages seems impossible to determine. That they are in Latapie's hand, as is the rest of the manuscript, and are as evenly spaced as the rest of the text—i.e., rather than cramped into margins—shows only that they were probably copied from an earlier manuscript on which the annotations were made, but does not aid in identifying their author. Henry's belief that these sections are the work of Rouelle *le cadet* and Darcet is based on the frequent references to these men within the bracketed passages and on Diderot's statement that Rouelle's courses were edited by his brother and son-in-law.[83] Such evidence seems to me quite inconclusive. Indeed, references to the younger Rouelle and Darcet in the text suggests that the editor was a student citing his teachers; the younger Rouelle, it should be remembered, assisted his brother and even replaced him on occasion, especially in the years 1766–68, and Darcet may also have done so.

The interfoliated notes, the third set of revisions in the Bordeaux manuscript, present problems which will not be treated here in any detail since these notes are distinct from the text and their content and dating are not vital for this study. However, the existence of the interleaved notes and the earlier interpolations raises the question of the possible purposes for which the Bordeaux manuscript was so heavily edited and annotated.

Rouelle's contemporaries testify that Rouelle *le cadet*, Jean Darcet, and Augustin Roux were editing the elder Rouelle's lectures in order to publish the textbook long planned but never completed by their teacher.[84] If these three men were in fact responsible for the annotations of the Bordeaux manuscript, there would be reason to suspect that this manuscript was the text destined for eventual publication. It is clear that the Bordeaux "Cours de chymie" is unique in the care lavished upon it and in the obvious efforts made to keep it up to date with progress in eighteenth-century chemistry. It was copied with considerable care and

---

[82] The context of the 1762 date, for example, clearly refers to Rouelle, beginning, "en 1762 il a fait. . . . " It is not always obvious that Rouelle's lectures were the source of the bracketed passages.

I am unable to account for the square brackets enclosing the section on the utility of chemistry in the Bordeaux MS. These brackets differ from those found elsewhere in the manuscript in that they are very faint (either lightly penciled or partially erased), do not contain the usual initial asterisk, appear to have been inserted between words as an afterthought, and are not the same shape as those brackets found throughout the manuscript.

[83] Henry, "Introduction à la chymie," p. 98, and "Le cours de chimie de Rouelle," p. 802; Diderot, "Œuvres," Vol. 3, p. 464.

[84] Grandjean de Fouchy, p. 148, and Grimm, Vol. 9, p. 109; also, Dorveaux, pp. 184–186.

embellishment in the beautiful hand of Latapie. That Latapie was the copyist suggests another link between the Bordeaux manuscript and the younger Rouelle and Darcet; Latapie could conceivably have been asked to act as copyist because he was already acquainted with Rouelle *le cadet* through the latter's courses and with Darcet during the period both he and Darcet spent at Bordeaux.[85] The latest date in the interfoliated notes, 1779, is in fact the year of the younger Rouelle's death and may mark the abandonment of the publication project when one of its instigators died.

A single objection may be raised against the suggestion that the Bordeaux manuscript was destined for publication: there is a lacuna in its text where a considerable number of Rouelle's experiments have been omitted with the omission left unrectified by the annotators.[86] The reason for this lacuna may simply be Diderot's failure to attend several of Rouelle's lectures, but he could hardly have missed the same set of lectures every year for three years. Perhaps Rouelle added these experiments to his courses after 1758, but then the annotators should certainly have filled the gap in this manuscript. A final possibility is that the editors exercised their privilege to omit a section—and it is a complete section that is missing—from their definitive version of Rouelle's lectures, although their motives seem obscure and such a procedure would surely have rendered the manuscript anything but definitive. If this curious lacuna does indeed show that the Bordeaux manuscript was not chosen for publication by Rouelle's disciples, there must have existed a manuscript in the Diderot tradition and similarly annotated which was being prepared for this purpose.[87]

The section missing from the Bordeaux "Cours de chymie" is supplied by two of the other manuscripts used in the preparation of this study.[88] The version found in the Clifton College "Cours de Chimie" does not differ significantly from that in MSS. 12303–12304 in the Bibliothèque nationale—with one notable exception. In the latter manuscript there is an indignant comment to the effect that P.-J. Macquer has "plagiarized" one of Rouelle's ideas; the same lines in the Clifton College manuscript have been carefully deleted so that all but a few words are illegible.[89]

[85] Latapie was the son of a surveyor employed at the Château de La Brède, was educated under Montesquieu's auspices, and became the secretary to Montesquieu's son. Darcet directed the education of this same son and was associated with Montesquieu until the latter's death in 1755. See M.-J.-J. Dizé, "Précis historique sur la vie et les travaux de Jean d'Arcet," Paris, 1802, pp. 6–9.

[86] The Bordeaux MS is continuously paginated, so that there is no possibility that these pages were simply lost.

[87] That this manuscript must have been in the Diderot tradition is evident from the note quoted below, p. 91. The Clifton College MS is in the Diderot tradition and was copied from a manuscript Roux intended to have published.

[88] Clifton College MS, pp. 910–985, and Bib. nat. MS 12303, pp. 240–332.

[89] Bib. nat. MS 12303, p. 262, and Clifton College MS, p. 929.

As I have noted elsewhere, Rouelle was famous for accusations of this kind; that there are no such references in the other manuscripts consulted testifies to the activities of the anonymous editors.

The Clifton College "Cours de Chimie" and the Duveen "Traité de Chymie" throw some light on the way in which Rouelle's courses were copied and circulated in manuscript form. The Duveen manuscript is an excellent example of one sold to a student who subsequently attended Rouelle's lectures and added data to the text of the original manuscript. Evidently derived from a manuscript in the Diderot tradition, there is at least one place which reveals that the Duveen manuscript dates from sometime after 1760 or 1761; the significant passage refers to a method employed by Rouelle in 1760.

> We put clean and pure olibanum in a glass vessel that we place in a sand bath. . . . We used in 1760 the open flame instead of the sand bath.[90]

On the flyleaf of the Clifton College manuscript is the owner's notation which I shall quote in full:

> This Course in chemistry is [taken] from that edited by Messrs. Roux and Darcet. It has been copied from the notebooks dictated by Mr. Roux. It cost me five golden louis for the workmanship of the scribe who copied it for me and who left in it a considerable number of errors all of which I have corrected in my own hand, so that this manuscript can be accepted as very exact and very trustworthy and makes a very precious work.

From this passage it is clear that the manuscript was bought as a reference work, perhaps by a person who had no intention of using it in conjunction with a course in chemistry. This conjecture is supported by the lack or internal evidence to indicate that any additions, other than the corrections mentioned by the owner, were made after the manuscript had been copied. Since Roux and Darcet probably began editing Rouelle's lectures after his retirement in 1768 (or after his death in 1770), it seems likely that the Clifton College manuscript was copied at least ten years after the lectures were delivered in 1754–57. This circumstance plus the fact that the section on *terres et pierres*—the section missing from the Bordeaux manuscript—was evidently sought out and appended to this "Cours de Chimie" suggest the high value placed upon Rouelle's lectures. Although his courses had not been published, they were in sufficient demand to make their circulation in manuscript form a costly necessity.

It should be clear from the foregoing discussion that my study of the organization and content of Rouelle's lectures has been limited by the

---

[90] Duveen MS, Vol. 1, p. 123; cf. Bordeaux MS, p. 224, which merely notes that either the sand bath or the open flame may be used.

nature of the material available to me. In most respects the manuscripts are substantially in agreement, but it should be remembered that they represent a relatively short period in Rouelle's career, though the period of his prime. For that interval it is possible to formulate accurate ideas about Rouelle's teachings, but the same material can reveal little of Rouelle's activities during the earlier and later years of his life.

The content of Rouelle's courses has been discussed recently in two little-known articles by the Swiss chemist and historian of science, Claude Secrétan.[91] These articles provide useful summaries of the subjects Rouelle treated, as well as numerous extracts from the Lausanne manuscript consulted by the author. Principally because M. Secrétan had at his disposal an incomplete Rouelle manuscript, lacking the entire treatment of inorganic substances (*règne minéral*), the same subject can be profitably reconsidered here. The following discussion will include the general plan of the lectures, a brief indication of the contents of each major section, and such conclusions about Rouelle's scientific method as may be drawn from the texts.

Rouelle's lectures open with some general remarks about the aims and methods of chemistry, followed by an introduction to chemical theory. The practice of beginning a series of lectures with theoretical principles and proceeding to particular experimental facts—established though it was in countless textbooks of chemistry—was probably a method imposed upon Rouelle by his colleague at the Jardin, Bourdelin.[92] If Bourdelin opened his lectures with a treatment of theory of the elements, Rouelle doubtless felt called upon to "correct" the professor by following with his own theory. There is good reason to suspect, however, that Rouelle would not have chosen this method under other circumstances, but rather the only alternative: facts first and theory last.[93] Rouelle's method of organization is stated briefly in his introduction to the *règne végétal*, which precedes both the *règne animal* and *règne minéral*.

M. Rouelle would have treated [the *règne animal*] first had he succeeded in finding the connection between the different phenomena which it presents. . . . [While] waiting for all [these phenomena] to be linked together, he will present vegetable analysis first as that in which all the parts harmonize best, and [which is] in the most suitable state to make known the true spirit of chemical analysis.[94]

[91] Above, note 7.

[92] This procedure had a long tradition that started with the earliest chemical lectures; cf. John Read, Chapter 6.

[93] He may have used this latter method in 1751, the date of Bib. nat. MS 4045; the table of contents of this manuscript lists the theoretical section after all the experiments. This is true of none of the Diderot manuscripts, and Diderot's testimony that the last two lectures every year were devoted to alchemy ("Oeuvres," Vol. 6, p. 409) checks with these manuscripts, showing that the order of subjects treated has not been altered by the editors.

[94] Bordeaux MS, p. 123.

The *règne animal* will follow, and then the *règne minéral*:

The examination of mineral substances of which the composition is much simpler, but more difficult to unfold will end this course. We will combine the different products [of this *règne*] with the substances taken from the two preceding realms, after having examined the phenomena which they present among themselves.[95]

Ideally, Rouelle would have liked to begin with the *règne animal*, the most complex, and to conclude with the *règne minéral*. That Rouelle preferred to treat the most complex substances first and the simplest last is implied by his ordering of the *règne minéral*, with gold, the most perfect and simplest of metals, ending the section.[96] A practical argument for this procedure is the relative instability of organic compounds, which makes them easier to decompose for the purposes of demonstration. But the evidence also implies that Rouelle favored an almost Baconian presentation of facts, to be followed by an attempt to show the relations between the facts. This method suggests that Rouelle would have preferred to present theory as a final generalization of experimental evidence.

If Rouelle did favor this facts-first, theory-last method, for two very sound reasons he did not use it. In the first place, as already noted, he probably had to follow the practice established at the Jardin and employed by Bourdelin. Secondly, Rouelle's primary function was teaching, and the better pedagogical method was to begin with theory. As Macquer justly remarked in stating his own philosophy of teaching, the facts-first, theory-last method is easier for the teacher, but a student presented with a mass of apparently unrelated facts is confused and fails to grasp the meaning of the experiments. It is difficult for the professor but far more profitable for the student if principles are discussed first and the facts then related to the established theories.[97] Rouelle's solution was to lecture first on chemical theory and then to present experimental data in a modified complex-to-simple order to show his students the proper method for arriving at generalizations and theories.

The bulk of Rouelle's lectures is a compilation of experiments on substances drawn from the three realms of nature. Since most of these experiments stem from other sources—French, German, and English— they are of little intrinsic interest or value in a study of Rouelle.[98]

[95] Bordeaux MS, p. 124.

[96] Most textbooks of chemistry of this period start with the *règne minéral* and treat gold, the perfect metal, first in that section. E.g., Nicolas Lémery, "Cours de chymie," and Paul-Jacques Malouin, "Traité de chimie," Paris, 1734.

[97] Macquer and Baumé, pp. 2–5.

[98] Rouelle performed many of the same experiments found in Lémery's "Cours de chymie." Careful comparison shows that there is in fact a surprisingly small proportion of experiments in Rouelle's lectures that cannot be found in Lémery's older textbook; most of these new experiments employ metals like zinc and cobalt or organic substances which Lémery did not use.

Occasionally, however, Rouelle varies his methods in some significant manner or illustrates his use of experiment in support of his chemical theories. The *règne végétal*, which consists of analyses of a variety of organic substances into their component resins, oils, acids, and alkalis, provides a noteworthy example of Rouelle's preoccupation with refinements of method. In his treatment of organic substances, Rouelle's method was singled out for praise by many of his contemporaries. By regulating the temperatures at which his experiments were conducted, he was able to obtain distillation products which would have been destroyed at high temperatures. He also classified and described the properties of the products thus obtained. In 1800 Fourcroy could no longer praise the obsolete theories of Rouelle, but he summarized Rouelle's achievements in organic chemistry with considerable admiration.

Rouelle who, in his courses, had already added much to Boerhaave's beautiful system of vegetable analysis, first distinguished with more care [than Boerhaave] the immediate components of vegetable matter, divided and characterized by their better known properties the different kinds of extracts; discovered the glutinous material in green leaves; compared gums and sugar to starch; published in his processes a more complete and especially more methodical outline of vegetable analysis than had been done until that time, and revived the hope of experimenters.[99]

Of greater significance for eighteenth-century chemical theory than Rouelle's method in organic analysis was his use of the work of Stephen Hales. It has recently been shown that Hales's *Vegetable Staticks* (1727) was unknown in France until after its appearance in French translation in 1735, and that, furthermore, it was through Rouelle's efforts that Hales's discoveries entered the mainstream of French chemistry.[100] Hales's experiments, which demonstrated that air is a constituent of many substances, were repeated by Rouelle in the course of his lectures. Not only did Rouelle adopt and improve upon Hales's apparatus for handling gases, but he also incorporated into his chemical theory the notion that air is one of the simple components of matter—a chemical element.[101]

Rouelle's treatment of the *règne animal*, which was later expanded by

[99] Fourcroy, Vol. 7, p. 39. The published "processes" probably refers to the outlines included in prospectuses issued for Rouelle's courses.

[100] Henry Guerlac, "Lavoisier and his Biographers," p. 59. See also the same author's "The Continental Reputation of Stephen Hales," *Archives Internationales d'histoire des sciences*, No. 15, 393–404 (1951), in which the influence of Hales on the Continent is established.

[101] For references to Hales and the experiments of Rouelle, Bordeaux MS, pp. 108, 206b, 207–208, and 990b. Also, Lavoisier, "Oeuvres," Vol. 2, p. 7. Rouelle's apparatus is reproduced in the "Encyclopédie méthodique," description in t. 66, p. 354 and t. 70–71, p. 16 of Explication des Planches; illustration in t. 70–71, Planche XVII, Fig. 45.

the younger Rouelle,[102] includes the analyses of milk, lymph, and urine, as well as the hard parts of animals. Admittedly more complex and less fully understood than the *règne végétal*, the *règne animal* contains at least one attempt to indicate a continuity among the three realms of nature. In his introduction to the study of insects, Rouelle makes an observation which was probably a shortened statement about more detailed work he was conducting in his laboratory.

Upon analysis all Insects generally yield some acid. In that they differ from other animals, and can be considered the transition from the *règne animal* to the *règne végétal*[103] which abounds in acid; just as cruciferous plants which give a great deal of volatile alkali, or gramineous plants which contain a very large quantity of mucous matter can be considered the transition from the *règne végétal* to the *règne animal*.[104]

Rouelle's work in animal chemistry is in no way emphasized in his lectures, but Diderot claimed that Rouelle, Macquer, and other chemists were experimenting widely in this area.[105] Their findings were not only used and admired by other scientists,[106] but were singularly impressive to Diderot who was well versed in contemporary science. After summarizing briefly some of the recent discoveries on the nature of organic matter, Diderot lets his imagination take over and records a series of *pensées* apparently inspired by the observation that there are no sharp discontinuities between the realms of nature.[107] He conjectures that the *règne animal* had its source in the *règne végétal*, which in turn is derived from the *règne minéral*; the latter probably came out of the "universal heterogeneous matter," the primary substance in Diderot's materialist philosophy. Carrying his conjectures to an extreme conclusion, Diderot speculates on the relation between the highly developed animal and the lowly machine: "What difference [is there] between a sensitive and living timepiece, and a timepiece of gold, iron, silver, and copper?"[108]

[102] Interfoliated pages of the Bordeaux MS, pp. 558a–558d, contain experiments on blood and bile performed by the younger Rouelle.

[103] Bordeaux MS (p. 554) and Nancy MS (p. 352) read *végétal*, but Duveen (Vol. 2, p. 39) and Clifton College (p. 431) read *minéral*, which is evidently not the intended meaning.

[104] Bordeaux MS, p. 554. Cf. *ibid.*, p. 489: "Mr. Rouelle commence son analyse animale par l'examen du Lait parce que c'est le premier de nos alimens, et celle de nos liqueurs qui diffère le moins des végétaux dont elle est tirée . . .," and Duveen MS, Vol. 1, p. 116: "les haricots en [i.e., phosphorus] donnent aussi, mais en si petite quantité qu'il se brule dans l'operation, ce qui fait regarder ces plantes par M. Rouelle comme les plus analogues au regne animal qui comme elles contient beaucoup d'alkali volatil et donne le phosphore, il les appelle plantes animales."

[105] Diderot, "Oeuvres," Vol. 9, p. 255.

[106] Fourcroy, Vol. 8, p. 83; Vol. 9, pp. 38, 50.

[107] Diderot, "Oeuvres," Vol. 9, pp. 256 ff.

[108] *Ibid.*, p. 265. Cf. passages on the differentiation of organic and inorganic substances in Bordeaux MS, p. 47, Duveen MS, Vol. 1, p. 10, and Clifton College MS, p. 13;

The use of Rouelle's experiments in organic chemistry for Diderot's philosophical speculations does not imply, of course, that Rouelle shared Diderot's views. This digression into the writings of the *philosophe* simply provides a striking example of the range of influence of some of Rouelle's work.

Most important for the light it sheds on Rouelle's theory of matter and the nature of chemical combination is the section of his lectures entitled *règne minéral*. By far the longest of the three major divisions of the course, the *règne minéral* includes experiments on bitumens, acids (sulfuric, nitric, and hydrochloric), the so-called demimetals, and metals. It is in this section that Rouelle discusses the composition of metals, the purification of metallic ores, the formation of salts, and the different classes of salts.[109] In the course of his observations of the location of mines and ore deposits, Rouelle was able to formulate a general theory of geological stratification. This theory, which is expounded in the introduction to the *règne minéral*, can only be summarized briefly here.

Rouelle's geological theory postulates two general strata, termed the *terre ancienne* and *terre nouvelle*, distinct from each other in their composition and the methods of their formation.[110] Sometime after 1763, probably influenced by the work of J. G. Lehmann and by the observations of his pupils, Rouelle modified this theory to include a third series of strata, referred to simply as "un travail intermédiaire."[111] Briefly described, the *terre ancienne* consists of massive, unstratified, granitic deposits; the intermediate strata contain bitumens and debris of various kinds; and the *terre nouvelle*, a complex and highly diversified structure, consists of horizontal layers deposited by slow sedimentation. The formation and deposition of fossils in the *terre nouvelle* is treated very perceptively, if all too briefly, in the manuscript lectures. Although Rouelle's discussion of volcanoes is not well developed, it is quite likely that Rouelle supplied

---

a similar series of passages is in Bordeaux, p. 481, Duveen, Vol. 2, p. 1, and Clifton College, p. 386. These passages, reminiscent of Diderot's ideas, should be checked against manuscripts outside the Diderot tradition since they differ from the vitalistic sentiment of Bordeaux, p. 559, and Clifton College, p. 438.

[109] Experiments on *sel marin* are in this section, Bordeaux MS, pp. 796–854, but a short summary of Rouelle's theory describing the three classes of salts occurs in the *règne végétal, ibid.*, pp. 329, 423–425.

[110] *Ibid.*, pp. 567–582.

[111] The details of this theory are given by Desmarest in his article "Rouelle," cited above, note 70. The dating of the three-strata theory is suggested by Lavoisier's adherence to the older theory (Lavoisier, "Oeuvres," Vol. 5, pp. 12, 226 ff., 236) after he studied with Rouelle in 1762–63 or 1763–64; Rouelle apparently had not yet formulated, or at least was not teaching, his three-strata theory.

Archibald Geikie, in "The Founders of Geology," 2nd ed., London, 1905, pp. 342–343, makes some use of Rouelle's geological ideas; his source is Desmarest's article.

I

the inspiration that led one of his most talented pupils, Nicholas Desmarest, to make the study of volcanoes his life's work.[112]

Rouelle's geological data came in great part from first-hand observations, many of them carried out during "plusiers voyages lithologiques . . . avec son digne ami Bernard de Jussieu,"[113] and later in his career from the field work of his pupils. As early as 1740, Rouelle is reported to have expounded in his lectures his ideas on the constitution of the earth.[114] Evidently his reputation as a geologist and mineralogist grew rapidly, attracting to his courses such men as Desmarest and Lavoisier as well as Nicolas Gobet and Antoine Monnet.[115]

The concluding portion of the course, and the subject of every year's final lectures, is a discussion of alchemy. As a disciple of Becher, who claimed to have known the secret of transmutation, Rouelle does not discredit alchemy entirely. He observes that, "It is to reason poorly to conclude that because it is impossible for us to reproduce a plant, an animal, that it is impossible to reproduce a metallic substance [which is] without organization and without life."[116]

Rouelle's failure to discredit alchemy should be understood in the light of his theory of the formation of metals, rather than as an inexplicable aberration of an otherwise sound scientific mind. In theory he considers it possible to combine the three principles of which metals consist in the correct proportions to produce gold.

We understand the inflammable principle and the vitrifiable principle. If we understood the mercurial principle as well [as the other two], we would perhaps succeed more easily in imitating the combination of gold.[117]

In practice, however, he cautions the would-be alchemist to avoid such costly and uncertain labors, unless there were a sure guide to success "to lead [one] in an operation which is only preserved by tradition."[118] Despite the good advice given to others, Diderot confides that Rouelle spent the last years of his life in alchemical studies.[119]

[112] Bordeaux MS, pp. 664–673. Rouelle's influence on Desmarest's choice of problems to be solved is suggested by Desmarest's remark that he will record "des différentes recherches dont j'ai commencé à m'occuper depuis 1763, relativement aux vues systématiques de Rouelle" (art. "Rouelle," p. 422).

[113] *Ibid.*, p. 410.

[114] Darcet, pp. 32–33.

[115] Monnet became Guettard's collaborator, succeeding Lavoisier, and aided in the completion of the "Atlas minéralogique de la France." Gobet has to his credit editions of the works of Bernard Palissy, Paris, 1777, and Jean Rey's "Essais," Paris, 1777, as well as "Les Anciens minéralogistes du royaume de France, avec des notes," 2 Vols., Paris, 1779.

[116] Duveen MS, Vol. 2, pp. 415–416.

[117] Bordeaux MS, pp. 1247–1248.

[118] *Ibid.*, p. 1258.

[119] Diderot, "Œuvres," Vol. 6, p. 409.

Some conclusions about Rouelle's scientific method can be drawn from his lectures. The available manuscripts are not particularly good guides in this respect, since they are written in a discursive style, with no attempts made to reproduce diagrams of equipment or tables of data.[120] A comparison of the manuscript discussions of the classification of salts with Rouelle's memoirs submitted to the Academy of Sciences provides evidence that his was not a wholly qualitative approach to chemistry. His careful regulation of the temperatures at which experiments were performed, his injunction about proper sealing of vessels to avoid loss of the reagents and products, and his use of an accurate balance all indicate awareness of the conditions needed for trustworthy experimental results.[121] Although the spirit of exactness which pervades his work is admirable, the limitations of Rouelle's methods are quite apparent upon closer examination. When, for example, vessels were sealed to guarantee accurate results, an air hole was left open to forestall breakage of the vessels under pressure. The sensitive balance was used to weigh reagents and liquid or solid, but not gaseous, products. Whether Rouelle fully understood the significance of evidence supplied by a balance is a question which merits more detailed discussion.[122]

The care with which Rouelle tried to perform his experiments arose from his conviction that the only sound theory was one based entirely on experiment. His method is epitomized by the remark which occurs invariably in the manuscripts: "La Chymie ne cherche pas de vains raisonnemens; elle ne cherche que des faits."[123] He impressed this dogma upon his pupils, one of whom, Monnet, completely mistook the spirit of Rouelle's motto and interpreted his teacher in this way: "Chemistry is only a collection of facts, most of them with no connection between them or independent of one another."[124] This clearly was not Rouelle's mean-

---

[120] Lemay's manuscript contains a reproduction of Rouelle's improved version of Hales's apparatus (Lemay, "Les cours de Guillaume-François Rouelle," p. 441). The Bordeaux MS does not even contain the exact proportions of reagents used in a particular experiment (p. 499), while Duveen MS, Vol. 2, pp. 9–10, does for the same experiment.

[121] The experimental conditions treated in Rouelle's lectures are: degrees of heat, Bordeaux MS, pp. 65 ff.; degrees of evaporation, *ibid.*, p. 430; the sealing of vessels, *ibid.*, pp. 172–173. On the value of thermometry, Rouelle is quoted in Secrétan, "Un aspect de la chimie prélavoisienne," p. 305; for Venel's dissenting view, *ibid.* Rouelle's chemical balance, used for a time by Lavoisier, is reproduced in Maurice Daumas, "Les Instruments Scientifiques aux XVIIe et XVIIIe Siècles," Paris, 1953, Plate 57, Figure 129; also, p. 292 and note 1.

[122] E.g., Rouelle argues that the increase in weight of the calx of antimony over the weight of the original metal is only apparent, and that "la pesanteur absolue etant toujours la même, il n'y a que la pesanteur specifique qui augmente sans doute parce que le volume diminue beaucoup plus que la substance réelle." Bordeaux MS, pp. 970–971.

[123] *Ibid.*, p. 36. Duveen MS, Vol. 1, p. 2 reads: "La Chymie ne s'appuye pas sur de simples raisonnements, elle veut des faits."

[124] Quoted in Pierre Duhem, "Le Mixte et la combination chimique," Paris, 1902, pp. 43–44, and Metzger, p. 89.

ing. Rouelle did not hesitate to construct general theories based on the findings of experiment, but strictly excluded speculation about a subject like the cause of chemical affinities; the latter was one of the "vains raisonnemens" he deplored.

To summarize what has been said about Rouelle's courses, Venel's description seems most appropriate.

> The courses which M. Rouelle has given at Paris for about twenty years, are, even in the opinion of strangers, among the best of this kind. The order in which particular objects are presented, the abundance and choice of examples, the care and exactitude with which operations are performed, the origin of and relation between the phenomena observed, the new luminous, broad insights suggested; the excellent manual precepts taught, and finally, the good, sound doctrine which sums up all the particular notions; all these advantages, I say, make the laboratory of this capable chemist such a good school, that one can in two courses, with ordinary dispositions, emerge sufficiently instructed, to deserve the title of distinguished amateur, or of artist able to engage successfully in chemical researches. This judgment is confirmed by the example of all the French chemists, for whom the first taste for *chemistry* followed the first courses of M. Rouelle.[125]

Venel has captured in this paragraph some of the excitement and enthusiasm that Rouelle could communicate to his students. The experiments, the illustrations abounding in the lectures, the "manual precepts taught"—all these mark the soundness of Rouelle's demonstrations. But it was his ability to unite the discrete facts by means of broad theories that distinguish Rouelle as a teacher and chemist rather than a mere laboratory assistant. Presented in the manner described by such pupils as Venel and Mercier, the doctrines taught by Rouelle were so widely accepted by contemporary chemists that the exact nature of these doctrines and their place in the history of the chemistry of the eighteenth century deserves further consideration and careful study.

MANUSCRIPTS OF ROUELLE'S LECTURES

Bordeaux MSS 564–565. "Cours de chymie de M. Rouelle, rédigé par M. Diderot et éclairci par plusieurs notes, divisé en neuf tomes." 1258 pp. in-12°. [Microfilm in possession of Professor Henry Guerlac, Cornell University.]

Bristol, England (Science Library, Clifton College). "Cours de Chimie, ou Leçons de Monsieur Rouelle, Recueillies pendant les années 1754, 1755. et rédigées en 1756. Revuës, et corrigées, en 1757, et 1758." 985 and 65 pp. [Microfilm in Cornell University Library.]

Nancy MS 265 (307). "Cours de Chimie." 772 pp. [Microfilm in Cornell University Library.]

---

[125] Venel, art. "Chymie," *Encyclopédie*, Vol. 3, p. 437.

New York City. Collection of Denis I. Duveen. "Traité de Chymie de Rouelle." 2 Vols., 207 and 421 pp.

Paris, Bibliothèque nationale, n.a.fr. 4043–4044. "Cours de Chimie, ou Leçons de Mr Rouelle, Recueillies pendant les années 1754, et 1755; Rédigées en 1756; revues et corrigées en 1757 et 1758." 2 Vols., 674 f°. [Fragment on microfilm in possession of Professor Henry Guerlac.]

Paris, Bibliothèque nationale, n.a.fr. 4045. "Cours de Chymie par M. Rouelle. 1751." 900 pp. [Fragment on microfilm in possession of Professor Henry Guerlac.]

Paris, Bibliothèque nationale, fr. 12303–12304. "Cours de chimie, ou leçons de M. Roüelle, démonstrateur au Jardin du Roy, récueillies en 1754 et 1755, . . . corrigées en 1757 et 1758." 1015 pp. [MS 12304, pp. 240–332 on microfilm in Cornell University Library.]

PUBLICATIONS BY AND ATTRIBUTED TO ROUELLE

Rouelle, G.-F. "Mémoire sur les Sels neutres, dans lequel on propose une division méthodique de ces Sels, qui facilite les moyens pour parvenir à la théorie de leur crystallisation," *Mémoires de l'Académie royale des sciences*, pp. 353–364 (1744).

Rouelle, G.-F. "Sur le Sel marin (Première partie.) De la crystallisation du Sel marin," *Mémoires de l'Académie royale des sciences*, pp. 57–79 (1745).

Rouelle, G.-F. "Sur l'inflammation de l'huile de Térébenthine par l'acide nitreux pur, suivant le procédé de Borrichius; Et sur l'inflammation de plusieurs huiles essentielles, & par expression avec le même acide, & conjointement avec l'acide vitriolique," *Mémoires de l'Académie royale des sciences*, pp. 34–56 (1747).

Rouelle, G.-F. "Sur les Embaumemens des Egyptiens, Premier Mémoire, . . . ," *Mémoires de l'Académie royale des sciences*, pp. 123–150 (1750).

Rouelle, G.-F. "Mémoire sur les Sels neutres, dans lequel on fait connoître deux nouvelles classes de Sels neutres, & l'on développe le phénomène singulier de l'excès d'acide dans ces sels," *Mémoires de l'Académie royale des sciences*, pp. 572–588 (1754).

Summary of 1750 memoir to the Academy of Sciences, in *Journal de médecine, chirurgie, pharmacie*, pp. 299–304 (1756).

"Analyse de l'eau minérale de . . . Passy," in *Analyses d'une eau minérale faites par MM. Rouelle, de l'Académie des Sciences, et Cadet, apoticaire major de l'Hôpital Royal des Invalides*, n. p., n. d.

Experiments published in 1760 and referred to in "Lettre de M. [Hilaire-Marin] Rouelle, Apothicaire de S.A.S. Monseigneur le Duc d'Orléans & Démonstrateur en Chymie au Jardin du Roi, &c. à l'Auteur de ce Recueil," *Observations sur la physique, sur l'histoire naturelle et sur les arts*, 2, 144–145 (juillet 1773).

"Examen analytique de l'eau du puits de l'Ecole royale militaire," unpublished memoir in Archives nationales ($0^1$1608, pièce 5).

Memoir communicated to *Bulletin de Pharmacie*, 4, 193, by Darcet after Rouelle's death.

A medical work, which cannot definitely be ascribed to the elder Rouelle, in Henry Saffory, "The inefficacy of all mercurial preparations in the cure of venereal and scorbutic disorders proved from reason and experience; with a dissertation on Mr. de Velnos's vegetable syrup, . . . and an accurate analysis

I

101 G.-F. ROUELLE: 18TH-CENTURY CHEMIST AND TEACHER

of that medicine, made by order of the Marshal Duke of Biron, by Messrs. Rouelle and Lacassaigne, professors of chymistry at Paris . . . ," 2nd ed., London, 1776.

Experiments on the destruction of diamonds and in collaboration with Jean Darcet mistakenly ascribed to elder Rouelle, but in fact the work of H.-M. Rouelle; listed in catalogue of Bibliothèque nationale and in J.-M. Quérard, "La France littéraire," 12 Vols., Paris, 1827–1864, Vol. 8, p. 173.

II

ROUELLE AND STAHL—THE PHLOGISTIC REVOLUTION IN FRANCE

WHEN Lavoisier began his work in chemistry, the phlogiston theory was the generally accepted chemical doctrine in France. As a unifying principle, this theory had much to recommend it; phlogiston was used primarily to explain combustion reactions, but also to account for physical and chemical properties ranging from the nature of color to the causes of fluidity and volatility. Despite its shortcomings, the phlogiston theory was so adaptable that to question its validity was to raise a host of difficult theoretical problems.

To say that the phlogiston theory alone dominated French chemistry is in reality to oversimplify the situation. Because phlogiston was an essential part of chemical theory and because the debates over its very existence raged for many years, phlogiston has come to epitomize and indeed to summarize in one word eighteenth-century chemistry before Lavoisier. This modern view would have surprised the chemist Guillaume-François Rouelle (1703–70), whose career encompassed the period before the polemical literature on phlogiston began to appear.[1] For Rouelle, phlogiston was but a single component of a more comprehensive theory: his "element-instrument theory."

Among his contemporaries Rouelle's reputation as a chemist was based primarily upon his outstanding ability to teach. But he was also called the "Restorer of French Chemistry"[2] and the person who had done most "to popularize chemistry in France in revealing to us for the first time the secrets of Germany, and in adding to it the soundest Theory and the surest experiment."[3] The German chemistry referred to is the work of Georg Ernst Stahl and his disciples, which remained relatively unknown in France for at least the first thirty years of the eighteenth century. While it is perhaps an exaggeration to say that Rouelle introduced a knowledge of Stahlian chemistry into France, he was certainly responsible for the popularization and eventual acceptance of a modified Stahlianism in the course of the century.

---

[1] For an account of Rouelle's life and work, see the author's "G.-F. Rouelle: An Eighteenth Century Teacher and Chemist," *Chymia*, **6**, 68–101 (1960).

[2] Johann Friedrich Meyer, *Essais de Chymie*, trans. P.-F. Dreux, 2 Vols., Paris, 1766, Vol. 1, "Avertissement du Traducteur," p. xvii.

[3] *Cours de chymie de M. Rouelle, rédigé par M. Diderot et éclairci par plusieurs notes*, 1258 pp., Bibliothèque de Bordeaux, MSS 564–565, p. 982b, note 1. Hereafter cited: Bordeaux MS.

His contemporaries were not fully aware that Rouelle was a critical disciple of Stahl and that, furthermore, his own theories both contradicted and elaborated on some of the doctrines of German Stahlianism. Rouelle added to Stahlianism his own thorough knowledge of the progress that had been made in chemistry, and he was thus able to formulate a new chemical theory which brought coherence to the work of his many predecessors. It will be the purpose of this article (1) to examine the element-instrument theory propounded by Rouelle, (2) to demonstrate the originality of Rouelle's work by a comparison with the doctrines of Stahl, and (3) to show that it was the revised Stahlianism of Rouelle which became the dominant theory in French chemistry in the 1750's. In the later years of the century, the attack on the phlogiston theory was accompanied by a simultaneous assault upon other features of the theory formulated by Rouelle. While a study of the origins of the chemical revolution cannot be undertaken here, a brief comparison of the methods and ideas of Rouelle and his most illustrious pupil, Lavoisier, should disclose some of the factors which weakened and eventually destroyed Rouelle's chemical system.

ROUELLE'S CHEMICAL THEORY

In the tradition of both the seventeenth and eighteenth centuries, the chemist was almost duty-bound to include in any statement of chemical theory some consideration of the nature of the ultimate particles of matter. Thus, Rouelle's staunch empiricism did not prevent his observing that

All sensible bodies, and even those which escape our senses, become the object of Chemistry. It makes known the latter, sometimes in displaying them through their effects, and sometimes in bringing them together to make them visible.[4]

The chemist is, in effect, extrapolating; he is drawing conclusions about ultimate particles from his knowledge of the properties of larger masses. Although these particles cannot be perceived individually, they can be described as "simple, homogeneous, indivisible, immutable and insensible, [and] more or less mobile, according to their different shape, their nature, their mass."[5] They are, moreover, hard and impenetrable, uniting by juxtaposition rather than by interpenetration. If ultimate particles do have varying shapes—and they probably do—these shapes are not known and are even unknowable.[6]

---

[4] Traité de Chymie de Rouelle, New York City (private collection of Denis I. Duveen), Vol. 1, p. 1. The manuscript consists of two volumes bound together and separately paginated. Hereafter cited: Duveen MS.

[5] Bordeaux MS, p. 43. This is actually Rouelle's definition of the particles of elements or principles.

[6] Cf. the views of Georg Ernst Stahl, Philosophical Principles of Universal Chemistry, trans. and ed. Peter Shaw, London, 1730, p. 67; Johann Juncker quoted in Hélène Metzger, Newton, Stahl, Boerhaave et la doctrine chimique, Paris, 1930, p. 105; and Hermann Boerhaave, A New Method of Chemistry, trans. Peter Shaw, 2nd edition, 2 Vols., London, 1741, Vol. 1, pp. 489–490.

Rouelle's conception of matter is better described as Newtonian than corpuscularian, despite his lack of emphasis on Newtonian mass and inertia. He simply implies that matter is inert and can be set in motion by the interplay of forces; and notably missing from his theory of matter is any vitalistic or materialistic idea of matter possessing energy—either the vitalist notion of the energy of an organized whole, or the materialist inner energy of individual particles. Taken in the context of his chemical lectures, the statement that matter is "more or less mobile" means nothing more than that some substances are more easily excited or volatilized than others, and that fluids appear to be in constant, insensible motion; Rouelle nowhere implies that mobility is somehow inherent in matter. Force, as used by Rouelle, signifies an existing relationship between bodies which enables them to be drawn together or to repel one another; force is an inherent property only in the sense that the effects of forces are universal characteristics of chemical reactions. Rouelle makes no attempt to explain the cause behind attractive and repellent forces, but, in the Newtonian manner, accepts the observed phenomenon of attraction as an unknown but essential property of matter.[7]

In his exposition of the primary substances formed by ultimate particles of matter, Rouelle follows closely the Stahlian system in which substances are organized into a hierarchy of principle, mixt, and compound (*principe, mixte, composé*), and increasingly complex orders.[8] Principles or elements are the simplest substances; these are formed directly from elementary particles, each such element consisting of an aggregate of homogeneous particles. There are four elements which are common to all the realms of nature: fire, air, water, and earth.[9] To these four a fifth, the mercurial principle posited by Becher, can be added, but Rouelle is uncertain of its existence.

It is impossible to isolate an element because the elements are not the immediate components of complex substances. Particles of elements unite to form mixts, with two, three, or even all four elements uniting in varying proportions to produce particles of mixts. This union of elements is maintained by an exceedingly strong force of adhesion, which can be broken by means of combinations: "to separate an element from a mixt one applies to the mixt another body, which adhering to this element detaches it from the mixt, and carries it away."[10] The element thus removed from a mixt combines again immediately with another substance and cannot be

[7] See below, pp. 82–83.

[8] Stahl, *Philosophical Principles*, esp. pp. 6–8. Unless otherwise noted, discussion of Rouelle's theory of matter and the elements is based upon the introductory theoretical section present in all the manuscripts consulted; in the Bordeaux MS, see pp. 34–94.

[9] Rouelle repudiates the *tria prima* and praises its refutation by Boyle, who, however, "a voulu substituer a cette doctrine celle de la philosophie corpusculaire qui n'est pas mieux fondée selon M. Rouelle" (Duveen MS, Vol. 1, p. 6).

[10] Bordeaux MS, pp. 44–45.

retained in its pure state. The metals are common mixts which, upon cal-cination, release combined fire (phlogiston) and leave a residue of earth; but the earth is impure and the fire escapes collection, so that the pure elements are never obtained. Thus, the mixts are the simplest substances the properties of which can be examined directly, while the elements can only be known through the properties they exhibit as they pass from one combination to another, escape into the atmosphere, or remain in an impure state after a chemical reaction.

Mixts of different kinds unite to form compounds, the "cohesive" bond uniting mixts being weaker than the "adhesive" bond combining ele-ments. Various compounds, analogous to modern chemical compounds, may combine to form supercompounds (*surcomposés*), an example of which is soap, a combination of an animal or vegetable oil and a fixed alkali.[11]

The foregoing discussion of elements, mixts, and compounds evidently has no meaning for individual particles of these substances, since these are not available to the senses. The only substances used in the chemical laboratory are large numbers or aggregates of particles, held together by a weak cohesive force.

While mixts, compounds, and supercompounds are convenient classifi-cations for the innumerable substances used by the chemist, the elements play a more important role in Rouelle's theory of matter. All chemical reactions can in fact be explained in terms of the chemical and mechanical behavior of the four elements. Each of the elements has, for Rouelle, a dual nature; each serves as both a component of matter and an instrument of chemical reaction.

In listing the four elements, Rouelle actually calls the first one fire or phlogiston, a designation which indicates its double function as an instru-ment and as a constituent of matter. The action of fire as a solvent or instrument results in changes of the physical state of the substance acted upon. In the rarefaction of air, for example, particles of fire insinuate themselves into the pores or interstices of an aggregate, separating the molecules from each other, but without changing the properties of the individual molecules.

The instrument fire is also responsible for chemical changes when its particles enter the pores of a mixt and disrupt the strong adhesive bonds. Since the union of elements to form mixts is very intimate, individual molecules of the mixt are sometimes completely dense and nonporous, and fire cannot penetrate to produce decomposition. But there is combined within every substance an amount of heat (or fire) which varies according to the density of the substance; the denser and more compact a body is, the less heat it contains and the more heat it will have to absorb before it will expand or enter into fusion. When fire is applied externally to a mixt, the

[11] *Ibid.*, p. 447. Rouelle rejects the still more complex orders in Stahl's hierarchy; see *ibid.*, pp. 45–46.

particles of fire already trapped within the mixt are set in motion in an effort to escape their confines. The greater the intimacy of union in a mixt, the more violent will be the struggle to escape. This, for example, accounts for the production of an explosion when nitre is subjected to heat.[12] In reactions of this kind the original mixt is decomposed and a new substance or substances formed. Here fire—both the applied fire and that within the mixt—has been the instrument of chemical change.

To distinguish the instrument fire from the element, Rouelle applies the name "phlogiston" to fire in its role as a constituent of matter. The combination of particles of fire within molecules of a mixt or compound forms the basis of the phlogiston theory, which was expounded by Stahl and adopted in modified form by Rouelle.[13] No systematic treatment of the phlogiston theory is presented by Rouelle, but phlogiston is referred to as the "material of fire," found combined with substances of all realms of nature and capable of being released by these substances during the process of combustion.

Phlogiston is not presented as the center and focus of Rouelle's chemical theory. In fact, there is no single place in his chemical lectures where the reader can find a statement of all the major points of the phlogistic doctrine. On the contrary, it is only after an examination of individual experiments and the explanations offered that a complete picture of Rouelle's views on phlogiston emerges. Rouelle states briefly that the chemical process involved in calcination is the release of phlogiston from a metal. To restore the metallic nature of a calx, a reducing agent like charcoal is employed as a source of the phlogiston which must recombine with the calx.[14] Rouelle cannot explain the augmented weight of the metallic calces, so noticeable for certain metals and of which the phlogistonists could present no reasonable interpretation. Stahl had ignored the problem, but his disciple Juncker made an attempt to solve it; and it is Juncker's solution which Rouelle adopts. Rejecting the possibility that the increased weight is a result of the addition of particles of fire to the metal, Rouelle argues that the increase in weight of the calx of antimony over the weight of the original metal is only apparent, and that

the absolute weight being always the same, it is only the specific weight which increases, doubtless because the volume [of the calx] decreases much more than the [amount of] real matter.[15]

[12] *Ibid.*, p. 760.

[13] Stahl, *Traité du Soufre*, trans. d'Holbach, Paris, 1766, pp. 55ff. The phlogiston theory first appeared in Stahl's *Zymotechnia fundamentalis seu fermentationis theoria generalis*, Halle, 1697. The properties of free fire and combined fire (phlogiston) are enumerated with admirable clarity by P.-J. Macquer, *Elements of the Theory and Practice of Chymistry*, trans. Andrew Reid, 2 Vols., London, 1758, Vol. 1, pp. 7–11.

[14] For phlogiston's conversion of iron into steel, Bordeaux MS, p. 1087.

[15] *Ibid.*, pp. 970f. Rouelle is cited on this point by d'Holbach, in Denis Diderot and Jean le Rond d'Alembert, *Encyclopédie, ou Dictionnaire raisonné des sciences, des arts et des*

One of the most curious aspects of the phlogiston theory is its use to explain color phenomena. Rouelle adduces several "proofs" of this theory which he uses to "refute" Newton's ideas about the nature of white light. Phlogiston, the principle of color, is declared to be absent from the calx of antimony, which is white or deprived of all color. This reasoning, fitting so neatly into the phlogiston theory of calcination, leads Rouelle to conclude that "the theory of Newton . . . is not correct, true though his experiments may be; for the whiteness that this great mathematician thought was composed of the seven primary colors is an absolute privation of all color."[16]

There are two aspects to Rouelle's discussion of the second element, air; he considers in turn its properties when it is free and its effect on substances with which it is in combination. Air itself consists of particles which, taken singly, are inelastic; on the other hand, an aggregate of such particles is elastic and subject to expansion and compression.

Most physicists believe that the particles of air are so many little elastic spirals, not noting the fact that if this were so, each molecule taken separately would be elastic, which experience shows to be false.[17]

The evidence offered to prove this property of air is the expansion of air when released from combination. In Rouelle's terminology, molecules of air released from combination form an aggregate which, being elastic, occupies more space than all the molecules taken individually. Fire causes air to expand to such a degree that the aggregation is destroyed and the elasticity lost; the reverse is true of cold. Carrying this idea to its conclusion, Rouelle asserts that in conditions of absolute cold, or in the absence of all fire, air would take a concrete form.

The role of air as an instrument in chemical reactions is expounded by

---

métiers, Paris, 1751–65, "Régule d'antimoine," Vol. XIV, p. 40. For Rouelle's rejection of Boyle's explanation of the augmented weight of a metallic calx, see Bordeaux MS, p. 970. Juncker's explanation, for which he was indebted to Kunckel, is quoted in Metzger, *op. cit.*, pp. 187–188: "Cette augmentation de poids vient d'une plus grande condensation des parties terrestres qui arrivent lors de la calcination. Les molécules occupant moins d'espace empêchent l'air de passer aussi librement entre elles; et conséquemment, comme la compression de l'air augmente, leur poids spécifique doit augmenter aussi. . . . Les substances les plus légères et qui occupent beaucoup d'espace, telles que la laine, le duvet, pèsent davantage quand on leur fait occuper un moindre espace. Les marchands d'étoffe de soie savent très bien cela."

For Macquer's cautious rejection of all theories on this subject, see his *Elemens de chymie-pratique*, 2 Vols., Paris, 1751, Vol. 1, pp. 307–308.

See also the excellent series of articles by J. R. Partington and Douglas McKie, "Historical Studies on the Phlogiston Theory," *Annals of Science*, 2, 361–404 (1937); 3, 1–58 and 337–371 (1938).

[16] Bordeaux MS, p. 978; also, *ibid.*, pp. 1121–1124 and 1128. For Senac's views, see below, note 45.

[17] Duveen MS, Vol. 1, p. 29.

Rouelle as part of his thesis that each of the four elements has a double function as nonparticipating instrument and as material reagent. Not only is the presence of free air recognized as a necessary condition for burning, a fact which Rouelle cannot understand or explain,[18] but Rouelle also describes the mechanical action of air in a distillation process. Alternate expansion and contraction of air in contact with the surface of a liquid to be distilled results in the following cycle: air expands due to the applied heat and leaves the distilling vessel; vapors carried by the air are then condensed in the recipient; and finally the decrease in pressure in the vessel causes air to rush in through a carefully prepared opening, and the process repeats itself.[19]

That air enters into chemical combination was established by the work of Stephen Hales, whose apparatus was improved upon by Rouelle in the course of his experiments.[20] However, the "air" obtained during the decomposition of any substance was likely to differ somewhat from other "airs" and was of necessity described according to its peculiar properties. In Rouelle's experiment to remove the air from a vegetable substance by distillation, the product is described as similar to the air of the atmosphere, but inflammable because it contains phlogiston.[21] Another such process, the object of which is to dissolve zinc in dilute hydrochloric acid, produces a *mélange* containing vapors which take fire when near a lighted candle, "which makes M. Rouelle say that the hydrochloric acid removes the phlogiston of zinc."[22]

Rouelle admits without comment the presence of the third element, water, as an ingredient in many substances. The very nature of the molecules of water make it capable of entry into combination to form solid substances. Opposing the atomist view that the fluidity of water can be explained by the oval shape of its particles, Rouelle remarks that, if such were the case, water could never solidify. Fluidity is the result of the presence in water of minute particles of fire (the principle of motion and

---

[18] Bordeaux MS, p. 174.

[19] *Ibid.*, pp. 131–132.

[20] Henry Guerlac has discussed Hales's influence on Rouelle in his "Lavoisier and His Biographers," *Isis*, **45**, 51–62 (1954). In his earlier article, "The Continental Reputation of Stephen Hales," *Archives Internationales d'Histoire des Sciences*, no. 15, 393–404 (1951), in which the influence of Hales in France was established, Rouelle was not mentioned. For references to Hales and the experiments of Rouelle, see Bordeaux MS, pp. 108, 206b, 207–208, and 990b. Also, A.-L. Lavoisier, *Œuvres de Lavoisier*, 6 Vols., Paris, 1862–1893, Vol. 2, p. 7. Rouelle's apparatus is reproduced in the *Encyclopédie méthodique*, Paris, 1782–1832: description in Vol. 66, p. 354 and Vol. 70–71, p. 16 of Explication des Planches; illustration in Vol. 70–71, Planche XVII, Fig. 45. Pierre Lemay's manuscript contains a reproduction of Rouelle's improved version of Hales's apparatus; see Lemay, "Les cours de Guillaume-François Rouelle," *Revue d'Histoire de la Pharmacie*, no. 123, 434–442 (Mars 1949).

[21] Bordeaux MS, pp. 207–208.

[22] *Ibid.*, p. 1003; for the reaction between iron and hydrochloric acid, *ibid.*, p. 1110.

fluidity), but when water loses most of its fire it also loses its accidental fluidity and becomes ice. Ice is the natural, pure state of water. If the molecules of pure water were visible, Rouelle conjectures that they would look earthy and proper for combination in solid substances.

As an instrument water fulfills important functions in solutions, crystallization, and combustion. Water acting as a solvent impinges on the particles of the solute, mechanically accelerating the process of solution. In his discussion of the role of water in crystal formation, Rouelle recognizes as a distinguishing property, useful in classification, the deliquescence and effervescence of salts. He also states clearly the difference between water of solution and water of crystallization, the latter being an essential part of crystal structure.[23] In combustion the expansion of water augments the flame and "is so much an instrument of the flame that substances which are totally deprived of it never give any [flame]: such is the state of perfect charcoal, that is, [charcoal which has been] well burned in free air."[24] But water also has the peculiar complementary property of extinguishing flame only when the burning substance is deprived of contact with air.

A consideration of the fourth element, earth, necessitates a statement of the ideas of Becher whose work influenced Rouelle through the writings of the Stahlian school.[25] Although Becher claimed he did not hold the *tria prima* theory of the Paracelsians, his doctrine of the three earths is entirely analogous to it. The three earths of Becher, however, are not three distinct elements, as are the spagyrical principles; Becher's earths are the principal classes of a single element. The first or vitrifiable earth occurs in its purest form in hard rocks, like flint or quartz, and may be compared with the Paracelsian salt principle. The second or sulfureous earth is the fatty and inflammable substance which Stahl named phlogiston, and which is evidently akin to the sulfur principle. Mercurial earth, the third in Becher's scheme, provides those properties peculiar to metals, giving them a "metallic essence" which distinguishes them from other types of substances. The existence of this latter earth was questioned by Becher's disciples and poses a serious problem for Rouelle. He suspects that phlogiston and mercurial earth are really identical. Having associated phlogiston with fire rather than with earth, however, Rouelle is not sure but that there may be only one kind of earth—the vitrifiable. He recognizes that seemingly different types of earths are formed after certain chemical reactions, but cannot decide whether or not these are impure manifesta-

[23] *Ibid.*, p. 428. For Rouelle's important memoirs on the classification of salts using crystal structure, see the author's "G.-F. Rouelle," p. 82.

[24] Bordeaux MS, pp. 469–470; this theme is developed at length, *ibid.*, pp. 466–470, 760, and 776.

[25] The theory of the three earths is discussed by Stahl, *Traité des sels*, trans. d'Holbach, Paris, 1771, pp. 6–7, and by Metzger, *op. cit.*, pp. 131–134.

tions of a single element. Whatever the solution, he concludes, earth is the principle which gives substances solidity, consistency, and stability.[26]

Rouelle's application of his element-instrument theory to the earth principle is not quite as clear as are his ideas about fire, air, and water. Having transferred Becher's second earth, phlogiston, to the fire principle, he is left with the vitrifiable and mercurial earths constituting the fourth element. About the vitrifiable earth there seems to be no confusion; this is the principle of stability and solidity and the basic constituent of most complex substances. Rouelle's description of earth as friable, porous, insipid, odorless, fixed, and immiscible with water, alcohol, and oils seems to apply only to the vitrifiable earth. As an instrument, earth of various appearances and properties—but all having the same fundamental make-up—serves not only in the construction of chemical vessels, but also as an aid in distillations.

Pure absorbent earth serves in chemistry to remove from the alkalis taken from plants or animals an empyreumatic oil which is united to them. It also serves to check the expansion of materials which would rise above the vessels, without that [earth] in distillation, like honey, manna, turpentine. . . . [Some earths] also serve in the distillation of nitre to prevent the molecules of nitric acid from reacting strongly with the glass. . . .[27]

The mercurial principle, in Rouelle's few statements about it, seems to have at least some of the properties attributed to phlogiston—the production of fluidity and volatility—in addition, of course, to giving metals their peculiar metallic essence.[28] Such properties place the mercurial principle in the class of instruments. The presence of this principle in metals also places it in the class of constituents of matter, and, thereby, makes the mercurial principle a candidate for inclusion in the element-instrument theory already applied to the four elements. That Rouelle occasionally made use of the mercurial principle, the existence of which he doubted, suggests that he did not flatly deny its existence because it fitted neatly into his theory and its exact status as a real substance had not been proved or disproved experimentally.[29] It is not clear, however, if Rouelle usually thought of this dubious substance as a fifth element or as simply a second type of earth.

[26] Bordeaux MS, p. 80.

[27] *Ibid.*, pp. 80–81.

[28] For the composition of metals, Bordeaux MS, pp. 1247–1248. For the similarity between phlogiston and the mercurial principle, *ibid.*, p. 864. Phlogiston and volatility, *ibid.*, pp. 339–340 and 540; mercurial principle and volatility, *ibid.*, pp. 819f. Phlogiston and fluidity, *ibid.*, p. 74; mercurial principle and fluidity, Duveen MS, Vol. 2, p. 210.

[29] Cf. Macquer, *Elements*, Vol. 1, pp. 45f. Macquer believes that there are "pretty strong reasons" for admitting the existence of the mercurial principle and goes on to say that the addition of phlogiston to sand, for example, does not produce a metal, which shows that metals must contain a distinctively metallic principle, the mercurial earth, in addition to phlogiston.

In order to explain fully the nature of chemical reactions and the mechanism involved, Rouelle supplements his element-instrument theory by a discussion of the action of solvents and the effects of affinities or attractions. Every reaction, whether it results in a physical or a chemical change, seems to begin with the separation of the particles of a solute by those of a solvent; only when this has occurred (or has at least begun to occur) can a chemical reaction on the corpuscular level take place. Inter-corpuscular changes are then determined by the attractive properties of the reagents.

In every chemical solution Rouelle considers the dissolving action to be reciprocal, solvent and solute acting on each other. Both agents should be called "dissolvents," but the name is usually applied only to the liquid which starts the process. There are two kinds of reactions which may occur in solutions; these are distinguished by the terms "resolution" and "disso-lution." Resolution describes a solution of salt in water, where the proper-ties of the salt remain unchanged and the removal of the solvent restores the salt to a form somewhat less solid but otherwise entirely similar to its original state. Dissolution, on the other hand, involves a chemical change in which the properties of solvent and solute are altered and their initial properties can be restored only by another chemical reaction.

In both types of processes the molecules of the solid substance are separated by those of the liquid. Since Rouelle refuses to reason from the figures of particles, he must find another way of explaining how some particles remain suspended in a solution, while others unite to form new substances or reunite to produce the original solute which precipitates out of solution. There must be some cause which produces the union of par-ticles, or the divided molecules would either separate out of solution according to their respective weights or remain mixed with the liquid solvent, rendering it opaque.

One can hardly fix the cause which begins the motion of dissolution; the solvent only possessed the movement of liquids, the solute was at rest, and with their mix-ture there was stirred up a heat which was neither in one nor in the other. This cause can only exist between the particles of the solvent and the particles of the solute, since if it depended upon the general agents air and fire, it would act indiscriminately on all bodies, instead of which it acts only on some. A certain solvent acts on this [solute], and does not act on that. . . .[30]

The cause of solution and precipitation, "or rather its effects,"[31] is called

[30] Duveen MS, Vol. 1, p. 40. The passage continues: "il faut donc recourir a une pro-prieté inhoerente." (Cf. Bordeaux MS, p. 88.) The implications of the term "inherent" seem inconsistent with the more commonly found expressions, reminiscent of Geoffroy's attitude, cited below and in notes 31 and 32.

[31] Bordeaux MS, p. 86.

affinity or rapport, the guiding principle of all chemical reactions. Describing the origin of the notion of affinity, Rouelle makes it clear that he considers affinity only a name for an observed phenomenon.

> Former chemists noticed that certain bodies [placed] at a certain distance [apart] attracted each other; they gave the cause of this phenomenon the name *sympathy* for which the moderns have substituted that of rapport or affinity: this attraction does not follow the law of the square of the distances as does Newtonian attraction, but that of the homogeneity of surfaces.[32]

The last lines quoted above imply that Rouelle has formulated some ideas about a law governing chemical affinities, but nowhere does he develop this statement further.[33]

For Rouelle, affinities are decidedly not in the realm of occult forces, although he does not fully understand how they operate. His discussions of affinities are restricted to descriptions of displacements observed in chemical reactions and tabulations of experimentally determined affinities. The first table of affinities, the compilation of E.-F. Geoffroy, appeared in 1718. It is entirely in accord with Rouelle's emphasis on the importance of experiment that he should have considered worthwhile the revision of this empirically useful tabulation.[34]

STAHL, BOERHAAVE, AND ROUELLE

How Rouelle obtained his knowledge of the works of Becher, Stahl,

[32] Duveen MS, Vol. 1, p. 40.

[33] The explanation advanced by Turgot, who was a Rouelle pupil, may be quite close to Rouelle's own views which are not developed in the manuscript lectures. In his article "Expansibilité" in the Diderot *Encyclopédie*, Turgot notes that attraction (affinity) between particles "ne serait sensible qu'à des distances très petites, et qu'elle serait infinie au point de contact; il est évident: 1° que l'adhérence résultante de cette attraction est en partie relative à l'étendue des surfaces par lesquelles les molécules attirées peuvent se toucher, puisque le nombre des points de contact est en raison des surfaces touchantes; 2° que moins le centre de gravité est éloigné des surfaces, plus l'adhésion est forte. En effet, cette attraction, qui est infinie au point de contact, ne peut jamais produire qu'une force finie, parce que la surface touchante n'est véritablement qu'un infiniment petit; la molécule entière est, par rapport à elle, un infini. . . ." A.-R.-J.Turgot, *Œuvres de Turgot et documents le concernant*, ed. Gustave Schelle, 5 Vols., Paris, 1913–1923, Vol. 1, p. 567.

[34] For Rouelle's use of affinities, see Bordeaux MS, pp. 418, 890, and 1056. His table of affinities is reproduced *ibid.*, unnumbered pages following p. 122; Duveen MS, pages appended to manuscript; J.-F. DeMachy, ed., *Recueil de dissertations physico-chymiques*, Amsterdam, 1774, Plate V; and *Encyclopédie*, Planches, Vol. 3, "Chimie," p. 1, Planche Iere. Rouelle's refusal to speculate about the causes of affinity has given him an undeserved reputation as a repudiator of the whole idea of affinity; cf. Pierre Duhem, *Le mixte et la combinaison chimique*, Paris, 1902, p. 43; Maurice Daumas, *Lavoisier*, n.p., 1941, p. 20; Metzger, *op. cit.*, p. 89; and R. Dujarric de la Rivière and Madeleine Chabrier, *La Vie et l'œuvre de Lavoisier d'après ses écrits*, Paris, 1959, p. 162. These writers produce no evidence from Rouelle's work, but usually group Rouelle with his pupils Venel and Monnet who did attack the doctrine of affinities. Dujarric makes Macquer the head of a rival school which subscribed to the notion of attractive forces.

and their disciples is a matter for conjecture. The source may have been J.-B. Senac's *Nouveau cours de chymie, suivant les principes de Newton et de Sthall* [*sic*], which appeared in 1723, but Rouelle also seems to have had some acquaintance with the achievements of Henckel, Juncker, Pott, and other German chemists and mineralogists. Latin editions of Stahl and Juncker, the two most important sources for Stahlian chemistry, were available to Rouelle, but Grimm raises some doubt that he was able to read Latin (and/or German) when he notes:

> Rouelle was a man of genius [but] without culture; . . . it is he who introduced the chemistry of Stahl [into France], and made known here that science about which we knew nothing, and that a number of great men have carried to a high degree of perfection in Germany. Rouelle did not know how to read them all; but his instinct was usually as sound as their science.[35]

Apart from Senac and the writings of the Germans themselves, a third possible source of Rouelle's knowledge of Stahl remains: the German pharmacist J.-G. Spitzley, with whom Rouelle had served his apprenticeship. However, not enough is known about Spitzley to make this line of speculation at all fruitful.

Whatever his sources—whether Stahl himself or one of his more or less orthodox German disciples—Rouelle was no timid follower of Stahlian thought, but unhesitatingly disagreed with, modified, and eliminated portions of his master's doctrines. The system of element, mixt, and compound is common to both Stahl and Rouelle, but Rouelle pared away much of Stahl's discussion of the more complex orders of chemical substances and strengthened the remaining theory by linking it as closely as possible to experiment. This process of fortifying the foundations of Stahlian theory was in fact accomplished by the addition of new features to Stahl's theory of the elements.

In order to determine the changes in theory introduced by Rouelle, it is necessary to examine Stahl's treatment of the same four substances called elements by Rouelle. Stahl's theory of the elements is summarized in the following passage:

> In general therefore we allow, with *Becher, Water* and *Earth* for the *immediate material Principles* of *Mixts*; and with him suppose this *Earth* to be of three kinds, *viz.* (1) *vitrifiable* or *fusible*, (2) *inflammable* in composition, and (3) *liquifiable* or *specifically mercurial* in *Metals*.[36]

Stahl has adopted the two-element theory of Becher and is including in his term "earth" the three earths of Becher.

In Stahlian chemistry fire and air are instruments of chemical change rather than elementary constituents of matter.

---

[35] Friedrich M. Grimm, *Correspondance littéraire, philosophique et critique par Grimm, Diderot, Raynal, Meister, Etc.*, ed. M. Tourneux, 16 Vols., Paris, 1877–1882, Vol. 9, p. 106.

[36] Stahl, *Philosophical Principles*, p. 8.

Fire or heat drives away the more moveable particles of the Concrete, and leaves all the rest immoveable by this means; but if applied in a violent degree, it not only forcibly carries off the more moveable parts, but along with them also the adjacent ones, that were otherwise immoveable; whereupon a *violent Combustion ensues.*

Air, on the other hand, can only wear and break off the finest particles of all; without being able considerably to move such as are more fix'd.[37]

Water too is given, in addition to its role as an element, an instrumental function as a solvent; particles of water are said to impinge upon adjacent particles of the solute, imparting motion to the latter.[38] For at least three of Rouelle's elements, then, their instrumental functions are common to the theories of both Stahl and Rouelle. One of the three, water, is also considered an element in both systems.

Stahl's description of the earth principle and its nature and properties presents a rather confusing picture. Although earth is "the Principle of Rest and Aggregation,"[39] motion and an instrumental function can be imparted to this passive principle. At least one of the three earths, phlogiston, is capable of motion, but only when it has been excited by the agency of fire.

It is important to observe that this fiery matter [phlogiston] left to itself, and without the assistance of air and moisture, is not found attenuated or volatile; but once it has been attenuated and volatilized by the motion of fire, and by contact with the open air, then it has a subtlety and a dilatation. . . .[40]

Even volatile phlogiston, which Stahl describes as "the true matter of fire, the real principle of its motion in all combustions,"[41] is still only one kind of earth and as such does not possess self-motion. When phlogiston is in combination, it is set in motion by the agency of fire and escapes from the substance with which it was combined; in this sense, both fire and phlogiston can be considered instruments of chemical decomposition.[42]

In summary, Stahl's system includes two elements, earth and water, and four instruments, earth, water, air, and fire.[43] Rouelle, on the other

---

[37] *Ibid.*, p. 45; cf. pp. 59ff. For Stahl's denial of air's chemical role, see his *Traité des sels*, pp. 349–356. Metzger, *op. cit.*, p. 113, asserts that although both Stahl and Juncker said air does not enter into mixts, they added cautiously that experiments designed to show the entry of air into chemical combination are not absolutely demonstrative.

[38] Stahl, *Philosophical Principles*, pp. 46f.

[39] *Ibid.*, p. 65.

[40] Stahl, *Traité du Soufre*, p. 56.

[41] *Ibid.*

[42] Cf. Rouelle, above, p. 77.

[43] Both Rouelle and Stahl actually list more than four instruments, since they include solvents and vessels. Rouelle calls the four elements "natural" instruments and solvents and vessels "artificial" ones; Bordeaux MS, pp. 50–51.

hand, was able to produce "evidence" that all four substances were both constituents of matter and instruments of chemical change. In order to do so, Rouelle had to introduce two changes into Stahlian theory.

One change in theory effected by Rouelle was stimulated by the work of Stephen Hales, who demonstrated conclusively that air enters into chemical combination. Rouelle repeated Hales's experiments in the course of his lectures and found it easy enough to conclude that air was both an instrument, as Stahl had claimed, and an element, which Stahl had specifically denied.

Rouelle's second change in Stahlian theory is a more startling and a subtler one. Stahl's phlogiston, a fatty earth, becomes associated in Rouelle's system with *fire* rather than with earth. When Rouelle refers to the inflammable principle which is a constituent of metals, he means corporeal fire, "the material of fire," or phlogiston, which is the name given to the instrument fire when it acts chemically. The properties of phlogiston, as enumerated by Stahl, remain essentially unchanged in Rouelle's theory; phlogiston is still the principle of odors and colors, the principle released in combustions, and an element which may be set in motion by the action of the instrument fire. But the only relation between phlogiston and earth in Rouelle's system is the same relation as obtains between air and earth and between water and earth: the particles of phlogiston, air, and water would all be "earthy"—i.e. solid—in appearance, if they were visible, and thus suitable for entrance into chemical combinations.

With the shift of phlogiston from an earthy to a fiery substance, Rouelle's theory remains with one kind of earth, the vitrifiable, and the mercurial principle, which may be a form of earth or a fifth element. Both the vitrifiable earth and the mercurial principle serve as elements and instruments for Rouelle, whereas in Stahl's theory they are ill defined.[44]

The originality of Rouelle's element-instrument theory is all the more striking when his chemical lectures are compared with the only professedly Stahlian work to appear in French before Rouelle started to teach. This book, Senac's *Nouveau cours de chymie, suivant les principes de Newton et de Sthall* (1723), is in some ways an improvement over orthodox Stahlianism, notably in the discussion of crystallization and in the use of Newton's work to explain color phenomena.[45] In an attempt to treat chemistry in a purely empirical manner, Senac further omits the Stahlian theory of principle-mixt-compound; however, his adherence to the experimental evidence is

[44] Antoine Baumé, Macquer's associate, evidently confused the ideas of Stahl with the modifications introduced by Rouelle when he claimed that Stahl recognized only the first two of Becher's three earths. See Baumé, *Manuel de chymie, ou exposé des opérations et des produits d'un Cours de Chymie*, Paris, 1763, p. 64.

[45] Jean-Baptiste Senac, *Nouveau cours de chymie, suivant les principes de Newton et de Sthall*, 2nd edition, 2 Vols., Paris, 1737, Vol. 1, pp. 250–252 and Vol. 2, pp. 181–183.

not consistently maintained, for Senac discusses at length the properties imparted to various substances by the shapes of their particulate components.[46]

In expounding his professedly Stahlian theory of the elements, Senac states that fire, water, and earth are the primary constituents of matter and that these three substances plus air comprise the major instruments of chemical operations.[47] From the preceding discussion of Stahl's theory, it is clear that fire has here been added to the roster of elements. In addition to this new feature, Senac's treatment of the earth principle also diverges from Stahlian orthodoxy.

Although Senac's general definition of the earth principle—a substance which is friable, porous, tasteless, and odorless—is consistent with Stahl's, he goes on to question the existence of the three types of earth posited by Becher and Stahl.[48] His doubts are similar to Rouelle's in that he suspects that the sulfureous principle, phlogiston, and the mercurial principle are identical; but Senac offers no solution to this problem and continues to use all three earths throughout his work. In the Stahlian tradition, Senac considers phlogiston a form of earth which he calls the "sulfureous principle," "oily material," or "inflammable principle." Phlogiston is the substance lost during calcination and regained in the reduction of a metallic calx.

Senac's treatment of fire is extraordinarily confused and confusing. After his initial declaration that fire is one of the three elements, he goes on to discuss only the other two. Immediately following the sections dealing with water and earth, he states that he will now treat "the materials which result [from them]: the principal ones are salts and sulfurs. Salt is a concretion of the three elements, fire, water, and earth."[49] In still another passage he declares: "The second earth . . . is the oily material which is found in plants and in animals. This oily material contains the inflammable principle."[50] Thus, phlogiston is clearly not the element fire Senac refers to but never defines. "The inflammable principle," phlogiston is thought to be a constituent of the Stahlian second earth.

The vagueness of Senac's discussions of the theory of fire can perhaps be somewhat clarified by reference to his explanations of chemical

[46] E.g., his discussion of the element earth which is "friable, parce qu'étant remplie de pores, ses parties ne se touchent que par leurs angles qui cedent facilement. Elle est insipide & sans odeur, parce que ses parties sont trop grossiéres pour ébranler les nerfs de la langue & du nez." *Ibid.*, Vol. 1, pp. 13–14.

[47] *Ibid.*, Vol. 1, p. 13 (elements); pp. 13–22 and 56 (earth); pp. 22–26 and 139–142 (water); pp. 55f., 131f., and 139 (air); and pp. 127–130 (instrument fire).

[48] *Ibid.*, Vol. 1, pp. 21–22.

[49] *Ibid.*, Vol. 1, p. 27. Despite constant use of *feu* and variant expressions, the discussion of "fire" included under Becher's second earth refers to phlogiston, an earthy principle (*ibid.*, Vol. 1, pp. 19–21).

[50] *Ibid.*, Vol. 1, p. 16.

reactions. Here, too, ambiguities and lack of consistent terminology hamper the reader. Senac talks, for example, of the universal presence of fire and of fiery particles (*parties ignées*) capable of penetrating all substances. But his examples, designed to show the omnipresence of fire, leave some doubt that the fiery substance under discussion is a chemical element rather than an instrument.

Damp hay takes fire, when it is gathered, [and] it is fermentation which disengages the particles of fire, and permits them to act; oils although cold are permeated by fiery particles: the nitre that one can obtain from plants, is only a true fire (*un véritable feu*); finally all that one obtains from animal substances is full of an inflammable material, phosphorus obtained from urine is a proof of this.[51]

It is evident from this passage that Senac is confusing the instrument fire with the earthy element phlogiston. The instrument fire is a particular material lodged in the interstices of the damp hay and the oils; when these particles are dislodged, they become capable of action. But are these particles the same as the "inflammable material" or phlogiston mentioned by Senac? Is phlogiston a form of fire or is it an earthy substance which is highly inflammable—i.e. fire-*producing*—but is not fire itself? Elsewhere, it will be recalled, Senac showed himself an adherent of the theory of the three earths, with phlogiston the fatty, sulfureous second earth.

Senac's discussion of the role of fire in calcination can perhaps shed additional light on his views of the nature of fire.

That fire divides and rarefies the particles of substances, this is shown by the augmentation of weight which occurs after calcination: tin and the other metals become heavier; . . . these surprising effects evidently prove that it is fire which augments the weight of these substances by its particles which insinuate themselves into the interstices. . . .] . . . [. . . While the absolute weight (*gravité absoluë*) increases, as I have shown, the specific gravity decreases; this occurs because the particles of the substances [having been] separated by the action of fire form a greater volume; moreover the fiery particles, lighter than those of the calcined material, scatter into their interstices, decrease by their lightness the specific gravity, and increase the absolute weight (*pesanteur absoluë*).[52]

The first paragraph is patterned on the views of Robert Boyle, while the second resembles those of Juncker and Rouelle. Here fire plays a part only in the physical effect which is detected by observing the increased weight of the metallic calces. That Senac does not at this point mention phlogiston's role in calcination lends support to the belief that he usually considers phlogiston a form of earth, for he cannot claim in the same breath that fire enters a metal to augment its weight while at the same time leaving the metal in the guise of phlogiston.[53] If phlogiston has any

[51] *Ibid.*, Vol. 1, p. 127.
[52] *Ibid.*, Vol. 1, pp. 160–161.
[53] Rouelle was able to avoid this difficulty by repudiating Boyle's theory, Bordeaux MS, p. 970.

function in calcination, it clearly must act as an earthy rather than a fiery principle.

Thus, despite Senac's statement to the contrary, there are only two elements—water and earth—posited in his *Nouveau cours de chymie*. In his theory of elements and instruments Senac is following the tradition of orthodox Stahlianism. Apparently unaware of the work of Stephen Hales, which was published in French translation just two years before the second edition of the *Nouveau cours de chymie*, Senac also follows Stahl in denying air any function as a chemical agent.

Finally, some consideration of Senac's discussions of chemical affinity will serve to show that in this area Rouelle made use of a tradition different from that of Senac. The title of Senac's work includes the name of Newton as well as that of Stahl, and it is in dealing with affinity that Senac makes a prominent display of his Newtonianism. He identifies chemical affinity with Newtonian attraction, observing that the force of attraction in this instance probably varies with some power of the distance greater than its square.[54] Then, possibly taking his cue from Newton's remark that "Bodies act one upon another by the Attractions of Gravity, Magnetism, and Electricity,"[55] Senac proceeds to confuse his terms and ideas in his customary fashion. The following are typical examples:

since the *rapports* or affinity of bodies depends in part on the tendency they have to unite, we are going to speak of the Magnetism that one observes in all of nature.[56]

and

. . . there is in nature a magnetism which brings bodies or their particles together; it is to this attractive force that it is necessary to attribute most of the phenomena which are most surprising in the composition or decomposition of bodies.[57]

Senac does not furnish tables of affinities, but simply lists, as Stahl and Newton did, the relative displacements of a few metals and acids.[58] The tradition of Geoffroy, continued by Rouelle, had no apparent influence on the author of the *Nouveau cours de chymie*.

The unique coherence and appeal of Rouelle's element-instrument theory are perhaps best shown by a comparison of his theory with the widely held views of his older contemporary, Hermann Boerhaave. The influence of Boerhaave actually extended to both theoretical and experimental chemistry. In his recent study, F. W. Gibbs ably defends the thesis that Boerhaave was considered the guiding spirit in the laboratory, while

---

[54] Senac, Vol. 1, p. 153.
[55] Isaac Newton, *Opticks*, New York, 1952, p. 376.
[56] Senac, Vol. 1, p. 74.
[57] *Ibid.*, Vol. 1, p. 76.
[58] *Ibid.*, Vol. 1, pp. 96ff., 117. Cf. Newton, pp. 380–381, and Stahl, *Philosophical Principles*, p. 40.

Stahl was, at the very same time, the great figure in systematic or "sublime and philosophical" chemistry.[59] While this distinction is an important one which will be referred to again, here I shall be concerned only with the replacement of Boerhaavian theory, and not method, by the revised Stahlianism of Rouelle.

Boerhaave's brilliant reputation, augmented by the publication of his chemical lectures, made Leyden a center of experimental science during his own lifetime and during the ascendancy of his pupils and successors. In 1753, fifteen years after Boerhaave's death, the influence of his *Elementa Chemiae* was still such that Venel complained that

the learned compilation of the famous Boerhaave on fire, [is] known, cited, and praised; while the superior views and unique things which Stahl has published on . . . [this subject], only exist for some chemists.[60]

When the French translation of Boerhaave's textbook appeared in 1754, J.-F. DeMachy, one of Rouelle's pupils, attempted to counterbalance its popularity by translating into French one of the major works of the Stahlian school, Juncker's *Conspectus chemiae*.[61]

Boerhaave formulated no theory of the chemical elements, although he was enough of an atomist to assert that all bodies are composed of elementary corpuscles, even if these had not yet been shown to be the end products of chemical analyses. He recognized the limits of analysis in the statement:

Fire may perhaps be exhibited pure and elementary, as it penetrates gold, and other of the most solid bodies: but no human art can produce the least drop of pure water, and much less of any of the rest, as air and earth, or the like.[62]

He described the universal presence of water as well as its action as a solvent or instrument of solution. A generic "earth" was also admitted by Boerhaave and its solidifying properties enumerated. Water and earth are not, however, given the status of elements, but are simply considered common constituents—whether elementary or complex—of most substances.

[59] F. W. Gibbs, *The Life and Work of Herman Boerhaave, With Particular Reference To His Influence In Chemistry*, unpublished University of London thesis, 1949, Part II, pp. 120–139. Gibbs makes use of a William Cullen manuscript to support his position (Part II, p. 137). Cullen's opinion of Stahl's works is very favorable on all counts—method, theory, and general utility—except presentation or style. See Leonard Dobbin, "A Cullen Chemical Manuscript of 1753," *Annals of Science*, **1**, 138–156 (1936).

[60] *Encyclopédie*, "Chymie," Vol. III, p. 408. The influence of Boerhaave and the Dutch physicists has been studied by Pierre Brunet, *Les Physiciens hollandais et la méthode expérimentale en France au XVIIIᵉ siècle*, Paris, 1926.

[61] DeMachy's translation bears the title *Eléments de chimie suivant les principes de Beccher et de Stahl*, 6 Vols., Paris, 1757. Cf. Metzger, *op. cit.*, p. 192, and F. W. Gibbs, *op. cit.*, Part II, pp. 139 and 236.

[62] Boerhaave, *op. cit.*, Vol. 1, p. 167.

The role of air in Boerhaave's chemistry changed considerably between the publication of the spurious 1724 edition of the *Elementa Chemiae* and the authorized edition of 1732.[63] The earlier work, repudiated by Boerhaave upon its appearance, describes the purely mechanical action of air as an instrument in the maintenance of fire.

All the use of air, then, in our vulgar fire, is to make a kind of *fornax* or vault around it, and thereby restrain and keep in the oily particles, and prevent their flying off too hastily.] . . . [

Air, therefore, acts no otherwise on fire, than as it confines, and keeps in the pabulum.[64]

In the 1732 edition of his textbook, Boerhaave makes use of the work of Hales to show that air enters into chemical combination; he also describes his own experiments to measure the quantity of air evolved in certain chemical reactions. His adoption of the position that air is a constituent of matter is summarized in the following passage:

Indeed, nearly all sorts of bodies, treated by fire, shew, that elastic air makes a considerable part in their composition; at least, that all known bodies do, by the force of fire, separate a fluid, elastic, compressible matter, that contracts with cold and expands with heat, which are the properties of elastic air; tho' this matter, when confined and bound down in bodies, does not produce the effects of air; but when once let loose, it has all the effects of air, and may again enter as an ingredient in the composition of other bodies.[65]

Boerhaave's much-admired "treatise on fire" presented theories of calcination and combustion which prevailed until replaced, in the middle of the eighteenth century, by the phlogiston theory.[66] As conceived of by Boerhaave, calcination is a physical rather than a chemical change and is characterized by the penetration of the metal involved by a subtle *materia ignis*. Reduction is, similarly, the removal of this tenuous material from the pores of the calx. The burning of limestone, producing quicklime, provides

---

[63] Boerhaave's changing ideas about air and his final position have been investigated very ably by Milton Kerker, "Hermann Boerhaave and the development of pneumatic chemistry," *Isis*, **46**, 36–49 (1955). For bibliographical treatment of Boerhaave's textbook, see Tenney L. Davis, "The Vicissitudes of Boerhaave's Textbook of Chemistry," *Isis*, **10**, 33–46 (1928); see Gibbs, *op. cit.*, Part III, pp. 3–73, for an excellent bibliography of the editions of Boerhaave's works.

[64] Boerhaave, *A New Method of Chemistry*, trans. P. Shaw and E. Chambers, London, 1727, p. 270. Hereafter, all citations will refer to the 1741 translation, which is based upon the authorized 1732 version of the *Elementa Chemiae*.

[65] Boerhaave, *op. cit.*, Vol. 1, p. 434.

[66] Boerhaave's theory of fire has been discussed by Metzger, *op. cit.*, pp. 209–245. The "treatise on fire" was actually the section of Boerhaave's textbook dealing with that subject; these chapters were dubbed a "treatise" because they constituted the most comprehensive and valuable treatment of fire then available and were frequently consulted independently of the remainder of the textbook.

a good example of the way in which the *materia ignis* functions: fire enters the pores of the limestone and is held there until the quicklime formed is slaked; the heat evolved in the slaking process is the *materia ignis*, which reappears when it is displaced by water.

Combustion, in contrast to calcination, is a chemical reaction in which a second fluid, called the *pabulum ignis*, or fuel of fire, contained within the combustible body, is consumed by fire.

Boerhaave's two-fluid theory of fire contains recognizable parallels to the phlogiston theory. For Boerhaave, the instrument fire is both a source of heat and a fluid (*materia ignis*), whose role in calcination is not unlike that of the subtle fiery matter of Boyle and Senac; and the pabulum or fuel of fire is a substance consumed during combustion but is not fire itself. Rouelle's theory eliminates Boerhaave's *materia ignis*, which is after all a form of the instrument fire, and changes the second fluid from a fuel to the chemical agent phlogiston, or elementary fire. What this revised Stahlian doctrine has achieved is some theoretical economy, but the superiority of one theory over the other is still not so striking that it would immediately convince a Boerhaavian to become a Stahlian. That the chemists of the eighteenth century were *not* at once converted to the phlogiston theory is evident from Venel's complaint that Boerhaave's treatise on fire was being read while Stahl's work was comparatively neglected. In order to become generally accessible, Stahl's theory needed clarification and popularization; and in order to become generally acceptable, the theory further needed coherence and systematization.

The sudden popularity of the phlogiston theory in the mid-eighteenth century is demonstrably a product of Rouelle's work, both as a teacher and advocate of the theory and as a modifier and adapter of Stahl's ideas. It is certainly true that the phlogistic explanation of calcination as a slow combustion effected an economy in chemical theory; a second and stronger appeal was made by the phlogiston theory to such writers as Eller and Gellert because it explained metallurgical reactions which had been largely neglected by Boerhaave.[67] Nevertheless, although Stahl's writings were not entirely unknown in the early part of the eighteenth century and although the phlogiston theory had been published as early as 1697, Stahlian chemistry attracted little or no attention for at least the first three decades of the century.

It is sometimes asserted that Stahlian influence in French chemistry dates from the work of Senac. This view was in fact advanced late in the eighteenth century by both Lavoisier and Fourcroy, who dated the

[67] Henry Guerlac, in "Some French Antecedents of the Chemical Revolution," *Chymia*, 5, 73–112 (1959), pp. 106–108, observes that the phlogiston theory "seemed to have originated with Beccher and also with Stahl in an attempt to interpret the manifold observations of the practical metallurgist." He cites Stahl, *Traité du Soufre*, pp. 108–110, and the writings of Eller and Gellert.

phlogistic revolution in France from 1723.[68] While it is difficult to evaluate exactly the importance of the *Nouveau cours de chymie* in the introduction of Stahlianism, more recent opinion that its role was minimal seems quite justifiable. In the writings of Rouelle's immediate predecessors, chemists like Louis Lémery and P.-J. Malouin, Stahl is cited solely for specific points of method or for important experiments, but not for theory.[69] Only in mid-century, during the peak of Rouelle's career, did Stahlianism begin to claim the adherence of French chemists.

The chronology of the acceptance of Rouelle's modified Stahlianism can be traced through the chemical writings of the 1750's and 1760's. One of the first of Rouelle's disciples, Macquer, presented the entire element-instrument theory in his textbook which was published in 1749 and which appeared in numerous subsequent editions. In 1760 there appeared a résumé of the current state of chemical knowledge; the anonymous author declared that his essay was based on Macquer's writings.[70] That the doctrines of Rouelle were not yet generally accepted in 1756 is suggested by Venel in his article "Feu" in the *Encyclopédie*; discussing the dual nature of fire, Venel notes that "the chemist, *at least the Stahlian chemist*, considers *fire* in two very different ways."[71] The implication is, of course, that the Stahl-Rouelle school of chemistry still had to face the opposition of the rival Boerhaavians.

By 1766 the element-instrument theory had become so firmly entrenched that Macquer could say in the article "Feu" in his *Dictionnaire de chymie*, "The Chemists consider fire, as well as the other elements, in two very different ways. . . ."[72] There was no longer any necessity to pin the label "Stahlian chemists" on the adherents of the element-instrument theory; Rouelle's modified Stahlian doctrines, also taught by his pupils Macquer, Venel, and Sage,[73] were being widely promulgated and had become the dominant chemical system of the period.

[68] Cf. Lavoisier, *Œuvres*, Vol. 3, p. 261; Fourcroy, "Chimie," in *Encyclopédie méthodique*, Vol. 3 of "Chimie, pharmacie et métallurgie," p. 302; and Venel, "Chymie," *Encyclopédie*, Vol. 3, p. 437. The evidence does not support this view of Senac's importance; see Metzger, *op. cit.*, p. 94, and Guerlac, "Some French Antecedents . . .," pp. 104–105. Guerlac cites Condorcet's opinion that Stahlian doctrines were not yet known in France in 1740 (*ibid.*, p. 105, note 109).

[69] P.-J. Malouin, *Traité de chimie*, Paris, 1734, pp. 19, 63, 75f., 109f., 163f., 170, 184, and 195; Louis Lémery, "Second mémoire sur le nitre," *Mémoires de l'Académie royale des sciences*, 122–146 (1717), pp. 123, 124, 132.

[70] *Histoire de l'Académie royale des sciences. Centième ou dernier Volume de la première Centurie. Contenant un Abregé Historique de chaque Science . . . jusqu'à l'année 1751. inclusivement*, 12°, Amsterdam, 1760, art. "Discours III. Sur la Chymie," p. 161.

[71] *Encyclopédie*, "Feu," Vol. 6, p. 609 (emphasis is mine).

[72] Macquer, *Dictionnaire de chymie*, 2 Vols., Paris, 1766, Vol. 1, p. 498. But cf. *Dictionnaire de chymie*, 2nd ed., 4 Vols. in-8°, Paris, 1778, art. "Principes," Vol. 3, p. 272, where phlogiston is called a fatty earth.

[73] Another of Rouelle's pupils, J.-B. Bucquet (1746–80), was professor of chemistry, Faculté de Médecine, Paris. He reputedly taught chemistry, mineralogy, and natural

94

A comparison of the theories of Stahl and Rouelle with those theories accepted in the 1760's reveals that the reigning doctrines were actually those of Rouelle, which had been built upon a substratum of the work of Stahl. However, even Rouelle's pupils referred to his teachings as the doctrines of Becher and Stahl, often to the complete exclusion of any acknowledgment of Rouelle's own important contributions to Stahlian chemistry. Venel, for example, although thoroughly aware of Rouelle's influence as a teacher, lauds Stahl who "has brought chemical doctrine to the point where it is today;" he further describes Stahl's *Specimen Becherianum* (1702) as "the code of *chemistry*, the Euclid of the chemists."[74] In his article "Principes" in the *Encyclopédie*, Venel does more than merely praise Stahl in the general terms just quoted.

The modern chemists have admitted rather generally for their first and inalterable *principles*, the four elements of the Peripatetics; fire which they call *phlogiston* with the Stahlians, air, water, and earth.[75]

When Venel states that the Stahlians identify phlogiston with fire, he is referring to Rouelle's theory, since Stahl actually thought of phlogiston as a form of earth. The same kind of error is found in Turgot's writings, when he asserts that Stahl "is really the first who perceived the role that air plays in nature as a principle of mixts;"[76] Stahl specifically denied the role of air as a constituent of matter, but Rouelle repeated for his pupils the experiments of Stephen Hales.

The above examples indicate clearly that the works of Stahl, couched in a barbaric mixture of German and Latin and difficult to understand even when translated into French or English, were probably not read by the generation of chemists taught by Rouelle. Rather than revealing a confusion between Stahl and Rouelle, the statements of Venel and Turgot show an apparent lack of awareness that Rouelle had added significantly to the doctrines of Stahl. A reasonable explanation for this curious state of affairs—discounting the petty motive of academic jealousies, which cer-

---

history in his course and continued the Rouelle tradition in his treatment of the *règne végétal*. See Condorcet, "Eloge de M. Bucquet," *Histoire de l'Académie royale des sciences*, pp. 60–76 (1780).

[74] *Encyclopédie*, "Chymie," Vol. 3, pp. 436, 434.

[75] *Ibid.*, "Principes," Vol. 13, p. 376; also, "Feu," Vol. 6, p. 609.

[76] Condorcet and Turgot, *Correspondance inédite de Condorcet et de Turgot, 1770–1779*, ed. Charles Henry, Paris, 1883, p. 111 (Turgot to Condorcet, 27 November 1772). See also, *ibid.*, p. 113. Turgot earlier cites the experiments of "Stales" as proof that air enters into chemical combination (p. 61); *Hales* is probably meant here. On pp. 111 and 113, however, Turgot writes *Stahl*. If this, too, should actually mean Hales (which is doubtful), the significance of the passage would still not be lost; Turgot has omitted Rouelle from the Stahl-Hales tradition.

tainly would not have affected Turgot—can be found in Rouelle's lectures. Rouelle claimed to be a disciple of Becher and Stahl, and his pupils, not having read the works of the German chemists, assumed that his teachings were the same as those of his confessed masters. This may account in large part for Macquer's failure to acknowledge his theories as those of his teacher, although he does praise Becher and Stahl highly.[77] Indeed, most references to Rouelle by his contemporaries suggest that he was little more than an outstanding teacher. Historians of chemistry have since shared the misconception of Rouelle's contemporaries and have applied the inaccurate term "Stahlian" to describe the chemistry of the mid-eighteenth century.

There are several factors which do much to explain Rouelle's success in having a relatively unknown doctrine accepted during his lifetime. First, and perhaps most significantly, Rouelle substituted for the verbiage and mystery of Stahl's doctrines a chemical system, the element-instrument theory, which was erected on an apparently solid foundation of experiment, and which had the additional virtues of economy and simplicity. Secondly, the phlogistic component of Rouelle's theory appealed to metallurgists and mineralogists. In the third place, the entire scheme of Stahlian-Rouellian chemical theory contained many features of currently accepted chemistry.[78] And finally, Rouelle taught the fundamentals of chemistry to a whole generation of chemists, most of whom modeled their own work on that of their teacher. Not only through his teaching and the missionary work of pupils like Macquer and Venel, but also by his encouragement of the translation of the writings of Stahl and his disciples did Rouelle effectively spread the modified Stahlian doctrines.[79]

---

[77] P.-J. Macquer and Antoine Baumé, *Plan d'un cours de chymie expérimentale et raisonnée, avec un discours historique sur la chymie*, Paris, 1757, pp. liii–lvii. In Macquer's case, academic jealousies may have been important, since he was doubtless aware that he was being publicly criticized by Rouelle.

Why Rouelle himself—a vain man who lived in fear of being plagiarized—did not assert the originality of his ideas is not clear. Perhaps Diderot alone among Rouelle's disciples had the ability and the interest to penetrate the ponderously Germanic Stahlian treatises, for he observes that Rouelle "a ajouté beaucoup d'idées neuves et utiles à la Doctrine de ses maîtres Stahl et Becher" (Bordeaux MS, p. 34). The various schools of thought about Rouelle's reputation are treated in more detail in the writer's "G.-F. Rouelle," pp. 70–71 and passim.

[78] Cf. the doctrines of Geoffroy in Lavoisier, *Œuvres*, Vol. 3, pp. 261–262; the revival of the four element theory, Macquer, *Dictionnaire*, 1766, Vol. 2, pp. 328–329; and Fourcroy, *Système des connaissances chimiques*, 10 Vols., Paris, 1800, Vol. 1, p. 51. Perhaps the most vital current in chemistry used by Rouelle was the doctrine of affinities, incorporated into Geoffroy's table of 1718.

[79] The complex subject of the flood of Stahlian translations which appeared at this time is discussed by Guerlac, "Some French Antecedents . . .," pp. 104–108, and the writer, "G.-F. Rouelle," pp. 80–81.

ROUELLE'S DOCTRINES AND THE NEW CHEMISTRY

Stahlian chemistry, as modified and taught by Rouelle, enjoyed a brief vogue; it was generally accepted in the 1760's and its influence declined in the 1780's. The system was destroyed primarily by the work of one of Rouelle's pupils, Lavoisier, even while it was still being supported—in modified versions—by other pupils. Their adherence to the old theories was explained by the phlogistonists as an unwillingness to substitute for the old system a new one which did not offer appreciably superior explanations for the same experimental facts. With the increasing numbers of new discoveries and the theories put forward by the rising generation of chemists, led by Lavoisier, the prevailing doctrines which Rouelle had advocated were weakened and eventually discredited.

Shortly after Rouelle's death in 1770, French chemistry came under the compelling influence of the British pneumatic chemists. Although the early work of Stephen Hales in this field was known and appreciated in France, it was only in 1772 or 1773 that first-hand knowledge of the experiments of Joseph Black was introduced. An abridged account of Black's experiments with magnesia alba had in fact been published in the *Journal de médecine, chirurgie, pharmacie* as early as 1758, but attracted no attention, probably because the abridgment featured a single footnote in which the authority of Rouelle was used to refute Black's arguments.[80] A second account of Black's experiments appeared in Johann Friedrich Meyer's *Essais de Chymie* (1766), a work intended to invalidate Black's conclusions.[81] Finally, after the death of Rouelle, Black's work of almost twenty years before became known in France. The isolation of carbon dioxide, the description of some of its properties, and the realization that this was a gas different from atmospheric air—these were the first steps in a long line of discoveries of various gases. A startling departure from Rouelle's theory, the identification of carbon dioxide and the subsequent developments in pneumatic chemistry undermined an essential aspect of the old doctrine.[82]

[80] "Expériences Sur la Magnésie; par M. Black, Docteur en Médecine à Edimbourg," *Journal de médecine, chirurgie, pharmacie*, **8**, 254–261 (March, 1758), pp. 259–260 and 260n. (Photostats of this article were supplied by Professor Guerlac.)

[81] Knowledge of Black's work in France, through the publication of 1758 and the writings of Meyer and Macbride, is treated in full by Henry Guerlac in his forthcoming monograph, *Lavoisier—The Crucial Year: The Background and Origin of his First Experiments on Combustion in 1772*. For Black's work in chemistry, see Guerlac's "Joseph Black and Fixed Air: A Bicentenary Retrospective, with Some New or Little Known Material," *Isis*, **48**, 124–151 (1957) and "Joseph Black and Fixed Air (Part II)," *Isis*, **48**, 433–456 (1957); also, Douglas McKie, "On Thos. Cochrane's MS. Notes of Black's Chemical Lectures, 1767-8," *Annals of Science*, **1**, 101–110 (1936).

[82] For the resistance of Baumé to the acceptance of "fixed air," see his *Chymie expérimentale et raisonnée*, 3 Vols., Paris, 1773, Vol. 3, pp. 693–698 (also reproduced in Lavoisier, *Œuvres*, Vol. 1, pp. 551–555); cf. Turgot in Condorcet and Turgot, *Correspondance*, pp. 111–113.

A second feature of post-Rouellian chemistry is the use of the quantitative methods of physics in the laboratory of the chemist. The methods were hardly new, having been employed very effectively by Boerhaave.[83] However, quantitative measurement as used by Rouelle had severe limitations, and Rouelle's school had come to dominate French chemistry. Renewed interest in Boerhaave is prominent in Lavoisier's work and helped to counterbalance the influence of Rouelle.

It is already evident from these brief considerations that the chemistry of Rouelle could have had only a limited influence on the new chemistry, since Rouelle did not and could not treat the problems raised by the pneumatic chemists. This limited influence was further circumscribed by a split in the allegiance of Rouelle's pupils, some of whom remained firmly loyal to the outmoded theories of their teacher, while others played a central role in establishing the new chemistry. If the rejection of Rouelle's entire system was a necessary step toward the acceptance of the new chemistry, then it would seem that Rouelle's influence was in fact limited to the diehard exponents of the old doctrines. The example of Macquer, however, who gradually accepted the new doctrines while he still advocated a modified phlogiston theory, indicates that Rouelle's theories and the new chemistry shared for a time the attention of the young chemists. A similar conservatism can be found in the writings of Lavoisier, who remained a lukewarm phlogistonist until he became convinced of both the inadequacies of the old theory and the superiority of his own ideas. Whether or not he did manage to eliminate entirely the doctines of revised Stahlianism from his own work has long been a subject for debate among historians of chemistry.[84] The question of the extent to which Rouelle's Stahlianism influenced Lavoisier will be the subject of the remainder of this article.

It is often asserted that Rouelle was a powerful force in shaping Lavoisier's scientific interests.[85] This statement, usually made with reference to Lavoisier's work in chemistry or to his initial interest in chemistry, is hardly an exact summary of the relationship between Rouelle and his famous pupil. Lavoisier did not begin to work in chemistry until at least five years after completing Rouelle's course (1763 or 1764), so it is highly

---

[83] See Milton Kerker, *passim.*

[84] For the relation of Lavoisier to Stahlian methods and theories, see Metzger, *La philosophie de la matière chez Lavoisier,* Paris, 1935, p. 35 and *passim.* See also the claim made by Wilhelm Ostwald, *L'Evolution d'une Science: La Chimie,* trans. Marcel Dufour, Paris, 1916, p. 21, that after the discovery and description of oxygen by Scheele and Priestley, Lavoisier could simply choose between the phlogiston theory and its "inverse"—i.e., Lavoisier could look at calcination as either the *loss* of phlogiston or the *gain* of oxygen. Daumas, *L'acte chimique,* Paris, 1946, p. 57, attacks the oversimplification of Ostwald.

[85] E.g., Hermann Kopp quoted in Claude Secrétan, "Un aspect de la chimie prélavoisienne (Le cours de G.-F. Rouelle)," *Mémoires de la Société vaudoise des Sciences naturelles,* Lausanne, 7, 219–444 (1943), p. 435, and Denis I. Duveen and Herbert S. Klickstein, *A Bibliography of the Works of Antoine Laurent Lavoisier 1743–1794,* London, 1954, pp. 1 and 4.

unlikely that Rouelle's lectures stimulated in him a lively interest in chemistry. In fact, examination of Lavoisier's references to Rouelle in his published works reveals that Rouelle is cited most frequently in geological and mineralogical rather than chemical contexts.[86] When it is also recalled that Lavoisier's writings and activities in the years immediately following his study with Rouelle were geological in nature, it would seem that Lavoisier attended Rouelle's lectures to learn the basic facts of geology and not chemistry.[87] Less concerned, then, with the theories of Stahlian chemistry than most of Rouelle's pupils, Lavoisier—it is reasonable to conclude—was relatively uncommitted to the principles of Stahlianism and possibly, for this reason, later found it easier to break with the Stahlian tradition than did his contemporaries.[88] That Lavoisier did not consider Rouelle the major source of his knowledge of chemistry is further confirmed by his constant references to the more readily available published writings of other chemists—notably Stahl, Boerhaave, and Macquer—when he speaks of the history of chemistry.[89]

Rouelle's influence upon Lavoisier cannot, however, be dismissed as entirely negligible when Lavoisier's chemical works are examined. Rouelle's teachings served as a source of many of the basic tools and facts in the repertory of the chemist and provided an example of a strictly empirical approach to the study of chemistry.

The "tools and facts," to be sure, consisted of the experimental data and results common to the tradition of eighteenth-century chemistry. The sheer mechanical transmission through Rouelle of the work of Stephen Hales furnished Lavoisier with a useful tool, the technique for collecting

[86] References to Rouelle in chemical contexts usually involve specific procedures and altogether minor points; Lavoisier, Œuvres, Vol. 1, pp. 92, 601, Vol. 2, pp. 7, 243. On the other hand Rouelle, Buffon, and Guettard are grouped by Lavoisier as the three greatest geologists of his day. Cf. Œuvres, Vol. 5, p. 226.

Rouelle's importance as Lavoisier's instructor in geology was first noted by Henry Guerlac, "A Note on Lavoisier's Scientific Education," Isis, **47**, 211–216 (1956).

[87] Lavoisier's early work in science has not been studied with the close attention it deserves. Some of the questions involved have been treated by A. N. Meldrum, "Lavoisier's early work in science, 1763-1771," Isis, **19**, 330–363 (1933) and **20**, 396–425 (1934), and "Lavoisier's Work on the Nature of Water and the Supposed Transmutation of Water into Earth (1768-1773)," Archeion, **14**, 246–247 (1932); Guerlac, "A Note on Lavoisier's Scientific Education"; and Pierre Courte, "Aperçu sur l'œuvre géologique de Lavoisier," Société Géologique du Nord, Annales, **69**, 369–375 (1949). Meldrum does not claim for Rouelle any great influence on the chemistry of Lavoisier, but neither does he mention at all Lavoisier's considerable reliance on Rouelle's geological methods and ideas; see "Lavoisier's early work in science," pp. 335 and 340–344.

[88] Besides Lavoisier, two of Rouelle's pupils independently questioned the phlogiston theory: Pierre Bayen and Turgot. See Meldrum, The Eighteenth Century Revolution in Science —The First Phase, Calcutta, 1930, pp. 43–44, and Condorcet and Turgot, Correspondance, pp. 59–63.

[89] Lavoisier, Œuvres, Vol. 1, pp. 7, 461, 464, Vol. 2, pp. 99–103, 623–655.

gases evolved in chemical reactions, and a significant fact, the entrance of air into chemical combination.[90]

In the areas of theory and method, there is little evidence to suggest that Lavoisier owed much to the teachings of Rouelle. Early in his chemical work, when Rouelle presumably would have been one of his sources of chemical theory, Lavoisier's choice of problems for research was significantly different from what it was to be shortly thereafter. Recent investigation by Henry Guerlac has revealed that one of Lavoisier's earliest papers in chemistry probably consisted of an attempt to write a system of the elements.[91] In preparation in 1772, this treatise was to have dealt with the four elements of Rouelle, and not with the empirically defined elementary substances of Lavoisier's *Traité élémentaire de chimie* (1789).[92] Two sections of this work were to have been concerned with fire ("feu élémentaire") and air, with the final portion probably a revised version of his famous paper of 1770 on the transmutation of water into earth.[93]

Quite suddenly Lavoisier stopped working on his system of the elements and never returned to it. None of his subsequent writings can be shown to have any direct relation to the teachings of Rouelle. Even Rouelle's much-admired motto, *Nihil est in intellectu quod non prius fuerit in sensu*,[94] became only a general tenet of method, while Rouelle's actual methods underwent considerable change and development when they were employed by Lavoisier. While Rouelle and Lavoisier both used and understood the importance of the chemical balance, Lavoisier added two components lacking in Rouelle's method: he extended the use of the balance to substances in the gaseous state, and he realized fully the significance of quantitative measurements. Where Rouelle had been interested in determining the quantities of materials used and obtained in reactions, Lavoisier noted that the results of weighing can reveal the kind of chemical reaction—one of addition or subtraction—which has occurred.

In the relative importance placed upon quantitative methods by

[90] *Ibid.*, Vol. 1, p. 601, Vol. 2, p. 7. Often Rouelle also supplied mineral samples for use in Lavoisier's experiments. *Ibid.*, Vol. 3, pp. 132–134 and 327.

[91] In his extraordinary article, "A Lost Memoir of Lavoisier," *Isis*, 50, 125–129 (1959), Professor Guerlac in effect deduced the existence of such a document. The "lost memoir" has since been discovered by M. René Fric and published—too late for use here—in the *Archives Internationales d'Histoire des Sciences*, no. 47, 137–168 (1960).

[92] Lavoisier, *Œuvres*, Vol. 1, p. 7: "si ... nous attachons au nom d'élémens ou de principes des corps l'idée du dernier terme auquel parvient l'analyse, toutes les substances que nous n'avons encore pu décomposer par aucun moyen, sont pour nous des élémens."

[93] *Ibid.*, Vol. 2, pp. 1–28. Cf. Rouelle's comments on the "successful" attempts of Digby and Boyle to convert water into earth: "la terre qu'ils trouvent chaque fois dans leurs vaisseaux etoit fournie par l'atmosphere" (Bordeaux MS, p. 78), and "M. Eller ... a cru aussi réduire toute l'eau en terre en la broiant dans un mortier de verre. La terre qu'on a obtenue par cette operation ridicule paroit plus venir du verre que de l'eau" (*ibid.*, p. 79).

[94] Lavoisier, *Œuvres*, Vol. 1, p. 246.

Rouelle and Lavoisier lies a key to the understanding of a major difference in outlook between pupil and teacher. Rouelle, and what was later called his school of chemists, distrusted and consciously avoided methods that smacked of the mathematical; they scorned Boyle and Boerhaave as chemists because their concern with the measurements of physics placed them in the class of *physiciens* rather than among the true chemists.[95] Lavoisier, on the other hand, used the methods of these early physical chemists most effectively and was praised for having at last achieved a union between chemistry and experimental physics.[96]

In the realm of chemical theory rather than method, there are at least two points of similarity which link the ideas of Rouelle and Lavoisier. As Tenney L. Davis once observed, Lavoisier's laudable definition of elementary substances, which has been hailed as a recognition of the true way to define an element, has its basis in the eighteenth-century empirical tradition, exemplified by Rouelle, Macquer, and Boerhaave.[97] Lavoisier's statement that an element is any substance which can suffer no further decomposition by laboratory means is simply an expression of current chemical practice, if not theory. While in theory Rouelle had considered the four elements the only truly simple substances, he knew that in fact he was conducting his experiments at the level of the mixt. The four elements of Rouelle were beyond the reach of the working chemist; they could not be obtained in a pure state and were thus not usable as laboratory reagents. In eliminating this theoretical substructure, Lavoisier was giving

---

[95] Bordeaux MS, pp. 29–30 and 34. The antimathematical feeling among Rouelle's disciples was described by Condorcet in a letter of October 1, 1772: "Je lis l'ouvrage de Romé [de l'Isle] sur la cristallisation. L'auteur proteste, à la tête de l'ouvrage, qu'il ne prétend pas que jamais la géométrie puisse expliquer ce phénomène; je ne sais s'il a le droit de faire les honneurs de la géométrie, mais il a mis cette phrase pour contenter l'école de Rouelle. Ils seraient bien fâchés qu'on puisse calculer quelques-unes de leurs opérations, et ils paraissent un peu tentés de la réputation d'être sorciers." Condorcet and Turgot, *Correspondance*, p. 98. Cf. Venel, art. "Chymie," *Encyclopédie*, Vol. 3, p. 436.

Daumas observes that the first partisans of pneumatic chemistry were physicists and mathematicians, and not chemists. Maurice Daumas, "La chimie dans l'*Encyclopédie* et dans l'*Encyclopédie méthodique*," *Revue d'histoire des sciences*, **4**, 334–343 (1951).

[96] Extract from the *Histoire de l'Académie royale des sciences* on Lavoisier's newly published (1774) *Opuscules physiques et chimiques*, in Lavoisier, *Œuvres*, Vol. 2, p. 96.

[97] Tenney L. Davis, "Boyle's Conception of Element Compared with that of Lavoisier," *Isis*, **16**, 82–91 (1931); also, Metzger, *Newton, Stahl, Boerhaave . . .*, pp. 128–129, and Emile Meyerson, *De l'explication dans les sciences*, 2 Vols., Paris, 1921, Vol. 1, p. 290. Marie Boas, in "Structure of Matter and Chemical Theory in the Seventeenth and Eighteenth Centuries," in *Critical Problems in the History of Science: Proceedings of the Institute for the History of Science at the University of Wisconsin, September 1–11, 1957*, ed. Marshall Clagett, Madison, Wis., 1959, pp. 499–514, also stresses this view and extends this understanding of the nature of chemical elements to many of the chemists of the late seventeenth century. She observes that the concern of chemists with elementary particles (physical principles) was limited to the realm of higher theory and could not be effectively linked to practice until Dalton's time.

expression to the prevailing practice, and bringing it to its logical conclusion.

The similarity between Rouelle's practices and Lavoisier's definition, however, should not be exaggerated. Lavoisier realized that chemical elements lose their properties and acquire new ones when in combination, while Rouelle suggested vaguely that the properties of a simple substance when in combination usually do resemble, albeit *de très loin,* those of the same substance in a free state.

A second point of theory in the new chemistry has called forth much comment about vestiges of Stahlianism in the work of Lavoisier. The caloric theory has provoked eighteenth-century Stahlians and recent historians of chemistry to identify Lavoisier as a Stahlian trying to disguise an old theory with new terminology.[98] There are in fact many points of resemblance between phlogiston and the subtle fluid Lavoisier named caloric: both are called the material of fire and are associated with the phenomena observed in combustions; both play roles in determining the states of bodies in which they are contained; and both present problems for quantitative measurement. Lavoisier's definitions of free and combined caloric may easily be applied to phlogiston.

Free caloric is that which is not united in any combination. Since we live in the midst of a system of bodies with which caloric has some adherence, it follows that we never obtain this principle in a state of absolute freedom.

Combined caloric is that which is bound within bodies by the force of affinity or attraction, and which constitutes a part of their substance, even of their solidity.[99]

These similarities between phlogiston and caloric are not identities. While the amount of caloric in a body determines its state as a gas, liquid, or solid, phlogiston only makes one substance relatively more volatile than another which contains less phlogiston. The presence of phlogiston makes a body capable of burning, but the escape of caloric merely accounts for the production of light and heat in a combustion. Thus, although phlogiston and caloric are products of combustion in both the Stahlian and the new chemical doctrines, respectively, phlogiston has the additional and inexplicable property of acting as the principle of combustions.[100]

[98] For the caloric theory, Lavoisier, *Œuvres,* Vol. 1, pp. 17–31. Its critics: Metzger, *La philosophie de la matière,* pp. 45f., and J. R. Partington and Douglas McKie, "Historical Studies on the Phlogiston Theory.—III. Light and Heat in Combustion," *Annals of Science,* **3,** 337–371 (1938), pp. 339–340.

[99] Lavoisier, *Œuvres,* Vol. 1, p. 28.

[100] Lavoisier was criticized for not accounting for the combustibility of substances; see Emile Meyerson, *Identité et Réalité,* Paris, 1908, p. 308.

For parallel uses made of oxygen and phlogiston, see Daumas, *L'acte chimique,* p. 56.

Caloric is in fact somewhat similar to both phlogiston and to Boerhaave's *materia ignis.* Lavoisier's debt to Boerhaave for features of the caloric theory is perhaps overstated by F. W. Gibbs, *op. cit.,* Part II, pp. 242–266. Cf. also, Metzger, *Newton, Stahl, Boerhaave . . .,* pp. 218 and 220f., and above, pp. 91–92.

Critics of the caloric theory justly point out that Lavoisier has reduced the number and altered some of the characteristics of phlogiston and has substituted the equally mysterious caloric. It has also been observed—and with equal justice—that until a more satisfactory explanation of the evolution of heat and light in combustion could be propounded, Lavoisier felt he needed a substance like caloric in order to explain fully the phenomena attending combustion. Admitting the necessity for positing the existence of caloric, it is still clear that the idea of caloric owes much to the phlogiston theory and is perhaps the one substantial link between the Stahlian doctrines and the chemistry of Lavoisier. That Lavoisier could have obtained his knowledge of the phlogiston theory from numerous sources other than Rouelle's lectures is quite obvious, and, indeed, in his discussion of the history of that theory Lavoisier frequently cites Stahl and omits completely all reference to Rouelle.[101] The general impression gained from reading Lavoisier's historical treatment of phlogiston is one of a chemist who has perhaps forgotten his early course in chemistry and who has had to look up the old doctrines in a textbook or reference work.

In conclusion, despite the few parallels that can be drawn between the chemistry of Lavoisier and that of Rouelle, it is all too clear that the major influences on the new chemistry did not stem from the prevailing doctrines of Rouelle. The crucial period for the survival of Rouelle's theories was the early 1770's, when the upsurge in pneumatic chemistry rendered Rouelle's modified Stahlianism insufficient to explain the wealth of new facts, and revealed the weaknesses in his method and theories. It is difficult to understand exactly how Lavoisier was able to see his way around the pitfalls of Stahlianism, while Macquer, also a chemist of the first rank, remained deeply committed to the ideas of the old school. There can be little doubt, however, that the chief sources of Lavoisier's chemistry are not to be found in the ideas of his teacher.

[101] Above, note 89.

Lavoisier's Geologic Activities, 1763-1792

HISTORIANS OF SCIENCE have long known that Antoine-Laurent Lavoisier was a geologist of distinction. Briefly summarized, the highlights of his geologic career are: a long field trip through the Vosges with naturalist Jean-Étienne Guettard in 1767, collaboration with Guettard in the preparation of a geologic atlas of France, and the completion, in 1789, of a justly famous memoir on sedimentary strata. These familiar facts, however, have been used by biographers to illuminate matters which have little to do with geology. Thus, the tour of the Vosges becomes an illustration of Lavoisier's early enthusiasm and versatility; his contributions to the atlas serve as an introduction to his subsequent dispute with Antoine Monnet; and his memoir on sedimentary strata is an example of what Lavoisier could accomplish in a science other than chemistry.

Apart from comments by biographers, the literature on Lavoisier's work in geology consists of three articles on his early career in science (1763–1771), two more on the memoir of 1789, and two bibliographical analyses of the geologic atlas. These important studies have been based almost exclusively upon the published Lavoisier documents, which in fact constitute less than one-half of his output in geology.[1] As a result, many questions that might be asked about this aspect of his career remain either wholly or partially unan-

*Vassar College. Much of the research for this article was done while the writer held a fellowship awarded by the American Association of University Women. The author is grateful for the aid and courtesy of the various libraries and private collectors cited below, but more particularly to Professor Henry Guerlac for his valuable criticisms and encouragement. The writer has been unable to examine one of the major collections of Lavoisier papers, the Archives de Chabrol, and there are in addition numerous relevant manuscripts described briefly in the Fichier Lavoisier at the Académie des sciences, Paris, but now inexplicably missing from the archives of the Académie.

[1] Henry Guerlac, "A Note on Lavoisier's Scientific Education," *Isis*, 1956, *47*:211–216.

A. N. Meldrum, "Lavoisier's Early Work in Science, 1763–1771," *Isis*, 1933, *19*:330–363, and *ibid.*, 1934, *20*:396–425. Pierre Comte, "Aperçu sur l'œuvre géologique de Lavoisier," *Annales de la société géologique du Nord*, 1949, *69*:369–375. A. V. Carozzi, "Lavoisier's Fundamental Contribution to Stratigraphy," *Ohio Journal of Science*, 1965, *65*:71–85. D. I. Duveen, H. S. Klickstein, *A Bibliography of the Works of Antoine Laurent Lavoisier 1743–1794* (London: Dawson and Weil, 1954), pp. 236–244, and the review of this volume by Guerlac, *Isis*, 1956, *47*:85–88. Duveen, *Supplement to a Bibliography of the Works of Antoine Laurent Lavoisier 1743–1794* (London: Dawsons of Pall Mall, 1965), pp. 129–132. Also É. J. A. d'Archiac de Saint Simon, *Géologie et paléontologie* (Paris: F. Savy, 1866), esp. p. 114.

swered. Some attention has been paid to Lavoisier's relationship with Guettard, but little has been done to disentangle their relationship to Antoine Monnet, who became director of the geologic survey in 1777. Furthermore, studies of the Guettard-Lavoisier collaboration have dealt only with such matters as the field trips taken jointly or independently and the authorship of notes, reports, memoirs, and maps; there has as yet been no study of the geologic ideas of either man. Discussions of Lavoisier's memoir of 1789 take into account neither the development of his ideas before that date nor any comparable work done by his contemporaries.

This article does not offer answers to the problems posed above, but it deals instead with two subjects which should logically precede consideration of the others: the nature of Lavoisier's activities and the extent of his interest in geology. A survey of his activities is intended to establish a chronological framework within which future study of more specialized topics can be conducted. Such a survey also provides material for a judgment of the extent of his interest in geology, a matter which has provoked some scholarly debate.

When Lavoisier ended his formal education in September 1763 with a Bachelor of Law degree, it was becoming apparent to his family and to Guettard, a friend of the elder Lavoisier, that the young man was increasingly interested in pursuing a career in science. He had already received some training in the mathematical sciences under La Caille, had accompanied Bernard de Jussieu on at least one botanical field trip, and was clearly being drawn to further study of the natural world. Guettard's impression, recorded at a later time, provides some clue to Lavoisier's state of mind when he was twenty-one years old:

> . . . jai employé les vacances de 1763 a parcourir une grande partie du Valois . . . jai été accompagné dans ce voyage par M. Lavoisier le fils qu'un gout naturel pour les Sciences porte a vouloir les connoitre toutes avant de Se fixer a une plutot qu'a une autre.[2]

By 1763 Guettard had probably suggested to Lavoisier that he assist in the preparation of a geologic atlas of France, a project which was conceived by Guettard, perhaps as early as 1746.[3] In order to be "un Mineralogiste aussi eclaire [sic] qu'on le peut etre," Guettard thought it necessary to learn enough chemistry to aid in the analysis and identification of rocks and minerals.[4] Happily, the rudiments of both chemistry and geology were easily available to Lavoisier in the celebrated *cours de chimie* offered by G.-F. Rouelle at the Jardin du roi and at his pharmaceutical shop on rue Jacob. Aware of

2 Guettard, "Memoire qui renferme des observations Mineralogiques faites dans plusieurs endroits des Provinces qui avoisinent la Champagne. Premiere Partie. du Valois," s.d., in MS *Lavoisier et Guettard: Géographie minéralogique de la France*, Vol. II, catchword "Valois," Lavoisier Papers, Cornell University. (Hereafter: LPCU.)

3 The date of Guettard's first mineralogical maps: "Mémoire et carte minéralogique Sur la nature & la situation des terreins qui traversent la France & l'Angleterre," *Mémoires de l'Académie royale des sciences*, 1746 (1751), pp. 363–392.

4 Guettard, *Memoire sur la maniere d'etudier La Mineralogie*, s.d. (after 1766), Muséum d'histoire naturelle, Paris, MS 2186, fol. 34r. (Hereafter: MHN.) Cf. *Œuvres de Lavoisier* (Paris: Imprimerie impériale, 1862–1893), Vol. III, p. 135.

Rouelle's interest in geology, Guettard may have suggested that Lavoisier attend one of Rouelle's courses, and the young man began to frequent Rouelle's pharmacy, perhaps in November of 1763.[5]

In addition to chemistry, Guettard had found his own early training in botany a useful introduction to a knowledge of physical geography. It may thus have been at his suggestion that Lavoisier began, in the summer of 1763, to accompany Bernard de Jussieu in his weekly "promenades philosophiques" to the environs of Paris. Every year, for about seven successive Wednesdays in June and July, Jussieu took his pupils to Sèvres, Meudon, St. Prix, and elsewhere, on field trips that often lasted all day. Since attendance at every excursion was not obligatory, Lavoisier may not have gone through the full series, but he did accompany Jussieu at least once in 1763 and twice in 1764.[6]

During these early years Lavoisier was also receiving less formal instruction from Guettard himself. On 17 September 1763, a few days after the award of his law degree, Lavoisier left Paris with Guettard and Clément Augez de Villers, Lavoisier's cousin, turning his annual vacation in Villers-Cotterets into a field trip. The excursion was a short and leisurely one, probably more memorable for discussions of French and Italian opera and the merits of Buffon's *Histoire naturelle* than for the botanizing done en route. In succeeding days, however, Lavoisier had some of his first lessons in examining strata, and he proved a good pupil when he noted that certain fossiliferous quarry beds "Sont Couchez horisontalement, Ce qui fait voir [qu'elles] n'ont point eté apporté par une revolution Subite, mais [qu'elles] ont eu La liberté d'y prendre Leur Situation naturelle."[7] This early expedition with Guettard was followed, in the years 1763–1765, by numerous geologic field trips, many of them to areas within a fifty-mile radius of Paris. Although he may have accompanied Guettard on some of these excursions, clear evidence is available only for two of Lavoisier's longer voyages during this period. In 1764 he and Guettard spent the month of September traversing part of Champagne to Mézières and Sedan, and in May 1765 they explored part of Normandy, going as far as the coast at Dieppe.[8]

The field trips of 1766 had as their express purpose the gathering of data for geologic maps, for it was early in that year that Guettard's projected atlas

[5] Guerlac, "A Note," *loc. cit.*, pp. 215–216. For contemporary references to Rouelle's interest in mineralogy and geology, see Faciot to Réaumur, 17 May 1753, Réaumur Papers, Académie des sciences, Paris, R.6.5, and Branlay to Guettard, s.d., MHN, MS 1996, #14.

[6] Notes dated 15 [June] 1763, 4 July 1764, and 25–26 July 1764, Lavoisier Papers, Académie des sciences, dossiers 424, 1388. (Hereafter: LPAS.) Announcements of Jussieu's excursions appeared regularly in the *Avantcoureur*; see esp. issues dated 2 July 1759 and 20 July 1761.

[7] Notes dated 17–23 Sept. 1763, LPAS 419. How long the excursion lasted is not clear, but Lavoisier was also in Villers-Cotterets in late October (*Œuvres de Lavoisier*, Vol. IV, pp. 1–7). Guettard's journal is entitled *Observa-*

tions faites dans le voyage de villers coterest [sic] en 1763 en 7ᵇʳᵉ 8ᵇʳᵉ 9ᵇʳᵉ, MHN, MS 2194, fols. 186v–193v.

[8] *Œuvres de Lavoisier*, Vol. V, pp. 1–11, 24–33; LPAS 417; *Journal de voyage Commencé le 8 7ᵇʳᵉ 1764 jusques et compris le 5 octobre de la [même] année*, LPAS 285; and Guettard, "Observations faites dans un Voyage Commencé de Compagnie avec Mʳ Lavoisiere [sic] fils le 8 7ᵇʳᵉ 1764...," in MS *Lavoisier et Guettard: Géographie minéralogique de la France*, Vol. II, catchword "Lorraine," LPCU. Lavoisier does not name his companion in Normandy, but part of the trip was spent in verifying Guettard's geologic theories, and Guettard is known to have visited Normandy fairly regularly since the 1740's. See also the Guerlac review, *loc cit.*

of France had finally been commissioned by Bertin, director of the Department of Mines. At least five voyages for the atlas took place between April and November, but earlier travels, too, had served the same purpose, because in June and July of 1766 Lavoisier and engraver Jean Louis Dupain-Triel were already at work completing maps of Vexin and Valois.[9] Among the expeditions of that year, at least three were in Guettard's company, but it would seem that Lavoisier was becoming the more enthusiastic and energetic of the two collaborators. In one instance, he preceded Guettard into the field, worked for almost two weeks on a map of Brie, and then wrote impatiently and gaily to Guettard who was still in Paris:

> . . . il y a bientost huit jours entiers qu'on vous attend à chaque instant, Cependant vous n'arrives pas . . . je Concluds donc que vous étes obligé, même en Conscience et Sous peine de peché au moins véniel de vous mettre incessamment en Campagne.[10]

The year's journeys—to Brie, Champagne, Orléanais, and much of the Ile de France—resulted in the completion of four geologic maps, the gathering of data for use in others, and the preparation of detailed memoirs and travel accounts.[11]

By the end of 1766, Lavoisier had emerged from apprenticeship to Guettard, as is shown by the development in his field notes of interests and techniques not fully shared by his mentor. The consistent theme in all his notes is the careful observation of the order of superimposed strata, information which Guettard also recorded; unlike Guettard, however, Lavoisier gradually began to use barometers for the measurement of the thickness and altitude of strata, and he sometimes noted dip and strike as well. Of these sets of data, Lavoisier gave most of his attention to stratigraphic succession and his barometric measurements, intending to use them for the construction of vertical type sections and, eventually, for a stratigraphic description of the whole of France. (Unfortunately, the latter project could not be realized because of difficulties in calibrating his barometers, but he did complete several type sections for the geologic atlas.) At the same time, Lavoisier had also become interested in the formation of littoral and pelagic beds and, in 1766, had begun to analyze this problem in a manner not far different from the exposition to be found in his memoir of 1789. Furthermore, he had begun to take the initiative in matters related to the atlas: conducting independent field work, supervising the labors of Dupain-Triel, and suggesting

[9] Notes dated 13 June 1766 and 22 July 1766, LPAS 415, 419. Lavoisier had visited these areas in 1765 and again early in 1766. For the voyages of 1766, see LPAS 412, 414, 416, 419, 421, 537–540, and Œuvres de Lavoisier, Vol. V, pp. 53–71, 93–142, 151–157.

[10] Œuvres de Lavoisier—Correspondance recueillie et annotée par René Fric (Paris: Albin Michel, 1955–1964), fasc. 1, p. 15; LPAS 421. Writing before the publication of Lavoisier's correspondence, Duveen and Klickstein, op. cit.,

p. 433, stated that there was "no record of Lavoisier's having travelled through this region [Brie] at that time," and they ascribed the published notes from this voyage to Guettard, Lavoisier having only made fair copies of them.

[11] Above, n. 9. The maps in question are numbers 40 bis (environs of Paris), 25, 26, 27; cf. Duveen and Klickstein, op. cit., pp. 240, 243–244, and Rappaport in Duveen, op. cit., p. 131.

and taking sole responsibility for the construction of vertical sections to accompany the maps.[12]

The first voyage for the atlas known to have received financial support from Bertin began in June 1767, when Guettard and Lavoisier embarked on their famous tour of the Vosges. The journey lasted more than four months during which they explored hundreds of square miles, often traveling under the most difficult conditions. The pleasures and pains of this expedition—visits with other well-known scientists, the purchase of books, the collection of mineralogical specimens, and the problems of transport, lodging, and communication with Paris — have all been described in some detail by Lavoisier's biographers, who have made good use of the numerous and fascinating letters written by Lavoisier to his father and aunt.[13] The same accounts also refer to the variety of Lavoisier's scientific activities during these months, but it is less often realized that Lavoisier assumed responsibility for certain activities either assigned to him or simply left to him by Guettard. As Guettard remarked in a letter to Bertin: "Monsieur Lavoisier . . . soccupe en outre a prendre les hauteurs des montagnes avec le Barometre, determine la pesanteur des eaux et me seconde infiniment dans le reste du travail."[14] Lavoisier described himself as "prodigieusement occupé" with his hydrometric analyses and his continual climbing of mountains "le barometre a la main."[15] Sometimes it was simpler to bottle a sample of water and postpone analysis until he returned to Paris. Although most of the results of this hydrometric activity were not published during Lavoisier's lifetime, they fulfilled one purpose of the voyage when they were used to reveal the nature of the terrains through which the analyzed streams and rivers had passed.[16] Less time-consuming was the job of taking periodic readings of thermometer and barometer, which he did both in the field and at least three times daily at his lodgings. One object here seems to have been meteorological and similar to that of Guettard in his own thermometric and barometric measurements made in Poland a few years earlier; but more important to Lavoisier was the calibration of his instruments so that his barometric readings would supply reliable and comparable data for the construction of vertical sections.[17]

[12] Cf. esp. Œuvres de Lavoisier, Vol. V, pp. 6, 28, 67–83, 109, 143–144, 205–206; LPAS 417, 419, 421; and Bibliothèque municipale de Clermont-Ferrand, MS 337, fols. 276–277. Guettard's papers and travel journals show that he used the barometer only for meteorological purposes (below, n. 17).

[13] Lavoisier's mineralogical collection is now at the Musée Lecoq, Clermont-Ferrand, and the catalogue among his papers, LPAS 150. His earlier collecting activities are mentioned in letters from Deperthuis to Lavoisier, 14 July 1765, and Pierre(?) to Lavoisier, 18 May 1765, MHN, MS 1929, IV. The latter communication has been incorrectly described by Guerlac, "A Note," loc. cit., p. 214, as written by Guettard and dated 1763, but Guettard is referred to in the letter as "M. Goutard."

[14] Guettard to [Bertin], Chaumont-en-Bassigny, 21 June 1767, MHN, MS 227, fols. 98–99.

[15] Œuvres de Lavoisier — Correspondance, fasc. 1, pp. 30, 38.

[16] Œuvres de Lavoisier, Vol. III, pp. 145–205; Guettard, Mémoires sur différentes parties de la physique, de l'histoire naturelle; des sciences et des arts, &c. (Paris: L. Prault, 1774–1786), Vol. II, pp. lxxv–lxxvi; and Duveen and Klickstein, op. cit., pp. 16–17, 387. The methods used by Lavoisier are discussed by Meldrum, op. cit., pp. 348–357.

[17] Above, n. 12, esp. Œuvres de Lavoisier, Vol. V, pp. 205–206. Cf. Guettard, "Observations météorologiques, faites à Varsovie, Pendant les années 1760, 1761 & 1762," Mém. Acad. R. Sci., 1762 (1764), pp. 402–430.

380

The travel journals kept by Lavoisier during his months in the Vosges contain records of meteorological and hydrometric data and descriptions of soil, river beds, and geologic outcrops. The latter information is similar to that noted by Guettard in his own journals. At times, when Guettard's descriptions include greater detail, Lavoisier seems to have entered many observations directly on leaves of the Cassini map of France, thus omitting much from his journals.[18] Since the primary purpose of the voyage was to gather data for maps, both travelers took every opportunity to record vertical sections, and long lists of observed strata fill many pages of their notebooks. Unlike Guettard, Lavoisier sometimes found it desirable to draw sketches of the terrain; these few pen-and-ink drawings are often diagrams of strata in which unconformities and the general direction of the beds are indicated.[19]

Within months after his tour of the Vosges, Lavoisier was elected to the Academy of Sciences and bought an interest in the Ferme générale, the company of financiers responsible for the collection of certain indirect taxes. In October 1768 he began the series of experiments which were to result in his famous memoir of 1770 on the supposed transmutation of water into earth. At the same time, he continued his hydrometric analyses, sorted his geologic observations, and completed maps of Alsace, Lorraine, and Franche-Comté. Although he seems to have undertaken no further voyages for purely geologic purposes, Lavoisier was now traveling for the Ferme and used these opportunities to gather additional data for maps. Thus, in 1769–1770 he visited Brie, Champagne, Picardy, and Flanders, combining his field work with such duties as examining the deployment of brigades set to catch bands of smugglers. The journeys and paper work of these years (1767–1770) resulted in the completion of additional maps for the atlas, bringing the total number to sixteen.[20]

The nature and extent of Lavoisier's geologic activities during the sixteen years after 1770 apparently underwent some change, if the striking decrease in the number of geologic notes and memoirs surviving from that period is a reliable guide. After 1772 he was, of course, increasingly preoccupied with chemistry, but the evidence also suggests that both he and Guettard continued to work on the atlas—although presumably less energetically than before—until 1777, the date of Antoine Monnet's appointment as director of the enterprise. In fact, in 1785 Lavoisier was to claim that, after the completion of the first sixteen maps,

18 MSS [Voyage des Vosges], II, 50r; III, 51v; IV, 13r; VIII, 20r–21r (LPAS, Tomes 1–8), and Douglas McKie, *Antoine Lavoisier: Scientist, Economist, Social Reformer* (New York: Collier Books, 1962), pp. 301–302. Guettard, MSS *Voyage de Loraine et des voges [sic],"* collection of Denis I. Duveen, and *Suite du voyage des vosges commencé le 14ᵉ juin 1767*, Guettard Papers, Académie des sciences.

19 MSS [Voyage des Vosges], IV, 54v; V, v[64]; VI, 47v. Similar drawings are also found

in his later travel journals, see below, nn. 20, 34.

20 MSS *Voyage en Flan[dres]*, 1769–1770, LPCU; *Voyage mineralogique de paris a champagne*, 1769, LPAS; and *Journal de voyage Commencé le 9 aoust 1770*, LPAS. Map numbers are given by Duveen and Klickstein, *op. cit.*, pp. 238–239, and further information about their publication by Rappaport in Duveen *loc. cit.*, pp. 130–132. *Cf.* also, Guettard's comments quoted in Duveen and Klickstein, *op. cit.*, p. 237.

. . . ayant été distrait de ce travail par des occupations d'un autre genre, il [Lavoisier] en avait négligé la suite; mais . . . ni M. Guettard, ni lui ne l'avaient entièrement perdue de vue . . . [et] ils n'avaient cessé de rassembler des matériaux [pour l'atlas].[21]

Lavoisier probably was exaggerating the extent of his collecting activities during the intervening period—he was addressing a gathering which he hoped would favor the reinvigoration and modification of the original project—but there is no reason to doubt that essentially he had not "entirely lost sight of" either the atlas or certain of his other geologic interests.

While there is no record of extensive travel by Lavoisier during the period 1770–1786, he was the author or co-author of four memoirs on hydrometry and three on mining and mineralogy. He also presented at least twenty reports to the Academy of Sciences on works dealing with these subjects and with natural history in general. In the same interval, he had Bertin's and his own mineralogical collections catalogued, attended a course of lectures on mineralogy, and, for at least a part of this period, continued to work toward the completion of new maps.[22]

The Guettard atlas in the two years after 1770 was beginning to founder for lack of continued financial support from Bertin, and both Lavoisier and Guettard made repeated efforts to keep the project going. These efforts took the forms of a memorandum to Bertin—in which Lavoisier suggested that the atlas be undertaken on a more modest and practicable scale — and several published appeals to provincial naturalists to aid the two geologists in gathering data. Although Bertin remained unable to help, many naturalists did respond, and it is likely that Lavoisier and Guettard, with secretarial assistance, spent some time in the 1770's sorting and entering on maps the observations received from the provinces. Indeed, in 1775 Lavoisier transmitted additional data to Dupain-Triel and was even concerned with possible ways to speed up work on the atlas.[23] When Bertin transferred control of the atlas to Monnet in 1777, Monnet expressed the hope that both Guettard and Lavoisier would continue to collaborate on the project. Guettard refused, while Lavoisier, about whom evidence is lacking, may have ceased to play any active role. That Lavoisier watched Monnet's progress with interest is more than likely, and, when a large set of maps was completed and published

---

21 Quoted in Henri Pigeonneau and Alfred de Foville, *L'Administration de l'agriculture au Contrôle général des finances (1785–1787): Procès-verbaux et rapports* (Paris: Guillaumin, 1882), pp. 68–69.

22 *Œuvres de Lavoisier*, Vol. II, pp. 29–37, 153–159, 234–247; Vol. III, p. 206–207; Vol. IV, *passim*; Vol. V, pp. 179–185. An undated MS *Cours de Minéralogie*, in Lavoisier's hand, contains a reference to chemist J.-B. Bucquet which could only have been written before his death in 1780 (LPAS 164, fol. 17r); later, in 1792, Lavoisier was to attend René-Just Haüy's course in crystallography.

23 *Œuvres de Lavoisier—Correspondance*, fasc. 2, pp. 485–486; *Œuvres de Lavoisier*, Vol. V, pp. 216–220; and Rappaport, *loc. cit.*, p. 130. At Cornell University there are 6 large foliovolumes of transcribed data presumably for use in the atlas: *Lavoisier et Guettard: Description Minéralogique de la France*, 2 vols., *Lavoisier et Guettard: Géographie minéralogique de la France*, 2 vols., *État des mines en France*, 2 vols. Among the Guettard papers, MHN, MSS 1326, 2187, 2194, there are dozens of sheets of data marked "transcrit" or "fait." The date at which all this copying was done cannot be determined.

by Monnet in 1780, Lavoisier was infuriated to find that his own contributions had barely been acknowledged.[24]

After 1777 Lavoisier continued to receive some data for the atlas from his provincial correspondents,[25] but it was not until 1785 that he saw an opportunity to resume work on a modified version of the original project. Vulcanologist Nicholas Desmarest proposed in 1785 to the newly created Comité d'administration de l'agriculture, of which Lavoisier was a member, that it consider his own plan for a geologic survey of France; since Desmarest's cartographic methods differed considerably from those of Guettard and Lavoisier, he insisted on the necessity of beginning afresh. Lavoisier protested jettisoning the work he and Guettard had begun and which he thought himself "obligé par des engagements d'honneur de continuer." Since Guettard's health was failing, Lavoisier further offered

> . . . de le suppléer et de faire exécuter sous ses yeux le dépouillement de ses [Guettard's] manuscrits et de ses mémoires imprimés; d'y joindre toutes les observations qui lui sont propres et qu'il a rassemblées dans les nombreux voyages qu'il a faits en France; de porter le tout sur la carte de Cassini, qui deviendra par là un manuscrit précieux qui appartiendra au gouvernement et qu'on pourra faire graver ensuite sur telle forme qu'on le jugera à propos.[26]

After long debate, the committee decided to endorse the general idea of a geologic survey, but these proceedings came to a halt on 31 July 1787, when it was announced that the Royal Treasury could not support such a venture.[27]

While these intermittent struggles to settle the fate of the atlas were taking place, there remained one area of geology pursued with vigor by Lavoisier before 1770 and oddly neglected by him until 1786: sedimentation and stratigraphy. His early studies of littoral and pelagic beds and his construction of type sections had taken place during field trips in connection with the atlas, but Lavoisier's ideas could not have been expressed fully on the maps alone, and in fact he had declared in 1765 and again in 1766 that he intended to continue his study of sedimentation until he could arrive at "un système . . . des changements arrivés au globe."[28] Yet this ambition not only seems to have been abandoned until 1786, but Lavoisier also made no effort to publish the results of his early field trips, apart from those observations included on the maps and several memoirs dealing with the relatively peripheral matters of hydrometry and the chemical analysis of minerals.[29]

In 1786, however, Lavoisier was again concerned with the problems he had

24 Œuvres de Lavoisier, Vol. V, pp. 220–221, and Duveen and Klickstein, op. cit., p. 238. Also, Bibliothèque municipale de Clermont-Ferrand, MS 719, fols. 39–42.

25 Œuvres de Lavoisier — Correspondance, fasc. 3, p. 686.

26 Pigeonneau and de Foville, op. cit., pp. 116–117. The original project was meanwhile floundering under Monnet's direction; he had published the Atlas et Description minéralogiques de la France . . . Première Partie (Paris: Didot l'aîné, 1780) but in the next 15 or more

years was able to complete only 4 new maps which were added to a 2nd edition of the atlas. See Rappaport, loc. cit., p. 131.

27 Pigeonneau and de Foville, op. cit., pp. 118, 408. Whether Lavoisier had given up entirely at this point is not clear; cf. Duveen and Klickstein, op. cit., p. 325, and Œuvres de Lavoisier, Vol. VI, pp. 252–255.

28 Œuvres de Lavoisier, Vol. III, pp. 109, 126.

29 Ibid., pp. 111–127, and above, n. 22.

begun to investigate in 1765 and 1766. On 5 September 1786 he presented to Condorcet, Perpetual Secretary of the Academy of Sciences, a memoir entitled "Observations generales Sur les Couches qu'ont ete deposées par la mer et Sur les Consequences quon peut tirer de leur disposition relativement a lanciennete du monde."[30] This memoir survives in an early précis and one draft—both slightly later in date than that shown to Condorcet—and the well-known version of 1789, and it is remarkable for its obvious reliance upon the observations and ideas expressed by its author twenty years earlier. The memoir presents some new data, but it is primarily a brilliant refashioning of material and ideas long in Lavoisier's possession.[31] Similarly, Lavoisier's early research provided data for a second memoir presented to the Academy in November 1792. In this modest "Mémoire sur la hauteur des montagnes des environs de Paris," published posthumously, Lavoisier emphasized the need for accurate barometric measurements of the kind he had attempted in the 1760's, and he went on to discuss his own barometric observations dating from 1771.[32] These memoirs were to have been followed by further publications, for Lavoisier intended to return to his youthful project of writing a "theory of the earth." As he put it in 1789, he meant not only to continue his work on sedimentation, but he would also enter into

> . . . des details très étendus sur les dégradations occasionnées par les eaux pluviales, sur la formation des ravines des vallées, sur le cours des eaux. Je m'efforcerai d'après les témoins qui nous restent de conclure quel étoit l'état de la surface du globe à différentes époques et de présenter quelques conjectures sur le tems où il a commencé à être habité par des animaux.[33]

These projected studies would probably have had the benefit of renewed field work, since Lavoisier was once more beginning to make extended journeys. In 1787 and 1788 he visited areas in Orléanais, Burgundy, and Normandy, and, as he had in 1769 and 1770, took the opportunity to record geologic observations whenever free of his other duties.[34] Lavoisier's work in geology had thus come full circle, as he returned in his last years to the activities and ideas of the 1760's.

At this juncture, it becomes possible to examine Edouard Grimaux's claim that "Lavoisier s'intéressa toute sa vie aux recherches géologiques."[35] Until about 1770 and after 1786, his very real interest in geology cannot be doubted. The period 1770–1786, however, presents at least three major

30 LPAS 335A consists of a single sheet bearing this title; the memoir itself may be among the missing documents of dossier 626.

31 The précis and draft (LPAS 335B, 335C) bear titles similar to that of the version published in Mém. Acad. R. Sci., 1789 (1793), pp. 351–371. The latter text, however, has suffered a few significant deletions found in the draft; it also includes a long footnote (p. 354) present in the draft but not published in Œuvres de Lavoisier, Vol. V, pp. 186–204.

32 Œuvres de Lavoisier, Vol. V, pp. 205–213. Early drafts of this memoir may be among the missing documents of LPAS 282, 338.

33 LPAS, 335C, fols. 17r–17v. This final paragraph was omitted from both of the published versions (above, n. 31) which conclude with references to future memoirs on pelagic and littoral formations. See also Œuvres de Lavoisier, Vol. V, p. 213.

34 LPAS, MSS le 4. 7bre 1787. Voyage D'orleans Pour L assemblée provinciale, untitled [Voyage en Bourgogne], 1787, and Voyage de cherbourg le 11 juin 1788.

35 Edouard Grimaux, Lavoisier 1743–1794 (2nd ed., Paris: F. Alcan, 1896), p. 26.

problems: (1) Why did Lavoisier temporarily abandon the study of sedimentation? (2) Why did he not publish earlier the materials dating from 1763 to 1771 and incorporated into his memoirs of 1789 and 1792? (3) Why did he resume fairly intensive work in geology in 1786? Several possible answers might be suggested, but none is wholly satisfactory. It may be, as A. N. Meldrum once observed, that Lavoisier's interest "dwindled" for a time; there are, however, so many geologic notes now known to be missing from among Lavoisier's papers that it seems hazardous to make any assertions of this kind. Or it may be, as Meldrum further suggested, that Lavoisier's "inspiration had failed" by 1772, and that he did not publish his early geologic work because he considered it of little value; [36] but this interpretation is questionable in view of the fact that Lavoisier's later geologic ideas, which he did publish, resemble to an extraordinary degree his achievements of the 1760's. Finally, it is possible that Lavoisier was so absorbed in chemistry that he could not devote much time to geology; but it can as easily be argued that Lavoisier was far busier after 1786 than he was before, and yet he found time to prepare two substantial memoirs and to plan further geologic research.

A less problematical, albeit unverifiable, solution to these questions involves the fortunes of the Guettard atlas after 1770. So much of Lavoisier's work in geology was directly connected with or grew out of this project that it should not be surprising to find the fate of the atlas influencing the course of his own activities. And, indeed, the years 1770–1777 which saw Guettard and Lavoisier struggling to keep the project from foundering, the appointment of Monnet, and the final disassociation of Guettard and Lavoisier from the enterprise, were also the years in which Lavoisier produced little of significance in geology. After 1777 there continues to be evidence only of miscellaneous and minor geologic work. In 1785, however, Lavoisier was faced with the opportunity to resume work on the atlas, only to have this possibility vanish two years later. This short-lived hope in 1785 may well have been the catalyst Lavoisier needed for the preparation in 1786 of a first draft of his memoir on sedimentary strata, and the failure of hope in 1787 may well have prompted him to undertake the revision and publication of that memoir. A second memoir, also based on his early research, followed in 1792. Lavoisier was thus publishing the materials accumulated during his voyages for the atlas, and publishing them only after it had become clear that he would be unable to complete the geologic survey of France. [37]

36 Meldrum, *op. cit.*, pp. 331, 420.
37 The major difficulty presented by this interpretation is that Lavoisier's memoirs as finally drafted do not require the support of geologic maps, but he may have originally intended to tie them closely to the maps of the atlas. Among Lavoisier's papers are documents resembling early drafts of memoirs and containing information to supplement that found on maps completed by 1770 (e.g., *Œuvres de Lavoisier*, Vol. V, pp. 67–71, 72–83, 91–92, 143–148).

IV

LAVOISIER'S THEORY OF THE EARTH

In the commonly understood meaning of the phrase, Antoine-Laurent Lavoisier did not write a 'theory of the earth'. That he intended to do so, however, is recorded in a passage probably written in 1766. There he announced a research programme, already begun, that would yield

> des connaissances exactes sur les anciennes limites de la mer, sur le lit qu'elle occupait, sur l'ancienne disposition des continents; en un mot *un système*, toujours guidé par des expériences et des observations sûres, *des changements arrivés au globe*.[1]

Although his ideas were soon to change, he promised in 1774 to make public the results of his geological research, and in 1788 he again declared his intention to continue his studies in order to arrive at conclusions about 'l'état de la surface du globe à différentes époques'.[2] Had he been able to carry out these projects, Lavoisier's theory of the earth—both the early theory and his mature one—would have differed markedly from those being written by his contemporaries, since Lavoisier displayed little interest in such topics as the formation of the earth's 'primitive core' and the role of volcanoes and of the Deluge. Instead, from the beginning of his career in geology, Lavoisier selected and emphasized problems of sedimentation and stratigraphy that could shed light on the evolution of the earth's crust.

Lavoisier's youthful theory of the earth is of particular interest for its combination of traditional and novel elements and for the inadequacies that led him to abandon it. At the end of 1766 he began to formulate a new theory, one based upon the principles later set forth in his famous memoir 'Observations générales sur les couches modernes horizontales qui ont été déposées par la mer et sur les conséquences qu'on peut tirer de leurs dispositions relativement à l'ancienneté du globe terrestre', completed by 1789 and published in 1793. This memoir occupies an anomalous place in both the history of geology and the biography of its author. Written in a period when Lavoisier is commonly said to have abandoned the geological studies of his youth, the memoir appears to have no immediate biographical context; and its brilliance has led

An earlier version of this study was presented (*in absentia*) to a meeting of the Asociación Argentina de Historia de la Ciencia, Córdoba, Argentina, 28 August 1970.

[1] 'Extrait de deux mémoires sur le gypse', in *Oeuvres de Lavoisier* (6 vols., Paris, 1862–93), iii. 109. The italics are mine. The second of the two memoirs on gypsum was read to the Académie des Sciences on 19 March 1766, and this abstract was probably written after that date but before October 1766, when Lavoisier modified his views on 'les anciennes limites de la mer'.

[2] *Opuscules physiques et chimiques* (Paris, 1774), in *Oeuvres de Lavoisier*, i. 441. I am indebted to Henry Guerlac for drawing my attention to this passage. See also R. Rappaport, 'Lavoisier's geologic activities, 1763–1792', *Isis*, lviii (1967), 383, and note 33.

historians to analyse it in isolation from its context in eighteenth-century geology. Within limits to be defined below, this study proposes to supply some of the biographical and geological background necessary to a fuller understanding of the memoir of 1789. In addition, it will be shown that this memoir once formed part of Lavoisier's unwritten theory of the earth.

Placing Lavoisier's work in its historical context necessarily involves some examination of his relationship to his predecessors and contemporaries. At least, this can and must be done for his early work, in the period 1763–6. His more mature ideas present different problems in that they probably originated in his own youthful studies, with minimal influence exerted by his teachers and other contemporaries.

Lavoisier was always reluctant to reveal the inspiration behind his ideas, but for his geological research he did once refer to the particular debt he owed to the writings of Buffon, the lectures of G. F. Rouelle, and conversations with J. E. Guettard.[3] While it is possible to find parallels between Lavoisier's ideas and those of still other contemporaries, such parallels would not constitute proof of indebtedness, and Lavoisier's own admission therefore provides the most useful point of departure. However, there remains some difficulty in determining precisely what Lavoisier owed to each of his mentors. The conversations with Guettard cannot now be reconstructed, and one can only assume that the ideas in Guettard's writings are those to which Lavoisier was exposed. Rouelle's geological teachings receive brief and unsatisfactory treatment in the manuscripts of his chemistry lectures, and ideas which contemporaries attributed to Rouelle are sometimes lacking entirely from the manuscripts. The geological and cosmological views in Buffon's *Histoire naturelle* may well have inspired Lavoisier; but Buffon's text is also rich in casual suggestions, and historians cannot be certain that Lavoisier noticed some of these asides. The existence of such lacunae does not render all analysis meaningless, but some questions about Lavoisier's early career must remain unanswered.

For the period after about 1770, Lavoisier's manuscripts yield little of geological interest, and this limits to some extent what the historian can do to trace the later development of his ideas. Nonetheless, two kinds of analysis are possible. The first is a comparison of Lavoisier's mature ideas with those of certain of his contemporaries; this method will be used very sparingly in this paper because mere similarity of ideas, in the absence of corroborative evidence, is inconclusive. For reasons that will appear below, it seems more instructive to treat Lavoisier's mature writings as an outgrowth of *his own* early work in geology. Indeed, it will be shown that his early notes can properly be used to supply the immediate background for ideas he was expressing twenty years later.

3 *Oeuvres de Lavoisier*, v. 226. Cf. H. Guerlac, 'A note on Lavoisier's scientific education', *Isis*, xlvii (1956), 211–16.

The first phase: 1763–6

These three years of what might be called Lavoisier's geological apprenticeship were marked by intensive field-work, the development of observational skills, and increasing use of such techniques as chemical analysis and barometric measurement. In 1763 he had read Buffon, had attended or was attending Rouelle's lectures, and was beginning to take field trips in the company of Guettard.[4] Not surprisingly, then, this was the period in which Lavoisier expressed geological views close to those of his teachers. Furthermore, he was using some of the special vocabulary developed by Rouelle and Guettard, and he occasionally referred to one or another of his teachers in a manner which suggests that he was working within and trying to verify portions of their geological theories.

Among the ideas shared by Guettard, Rouelle, and Buffon—and other geologists of the period—was the belief that the earth's crust consists of two major groups of formations: the unfossiliferous, largely granitic masses which Rouelle called the *terre ancienne,* and the more recent, fossiliferous, sedimentary *terre nouvelle.* (Volcanic and alluvial terrains were considered to be relatively less important in the overall history of the earth.) Sedimentary formations had been deposited during an era when the sea covered all or most of the present continents, except perhaps for the peaks of the highest mountain ranges; this era had ended when the waters had somehow retreated and exposed the present land masses. Such occurrences as the appearance from time to time of new islands did not impair the general picture adopted by Lavoisier's teachers, and, with the possible exception of Buffon, they assumed that the sea had only once covered the land. In short, the entire *terre nouvelle* was the product of continuous deposition (perhaps interrupted occasionally in some localities) in a deep and stationary sea.[5]

Lavoisier's teachers all discussed the dynamics of sedimentation and distinguished between what their pupil later called pelagic and littoral beds.[6] But they did not explain clearly the relationship of this subject to their general theory of the earth, and so they left wholly or partially unanswered the question: if the *terre nouvelle* was laid down on the ocean floor, why then do some of its strata appear to be *littoral* in origin? The question did not arise in connexion with present shorelines, and Buffon was able to describe at length their gradual elevation or

[4] Details of his activities during this period are in Rappaport, op. cit. (2).
[5] Buffon, 'Théorie de la terre', in *Oeuvres complètes de Buffon,* ed. J.-L. de Lanessan (14 vols., Paris, 1884–5), i. 34–66, and 'Cours de chymie de M. Rouelle, rédigé par M. Diderot et éclairci par plusieurs notes', Bibliothèque Municipale de Bordeaux, MSS. 564–5, pp. 567–82. Guettard never emphasized a general theory of the earth but at least once he expressed views close to those of Buffon and Rouelle and claimed to have held them for some years (Muséum d'Histoire Naturelle, Paris, MS. 2193, ff. 18–20).
[6] The term 'pelagic' had a broader meaning in the eighteenth century than at present, signifying anything to do with the open seas, at any depth; and it will be used in that sense throughout this study. See also note 34.

250

degradation as the sea engages in a perpetual shift from East to West around the globe. But littoral deposits far from modern seas posed a problem of interpretation that even the ingenious Buffon found difficult of solution. In fact, he offered two mutually exclusive explanations. On one occasion Buffon suggested that alternating littoral and pelagic formations could be attributed to the fact that the seas 'ont . . . couvert et peuvent encore couvrir successivement toutes les parties des continents terrestres'.[7] More often, however, his analysis resembles the theory put forward earlier by Benoît de Maillet in *Telliamed* (1748). Believing that the ocean floor can be agitated by strong currents, Buffon concluded that most of the earth's mountains had been built up by the combined action of sedimentation and erosion taking place on the ocean floor; hence, the present land masses consist of sedimentary formations which are all deep-water in origin.[8]

Unlike Buffon, Rouelle believed that disturbances extend to a depth of only ten or twelve feet below the surface of the sea; logically, therefore, littoral formations could not have originated in deep water. The evidence that Rouelle attempted to explain this problem is slender and indirect, and it seems he may have believed that littoral formations (but not single beds) mark the boundaries of former sea basins.[9] If this statement accurately represents his ideas, then Rouelle clearly had a more sophisticated view than either Guettard or Buffon of a once complex distribution of sea basins and land masses. But neither Rouelle nor his contemporaries offered a solution for the problem of single, apparently 'anomalous', littoral beds that 'interrupt' formations otherwise pelagic in origin.

Each of Lavoisier's teachers added distinctive touches to the broad theory which was shared by all, and the most significant of these additions for the present study are Guettard's concept of the geological *bande* and Rouelle's definition of the *tractus*. *Bandes* were divisions of the earth's superficial crust into regions characterized by the predominance of certain rocks and minerals. For France, Guettard outlined a *Bande sableuse* with its centre near Paris and consisting mainly of sandstones and limestones; the *Bande marneuse* circumscribed the *sableuse* and was in its turn surrounded by the *Bande schisteuse ou métallique*. The scheme was a lithological one, taking no account of the relative age of formations or of palaeontology.[10]

[7] Buffon, op. cit. (5), i. 56.

[8] Ibid., i. 34–66, 104 ff. Cf. Guettard, 'Mémoire sur les poudingues. Seconde partie, *Mémoires de l'Académie Royale des Sciences, 1753* (1757), pp. 181–2. Cf. Benoît de Maillet, *Telliamed*, trans. and ed. Albert V. Carozzi (Urbana, Illinois, 1968), especially pp. 65–72, and Carozzi's comments on Buffon on p. 4.

[9] Rouelle cited by Lavoisier; see below, p. 252.

[10] Guettard, 'Mémoire et carte minéralogique sur la nature & la situation des terreins qui traversent la France & l'Angleterre', *Mém. Acad. R. Sci. 1746* (1751), pp. 363–92. See also R. Rappaport, 'The Geological Atlas of Guettard, Lavoisier, and Monnet: conflicting views of the nature of geology', in C. J. Schneer (ed.). *Toward a history of geology* (Cambridge, Mass., 1969), pp. 272–87.

Rouelle's *tractus* was a division of sedimentary strata on the basis of fossil populations. France was marked off into a number of 'cantons',

> déterminés par le coquillage qui s'y trouve le plus abondamment, et il [Rouelle] a appellé *centre* de ce canton ou de ce *tractus* . . . le lieu où ce coquillage se trouve en plus grande quantité. De là jusqu'à l'extrémité de ce canton ce coquillage vient de plus en plus rare; mais aussi à mesure qu'on s'éloigne de ce centre on trouve de nouveaux coquillages dont le nombre augmente continuellement; et on parvient enfin . . . [au] centre d'un nouveau *Tractus*.[11]

Rouelle's system was based on the idea, not uncommon at the time, that the ocean floor of the past resembles that of the present and that, therefore, fossil populations in the strata, like their living analogues, are grouped together as colonies in which particular species predominate. There is no evidence that he assigned different species to different periods of geological time, and in his neglect of time his approach was not unlike that of Guettard.

During the years 1763–6 Lavoisier adopted significant elements of the theory of the earth expounded by his teachers, while, at the same time, he modified and eventually abandoned those concepts attached to the terms *bande* and *tractus*. The latter process of modification appears most clearly in Lavoisier's reference, in 1765, to 'cette partie basse du globe qui a été couverte par les eaux de la mer, et que quelques naturalistes ont appelée la bande ou le *tractus* calcaire'.[12] In a single sentence Lavoisier has equated three quite different ideas. 'Cette partie basse' was the phrase he commonly used for all the sedimentary formations that comprise the *terre nouvelle*, while neither *bande* nor *tractus* had ever possessed so broad a meaning. In fact, Lavoisier has so confused these special terms that the only clear idea to emerge from his statement is a distinction between sedimentary formations and older masses. Similarly, he referred to ore-bearing strata as belonging to Guettard's *Bande schisteuse* and Rouelle's *terre ancienne*; both phrases are here used correctly, but Guettard's lithological concept and Rouelle's stratigraphic one are not equivalent.[13] So, while employing the vocabulary of two of his teachers, Lavoisier actually dispensed with some of the variations and refinements in their ideas and retained the simpler stratigraphic scheme which they shared with Buffon.

The key idea that Lavoisier adopted without modification was his teachers' assumption that there had been only one era of deposition in stationary seas. Hence, when in 1766 he announced his intention of preparing a theory of the earth, he referred to 'le lit' formerly occupied

---

[11] Rouelle, op. cit. (5), p. 574. On occasion Rouelle also used *tractus* lithologically, for example in *tractus calcaire* (ibid., p. 572).
[12] *Oeuvres de Lavoisier*, iii. 126.
[13] Ibid., iii. 135.

252

by the sea. Elsewhere he described sedimentary formations as occupying that part of the earth's crust 'où la mer paraît avoir séjourné pendant des siècles'.[14] At the same time he came to doubt that all sediments could have originated on the ocean floor. In an early (undated) note no such doubt is apparent, and he produced diagrams which illustrate Buffon's theory of mountain building.[15] In 1765, however, he explicitly criticized Buffon's theory on the grounds that littoral and pelagic formations should not be explained by the same mechanism of deposition.[16] This noteworthy change in Lavoisier's thinking may well have been the result of Rouelle's teaching, as is implied in a note written in that same year, 1765. Here Lavoisier remarks upon strata of coarse limestone and sandstone superimposed upon much finer sandstone; the position of these beds suggested deposition under the same conditions and during a single epoch, but their lithology suggested respectively littoral and pelagic origins. More careful observation, he goes on to say, convinced him that the fine-grained sandstone did not actually pass under the other strata but seemed to be 'appliqué Contre [eux]'. He thus interpreted the littoral deposits as revealing the former limits of the sea basin, and he concluded triumphantly that these observations served as 'confirmation aux idees de m^r rouelle et aux miennes'.[17] This brief episode not only illustrates Lavoisier's departure from the ideas of Buffon, but it also explains the way in which he was using pelagic-littoral distinctions in his research programme to chart precisely 'les anciennes limites de la mer'.

By 1765 Lavoisier's theory of the earth closely resembled Rouelle's in its major tenets: a single era of deposition, a complex distribution of sea basins and land masses, and important distinctions drawn between pelagic and littoral formations. It has already been noted that a theory of this kind left unexplained those littoral beds which seem to 'interrupt' pelagic formations. This problem did not go unnoticed by Lavoisier who, in field notes of June 1765, remarked upon a bed of sandstone which seemed to be 'out of place', or, as he put it, 'vraisemblablement accidentel car il est enclavé dans les bancs Calcaires'. By invoking the then common device of the geological 'accident'—which might be defined as a negligible break in the otherwise predictable course of nature—Lavoisier was for the moment able to ignore an anomaly. At this stage in his thinking, Lavoisier could interpret certain littoral formations as the remnants of

---

[14] Ibid., iii. 137

[15] Ibid., v. 12–13.

[16] Lavoisier Papers, Académie des Sciences, dossier 420 (notes dated 1765, on the route from Paris to Senlis).

[17] Lavoisier Papers (16), dossier 423 (25 March 1764). To judge from other notes in this dossier, the date should actually read 1765. It is possible, too, that Lavoisier had read the work of one of Buffon's severest critics: R. E. Raspe, *Specimen historiae naturalis globi terraquei* (Amsterdam, 1763). This volume was in Lavoisier's library. Of the several catalogues of the library, I have consulted the 'Catalogue des livres de la bibliothèque de Madame la Comtesse de Rumford' (2 vols.), in the Lavoisier papers, Cornell University Library; see especially i. 62–80.

ancient shorelines, but the occasional littoral bed was merely an 'accident' that need not be taken into account in a general theory of the earth.[18]

Pelagic-littoral problems were to increase in importance for Lavoisier towards the end of 1766, while before that date his attention was as often directed towards two other activities fundamental to his theories: his barometric measurements and the construction of what he called *coupes générales*, or general sections. Lavoisier used the barometer to measure the thickness and altitude of strata, his aim being to identify precisely and then trace the lateral extent of each bed. At a time when the stratigraphic significance of fossils was not recognized, identifying similar strata in two outcrops was a matter of lithology and position; and Lavoisier was using the barometer for a more accurate determination of position. Although this technique was probably learned from Rouelle, the potential usefulness of such measurements was not immediately obvious to Lavoisier; he was not yet using a barometer habitually in 1764, customarily carried one or two barometers by 1766, and in 1767 described himself as climbing mountains 'le barometre a [*sic*] la main'.[19] In his desire to study the continuity of sedimentary formations Lavoisier was doing, systematically and with precision, what two of his teachers had advocated but scarcely practised. There is no evidence that Rouelle used the barometer with any regularity, although demonstrating the continuity of sedimentary strata was one of his principal concerns. The latter was of even greater interest to Guettard, since he continually sought to refine his system of *bandes*; but barometric measurement was used by Guettard only for meteorological purposes. By contrast, Lavoisier came to believe that precise measurement would provide a firm foundation for his theory of the earth. The research programme he described as already begun in 1766 was, in fact, a series of such measurements.

By the early months of 1766 Lavoisier was using his barometric data to construct general sections. These composite, idealized sections were put together after examination of the various outcrops visible in any given locality.[20] While each section represented the general order of strata in one locality, the construction of such sections for the several regions of France would eventually yield a broader picture of geological change. Here too Lavoisier's ideas show a certain resemblance to those of his teachers. Both Guettard and Buffon had commented earlier on a general order discernible in sedimentary strata, Guettard then using as his example the environs of Paris and Buffon describing what he considered

---

[18] Lavoisier Papers (16), dossier 420 (3 June 1765). This kind of explanation was (and is) not unusual; cf. Guettard (op. cit. [10]), who admits the existence of local exceptions to his system of *bandes*, but says they do not invalidate the scheme he is proposing.

[19] Rappaport, op. cit. (2), 378–9, and op. cit. (10), 281. Buffon noted briefly that the barometer can be used to measure accurately the height of mountains; see *Oeuvres complètes de Buffon*, i. 136.

[20] *Oeuvres de Lavoisier*, v. 75. See also Lavoisier Papers (16), dossiers 415, 419, and Rappaport, op. cit. (10), 282.

254

to be the universal order of all strata composed primarily of vitrifiable materials. Closer reading, however, reveals that Guettard was simply adopting this device as a way of making rapid generalizations, while Buffon, typically, was seeking to formulate global laws. Lavoisier was doing neither but was trying to show that orders of deposition can be established for particular regions. In addition, Lavoisier, far more than his teachers, stressed the geological importance of the general section; indeed, it was Lavoisier who insisted that general sections be included on the geological maps that he was preparing in collaboration with Guettard, while Guettard did not fully realize why the matter should be so significant.[21]

Maturity: 1766 and 1789

By the autumn of 1766 Lavoisier was equipped with a geological theory that he was testing in the field, and his skills had developed to such an extent that he was doing independent and original work in geological cartography. But, despite some modifications in the ideas he had inherited, the general framework he was using remained essentially that of his teachers. This state of affairs continued until October 1766, when he announced: 'Je Commence de Ce moment a Considerer toutes ces choses Sous un aspect tres different'.[22] The occasion of this remark was a geological excursion to Brie, and 'ces choses' is a reference to pelagic-littoral problems. His observations, he goes on to say, seemed suddenly to require profound changes in his theory of the earth. These changes were to affect all his subsequent thinking in geology and were ultimately to supply the basis for his memoir of 1789.

While in Brie in 1766, Lavoisier once again noticed a series of beds alternately littoral and pelagic, but this time he could not resort to the device of the geological 'accident'. As he put it, to find limestone resting upon sandstone now struck him as a *common* rather than a rare phenomenon and therefore one that required attention. He thus began to consider the possibility that 'les pierres Calcaires reposent non pas Sur la terre primitive . . ., Sur l'ancienne terre, mais Sur des matierres que la mer a elle meme deposée lorsquelle [sic] gagnoit Sur les terres'. The latter 'matierres', littoral deposits, entered deeper water as the seas advanced. With time and tranquillity, pelagic beds were deposited horizontally atop the littoral until the seas began to subside, restoring those conditions under which littoral beds had earlier been formed. Thus, he concluded,

---

[21] See *Oeuvres complètes de Buffon*, i. especially 229–30, and Guettard, 'Description minéralogique des environs de Paris', *Mém. Acad. R. Sci. 1756* (1762), pp. 225–7. Disagreement between Guettard and Lavoisier on the subject of map-making is discussed in Rappaport, op. cit. (10), especially 278–9.

[22] Lavoisier Papers (16), dossier 421: 'Observations mineralogiques faittes en 8bre 1766 dans [une] partie du vallois du Soissonnois de la champagne et de la Brie', ff. 1v–2r.

we can discern 'trois formations bien differentes pour la datte dans les bancs que nous habitons', and by going deeper within the earth's crust we might find three more.[23]

Lavoisier had clearly arrived at a rather different conception of the evolution of the sedimentary crust than was possessed by his teachers. Indeed, by implication at least, he was also raising questions about the possibly sedimentary origins of portions of the *terre ancienne*. Moving decisively away from the idea of a single era of deposition in stationary seas, he postulated a succession of ages characterized by a cyclically advancing and subsiding ocean.

How did Lavoisier arrive at these revolutionary ideas? The possible influence of Buffon must be discounted. Lavoisier may well have noticed Buffon's brief comments on the present land masses having been successively covered and then exposed by the sea; but Lavoisier had read the first volumes of the *Histoire naturelle* by 1763, and yet had paid no attention to an idea that he was to find so fruitful three years later. Lavoisier's own explanation, that his observations in Brie provided the needed inspiration, is probably as close to a satisfactory analysis as we can come. It is significant that he remarked not on the novelty of his observations but on the fact that such phenomena were more common than he had realized. One might well ask, when does a particular phenomenon cease to be rare and begin to seem common? To attempt an answer would require an excursion into philosophy and psychology for which evidence, in Lavoisier's case, is lacking.

It is at this critical stage in Lavoisier's development that the documents become sparse and unsatisfactory. As he stated in the journal of his voyage to Brie, his new theory would benefit from additional field-work, and for the years 1767–71 there are records of geological trips to Alsace and Lorraine, Champagne, Flanders, and Picardy. However, his well-known journey with Guettard to Alsace and Lorraine in 1767 did little to support his theory because he examined areas so disturbed geologically that he was unable to establish the general order of the strata.[24] Other terrains explored during this period could have provided useful data, but Lavoisier's travel journals give no indication of the ways in which his theory was being developed or confirmed. The journals, in fact, contain field observations similar to those Lavoisier had been making for years, as well as his usual barometric measurements. After 1771 there is a dwindling record of field-work consisting of voyages to Brittany in 1778 and to Orléanais, Burgundy, and Normandy in 1787 and 1788. Like certain of the earlier voyages, these were not primarily geological

---

[23] Ibid., ff. 2r, 2v, 3r.

[24] This is evident from an examination of the maps based on this voyage; the relevant map numbers and titles are given in D. I. Duveen, *Supplement to a bibliography of the works of Antoine Laurent Lavoisier 1743–1794* (London, 1965), pp. 131–2.

256

in purpose, and Lavoisier's notes, with the exception of some observations at the coast near Cherbourg, are singularly unrevealing.[25]

These lacunae are less serious than one might suppose. Examination of Lavoisier's notes of the 1760s shows that they are sufficiently detailed and precise to have been of value to him twenty years later. In fact, his few examples of field observations in the memoir of 1789 are virtually all references to his earliest voyages: thus, repeated allusions to coastal deposits in Normandy do not stem from his visit there in 1788—his memoir was then already drafted and the results of that voyage were placed in a footnote—but rather to knowledge acquired during a field trip in 1765. Furthermore, Lavoisier's mature ideas were based only in part on field-work, while elements were also drawn from laboratory analysis; for example, when in 1789 Lavoisier remarked on the relative purity of clayey or calcareous materials in particular sedimentary formations, some of his information came from the analysis of specimens in his own mineralogical collection and that of the Académie des Sciences.[26] After 1766, therefore, it was entirely possible for Lavoisier to work out the details of his theory without additional observations in the field. His real need was instead to study and reinterpret the data he already possessed.

Lavoisier's memoir of 1789 can now be examined as his mature statement of observations and ideas dating from his youth. The burden of the memoir is to demonstrate that sedimentary formations are the work of an alternately advancing and subsiding ocean. To this end he presented a detailed theoretical analysis of the deposition of pelagic and littoral strata illustrated by a series of diagrams showing the process occurring under ideal conditions. Assuming, he continued,

> que la mer ait eu un mouvement d'oscillation très lent, une espèce de flux et de reflux, dont le mouvement se soit exécuté dans une période de plusieurs centaines de milliers d'années, et qui se soit répété déjà un certain nombre de fois, il doit en résulter qu'en faisant une coupe de bancs horizontaux . . . cette coupe doit présenter une alternative de bancs littoraux et de bancs pélagiens.

And one should then be able to judge from the number of beds 'le nombre d'excursions que la mer a faites'. To show that observations in the field do support these contentions, Lavoisier constructed a series of vertical sections for areas near Villers-Cotterets, Meudon, and St Gobain.[27]

[25] The journals are at the Académie des Sciences and at Cornell University Library. For a discussion of the period after 1770 see Rappaport, op. cit. (2), 380–4.

[26] Oeuvres de Lavoisier, v. 187, 203. Cf. a minute of a letter from Lavoisier to the Académie (undated but written after Guettard's death on 6 January 1786), Lavoisier Papers, Cornell University Library, MS. 29. In this letter Lavoisier emphasizes that specimens he wants to analyse must be clearly identified as to place of origin.

[27] Oeuvres de Lavoisier, v. 198 and Plates. He also noted that the superior beds can be destroyed by a subsiding ocean and that the number of beds therefore may be fewer than the number of 'excursions' of the sea.

It is clear that the key idea in this memoir had been fully formulated by Lavoisier in 1766, and the vertical sections, too, represent regions that Lavoisier knew well and had mapped in detail by 1766.[28] Furthermore, elements in his theoretical analysis of the configuration of pelagic and littoral beds had long since been worked out, and it is a simple matter to select passages from the memoir of 1789 and to find clear parallels in notes dating from 1765 and 1766 (see Appendix). Preliminary versions of the diagrams used in 1789 still exist, and it is possible that these too were prepared as early as 1770; however, the dating of these sketches remains problematical.[29] In short, there is no doubt that Lavoisier possessed significant parts of his analysis of pelagic and littoral beds quite early in his career and that late in 1766 he developed the theory which was ultimately incorporated into his published memoir.

A second point of major importance to Lavoisier, as the full title of his memoir reveals, was his desire to provide evidence for the antiquity of the earth. In a preliminary précis of the memoir, he had in fact remarked that 'tout atteste que le globe que nous habitons est dune anciennete pour ainsy dire incalculable'.[30] Modifying this statement for publication, he simply observed that the slowness of marine oscillations could be measured by 'plusieurs centaines de milliers d'années'. That he had in mind some further extrapolation is implied in his comment that the expression *terre ancienne* is not especially precise, and that these so-called primitive masses are probably 'un composé de bancs littoraux beaucoup plus anciennement formés'. This lengthy evolution of the earth's crust he planned to investigate in a series of memoirs that would treat, among other topics, the condition of the crust at various stages in the past and the development of life on its surface.[31]

As noted earlier, Lavoisier's suspicion that part of the *terre ancienne* might be sedimentary in origin dated from 1766. The rest of his evolutionary analysis and his views on geological time, however, have no clear precedent in his youthful notes, although the latter do contain some suggestive comments. Such, for example, is his statement in 1766, quoted above, that the alternating formations observed in Brie were 'bien differentes pour la datte . . .', and that further examination might show such strata recurring at greater depths beneath the earth's surface. Both in the early notes and in the mature memoir Lavoisier often referred loosely to periods of 'centuries'; while in 1789 it is clear that he had a

---

[28] Ibid., v. 48–52, 72–83. The corresponding maps are dated 1766 and numbered 26 and 40bis, the former in the *Atlas et description minéralogiques de la France* (Paris, 1780), the latter in the rare second edition of the *Atlas* and reproduced in Rappaport, op. cit. (10), 280.

[29] Muséum d'Histoire Naturelle, Paris, MS. 237, folder 6. An early date is suggested by the fact that the sketches are accompanied by a separate note dated 1770 and are in a folder labelled by Lavoisier: 'Histoire naturelle memoire/generale/Tractus Calcaire. (Coupes géologiques.)'. The term *tractus* is characteristic of his youthful notes. Evidence for a later date includes the use of the word 'géologie', which was not widespread in France before about 1780.

[30] Lavoisier Papers (16), dossier 335B, f. 2v.

[31] *Oeuvres de Lavoisier*, v. 201–4.

longer time scale in mind, there is no way of judging whether the same can be said about his thinking in the 1760s. If his ideas on the time scale did undergo some development between 1766 and 1789, there is no evidence permitting us to explain when and why this happened, although a few conjectures can be hazarded. Lavoisier's library is known to have contained at least two works which emphasize a great extension of the time scale: Buffon's *Les époques de la nature* (1778) and J. T. Needham's *Théorie de la terre* (1769).[32] In fact, bold statements about the age of the earth were becoming increasingly common, and it would have been difficult for Lavoisier to have avoided exposure to such ideas. In any case, his theory certainly required long periods of time to accomplish major and repeated displacements of land and sea, and Lavoisier, unlike Needham and Guettard, was not a man to resist the implications of his own ideas. What made Lavoisier's memoir unusual for its day was not the fact that he adopted a long time scale, but rather his use of marine oscillations as the basic evidence in support of the new geochronology.

Although Lavoisier's reliance upon ideas formulated or outlined in 1766 was considerable, the memoir of 1789 does possess some features which are not to be found in his early work and whose origin is not readily identifiable. Among these, for example, are his brief comments on the gradual development of life on earth—this idea was not uncommon in the eighteenth century—and his remark on the equilibrium of coastlines that would prevail were it not for gradual shifts in the position of the sea basin.[33] Noteworthy too is the memoir's increased sophistication and richness in analysis (see Appendix). There are, furthermore, stylistic changes which cannot be dismissed on the grounds that a polished memoir would naturally differ from a series of field notes. Not only have *bande* and *tractus* vanished, but there is the adoption and consistent use of the terms 'pelagic' and 'littoral', chosen to avoid circuitous phrases.[34]

---

[32] See J. T. Needham, *Recherches physiques & métaphysiques sur la nature & la religion, & une nouvelle théorie de la terre*, in Lazzaro Spallanzani, *Nouvelles recherches sur les découvertes microscopiques, et la génération des corps organisés* (2 vols. in 1, London and Paris, 1769), ii. especially 82, 102–3, where Needham uses sedimentation rates as evidence for an extended time scale. Lavoisier's copy of this work is at Cornell University and, although used in appearance, contains no marginalia.

[33] The latter point, often emphasized by historians of geology, has here been subordinated to those ideas which Lavoisier himself considered most important and treated at length. Lavoisier emphatically attributed to the mathematician Gaspard Monge his ideas on the development of life on earth; see *Oeuvres de Lavoisier*, v. 202. Monge is not known to have had any competence in geology, but some evidence for his interest in natural history and his participation in conversations in Lavoisier's laboratory can be found in Paul V. Aubry, *Monge: le savant ami de Napoléon Bonaparte, 1746–1818* (Paris, 1954), pp. 30, 39, and René Taton, *L'oeuvre scientifique de Monge* (Paris, 1951), p. 324.

[34] Lavoisier Papers (16), dossier 335C, f. 4v, and *Oeuvres de Lavoisier*, v. 189. These terms occur here for the first time in Lavoisier's writings. In the published memoir he claimed to be following Rouelle's usage, but this is not borne out by the Rouelle manuscripts. Buffon is the only one of Lavoisier's teachers known to have used these terms and then in only one passage: *Oeuvres complètes de Buffon*, i. 128. But the terms are Latinisms that Lavoisier could have found in other works, such as R. E. Raspe, *An account of some German volcanos, and their productions* (London, 1776), pp. 15, 17. Raspe's book was in Lavoisier's library.

If the words are of minor importance when compared with the ideas, this new vocabulary remains a significant symbol of the assurance with which Lavoisier expressed himself in 1789, and their use gives his memoir an air of professional competence lacking in the earlier notes.

Having laid the foundations for a theory of the earth, Lavoisier was able to complete only one more geological memoir, his brief 'Mémoire sur la hauteur des montagnes des environs de Paris', presented to the Académie des Sciences in 1792 and published posthumously. Using as examples barometric measurements dating from 1771, he advocated the systematic application of this technique as the basis for investigating the geological history of France. These had been his aims in the period 1766–70 when he was most active in the preparation of the geological atlas of France, and despite the fact that much of his early data would have to be discarded—he had been unable to correct for variations in his instruments—he announced his intention to resume these observations.[35] While his data were drawn from France, it is evident here, as in the memoir of 1789, that Lavoisier meant to arrive at more universal conclusions about the structure of the earth's crust. In view of the solidity and brilliance of Lavoisier's achievements, historians can only regret that he did not live to complete this and other projects.

APPENDIX

The passages in the left column all date from 1765 and 1766 and are drawn from the Lavoisier Papers (cited in note 16), dossiers 417 and 421. Those in the right column are taken from the memoir of 1789, printed in *Oeuvres de Lavoisier* (cited in note 1), v. 191, 192, 194–5.

I. Les Cotes de dieppe Sont toutes garnies de Galets et voici ce quon observe; les plus gros et les plus arondis, ceux qui Sont dans le bas ressemblent presques a du gravier de Rivierre enfin encore plus bas dans lendroit qui nest decouvert que par les basses marées tout est Sable . . .

La mer ote donc aux cailloux leurs angles, les parties plus legeres detachees des Cailloux demeurent quelques instants flottante dans les eaux ce qui les entraine peu a peu dans la partie plus basse de Sorte que peu a peu a mesure quon avanceroit et qu'on S'eloigneroit de la Cote, on devroit trouver un Sable plus fin, ce qui est a examiner.

Les matières les plus grossières, telles que les galets, doivent occuper les parties le plus élevées, et former la limite de la haute mer. Plus bas doivent se ranger les sables grossiers qui ne sont eux-mêmes que des galets plus atténués; au-dessous, dans les parties où la mer est moins tumultueuse et les mouvements moins violents, doivent se déposer les sables fins; enfin, les matières les plus légères, les plus divisées . . . doivent demeurer longtemps suspendues; elles ne doivent se déposer qu'à une distance assez grande de la côte, et à une profondeur telle que le mouvement de la mer soit presque nul.

[35] *Oeuvres de Lavoisier*, v. 206, 213.

II. les bancs formes a la Cote doivent S'abaisser de niveau a mesure quils approchent de la mer . . . les autres [i.e., pelagic beds] peuvent bien prendre a la verité une petite inclinaison lorsqu'ils Commencent a approcher de la Cote, mais quen tirant vers la plaine mer ils ont du remplir les creux et retablir le niveau.

III. [As the seas advanced and littoral deposits entered deep, tranquil water,] alors les Coquilliage[s] ont Commencé a pouvoir y vivre, les premiers qui y on[t] vecu ne Se trouvant pas encore fort eloignés de la Cote le Sable ou la glaise Sur les quels ils etoient poses a du recevoir asses d'agitation pour Se meler avec eux dela il a du arriver que les premiers bancs etoient de pierres Calcaires Sableuses.

Tous les bancs . . . qui se sont formés ainsi à la côte sont ceux que j'ai nommés bancs *littoraux*; enfin, on voit . . . le commencement des bancs calcaires . . . formés en pleine mer des bancs que j'ai nommés *pélagiens* qui se continuent en s'approchant de plus en plus de la ligne horizontale . . .

. . . c'est alors que les animaux de la mer, ceux mêmes qui sont revêtus d'enveloppes fragiles et qui craignent le mouvement, ont commencé à s'y établir . . . Mais les animaux . . . ont dû être incommodés quelquefois par les très grands mouvements de la mer, et . . . aussi des portions de sable fin ont pu demeurer assez longtemps suspendues dans l'eau, pour parvenir jusqu'à eux: les premières coquilles doivent donc encore aujourd'hui se trouver mêlées d'une portion de sable de même espèce que celui sur lequel elles reposent . . .

V

The Geological Atlas of Guettard, Lavoisier, and Monnet:
Conflicting Views of the Nature of Geology

The *Atlas et Description minéralogiques de la France* (1780) was once described by Sir Archibald Geikie as the first of the national geological surveys. While it is true that there was, at least in scale, no comparable work until the nineteenth century, Geikie's characterization is nonetheless misleading, suggesting as it does that this early French survey was a direct ancestor of the massive productions of the next generations. Such a comparison is, in fact, no compliment to Guettard, Lavoisier, and Monnet—although Geikie clearly intended it to be—because their *Atlas* falls far short of the revolutionary ideas and methods developed and used so effectively by William Smith, Georges Cuvier and Alexandre Brongniart, and their successors.

To understand and appreciate the Guettard *Atlas*, it is more appropriate to study these maps in their own eighteenth-century context; that context might be summarized briefly as, first, a lack of established cartographic methods to guide Guettard and his collaborators, and second, a certain degree of confusion and disagreement about the aims and even the subject matter of geology. In other words, when the various eighteenth-century maps and the explanatory memoirs published with them are examined in detail, it is clear that geological cartography in this period was a personal affair, largely dependent upon the views of the individual cartographers. There are, to be sure, some problems, ideas, and methods shared by many of these early map-makers, but to compare the differences rather than to chart the similarities is an exercise that can shed light on the development of geological thought in a period of rapid change and controversy. As the collaborative work of three geologists, the Guettard *Atlas* offers a particularly good example

V

of conflicting ideas about the nature of geological science and the purpose of geological maps.

The construction of the *Atlas* began as early as 1746, when
Guettard produced his first "mineralogical map" of France. Intended
as a preliminary survey, this map was later modified in detail over a
period of almost 40 years as Guettard continued to gather information
about the location of mineral deposits and rock formations. In 1766,
Guettard's work came to the attention of Bertin, Minister and Secretary
of State, whose own concern with the efficient use of French natural
resources made him look with favor on the project which, until that
time, Guettard had carried out alone. The financial support arranged
by Bertin and the collaboration of Lavoisier began almost simultaneously, and while the former soon ended, Lavoisier continued to
work with Guettard for about 12 years, until 1777. At that date, with
16 quadrangles completed and perhaps an equal number close to completion — out of a projected total of 214 — Antoine Monnet took over
the direction of the survey and in 1780 published the set of 31 maps
known as the *Atlas*. Additional maps were to appear some years later,
bringing the final total to 45 completed quadrangles.[1]

Although more might be said about these events, three points in
this brief chronicle are of particular significance here. In the first place,
Guettard is obviously the key figure, since it was he who originated,
designed, and brought to completion the preliminary map of France;
for the next 20 years, he gathered data that were to be useful in preparing the more detailed quadrangles; and it was his pioneer work that
attracted Bertin's attention. A second point is the early date at which
Guettard began his work, before the publication of such classics as
Buffon's "Théorie de la terre" (1749) and Desmarest's geological map
of Auvergne (1774); some features of the *Atlas* are therefore more
easily understood as the product of an earlier scientific generation to
which Guettard belonged. Finally, there are the very different men
involved; if one were to apply a label to each, Guettard might best be
called a naturalist, Lavoisier a stratigrapher, and Monnet a mineralogist.

---

[1] Denis I. Duveen and Herbert S. Klickstein, *A Bibliography of the Works of
Antoine Laurent Lavoisier 1743–1794* (London: Dawson and Weil, 1954), pp. 236–
244, with additions and corrections in Duveen, *Supplement to a Bibliography of
the Works of Antoine Laurent Lavoisier 1743–1794* (London: Dawsons of Pall
Mall, 1965), pp. 129–132. In addition, Guettard prepared five quadrangles describing the province of Dauphiné; these were not published and cannot now be located.
See Archives nationales, F^{14}1312, and E. J. A. d'Archiac, *Introduction à l'étude
de la paléontologie stratigraphique* (2 vols.; Paris, 1864), Vol. I, p. 294.

They differed widely in professional training, interests, and ability, and the fact that they were engaged in an enterprise without precedent gave each one the hope that his own ideas might guide the survey. At the same time, whatever their hopes, Lavoisier and Monnet were able only to modify to a greater or lesser degree the project as conceived by Guettard and sanctioned by Bertin.

To analyze the maps of the *Atlas*, it is essential to begin by examining the state of geology in the 1740's and the ideas of Guettard. Geological cartography at that date can immediately be dismissed as virtually nonexistent. There is no evidence that Guettard then knew either the early hydrographic map by the Abbé Coulon or the "chorographic chart" by Christopher Packe, and certainly neither of these looks like a possible model for Guettard's distinctive mineralogical map.[2] If we turn to the geological literature before 1746, it is clear, too, that there existed no generally accepted principles, theories, or defined subject matter that Guettard might have been expected to adopt as the basis for his cartographic work. The pioneer treatise of Steno seems, in retrospect, to mark the start of modern geology, but in Guettard's youth, Steno, insofar as his work was known, was only one of several authors whose writings seemed to pull in different directions. It would have taken a mind of some brilliance and daring to attempt to transform into a coherent discipline the achievements of Steno, the speculations of Leibnitz, the paleontological memoirs of Réaumur and Jussieu, and the natural theology of the Abbé Pluche and John Woodward. In 1746, however, Buffon's great synthesis was still three years away.

After the publication of Buffon's "Théorie de la terre," there was discernible in France a new geological climate of opinion that, in some ways at least, is directly traceable to the influence of Buffon. Whatever the merits of Buffon's treatise, it was undeniably a best-seller and it succeeded, too, in bringing together, dramatically and eloquently, ideas that had never quite been combined in a single synthesis and that might, in certain cases, have remained neglected. Although Buffon's synthesis did not go unchallenged, it did provide geologists with a common fund of information and ideas and a common point of reference in the ensuing debates. Indeed, as Jacques Roger has shown, Buffon's synthesis brought to a close one era in the history of geology and opened another during which geologists paid more attention to the solution of specific

---

[2] I have been unable to locate a copy of Coulon's map, but it is described briefly by Archiac, *Géologie et paléontologie* (Paris, 1866), p. 106. Packe's map, published in 1743, is reproduced by E. M. T. Campbell, "An English Philosophico-Chorographical Chart," *Imago Mundi*, 6 (1949), 79–84.

V

problems than to the broader issues involved in a "theory of the earth." As a result, Buffon's subsequent masterpiece, *Les Epoques de la nature* (1778), was published some 20 years too late and possessed "un certain air désuet, et comme un parfum de passé."[3]

Among the issues treated by Buffon, those most relevant here are the questions of the extent of the geological time scale and the general structure of the earth's crust. Buffon described the earth's crust as consisting of a primitive — to him, igneous — core surrounded by a series of sedimentary formations. Explicit, too, was the idea that the deposition of sedimentary strata had required a long time and that to chart the history of successive formations would serve to reconstruct the history of both the earth's surface and the forms of life that had inhabited it. Whether or not Buffon's contemporaries took these ideas from his "Théorie de la terre" or from some other source is less important than the fact that the ideas gained wide currency after 1749. Geologists might (and did) disagree about the nature of the primitive core of the earth, about the simple dictinction of Primary and Secondary formations, and about the precise extent of the time scale; in general, however, they could no longer ignore the broad proposition that the earth's crust had been formed in stages, so to speak, and over a period of time.[4]

If the foregoing generalizations about "pre-Buffon" and "post-Buffon" geology are at all valid, one would surely expect geological map-makers after 1749 to reveal some interest in the general stratigraphic structure of the earth's crust. In very different ways, this is certainly true of the maps and memoirs of J. F. W. Charpentier, Nicolas Desmarest, and J.-L. Giraud-Soulavie, to name only three of the later cartographers. Significantly, however, this was true neither of Guettard's first map nor of the revised versions he subsequently produced. Ignoring for a moment the shaded regions on the map (Fig. 1), Guettard has simply dotted the map's surface with symbols representing rock formations and mineral deposits.[5] There is no geological purpose apparent here, no indication of stratigraphic levels, no tracing of the extent of any one formation. Given the geological confusion of his youth, Guettard was at liberty to devise his own cartographic methods

---

[3] Buffon, *Les Epoques de la nature*, ed. Jacques Roger (*Mémoires du Muséum national d'histoire naturelle*, Série C, Tome X, Paris, 1962), Introduction, esp. pp. lix–lx, cxxix–cxliv.

[4] R. Rappaport, "Problems and Sources in the History of Geology, 1749–1810," *History of Science*, 3 (1964), esp. pp. 65–66, 68–69.

[5] Copies of this map are to be found at the Bibliothèque nationale (cf. Duveen and Klickstein, *Bibliography*, p. 240), Ecole des mines, and Academy of Natural Sciences of Philadelphia.

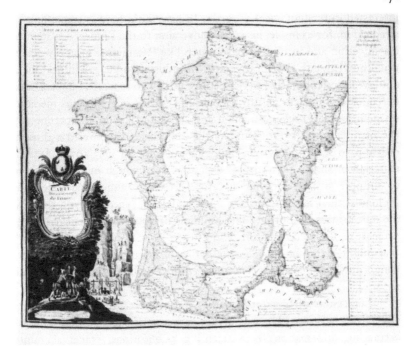

FIG. I.   Guettard's "Carte Minéralogique de France," 1784.

and to express on this map his personal views about the nature of geological science.

Guettard's geological ideas are a peculiar blend of his scientific education, religious convictions, broad interests as a naturalist, and a wide streak of utilitarianism. His formal education was in medicine, in which he received his degree in 1742, but his stronger inclination was toward botany. Early field trips with his grandfather, a pharmacist and amateur botanist, had been followed by friendships with the Jussieu brothers at the Jardin du Roi and with Réaumur, then owner of one of the finest natural history collections in France. Guettard became curator of Réaumur's collection, research assistant to Réaumur, and, gradually, a versatile naturalist in his own right. By 1747 he was "médecin-botaniste" to Louis, Duc d'Orléans, and from that date had a chemical laboratory at his disposal, a growing natural history collection to care for, and the beginnings of an international scientific correspondence of considerable scope and volume.

Throughout his career, Guettard showed himself to be a meticulous observer of nature, cautious in his reactions to new ideas, and especially

V

interested in technological progress. In his youth, he was something of a radical in, for example, his early conversion to the Linnaean system of classification and in his desire, so much like Réaumur's, to apply scientific knowledge to the improvement of industrial and agricultural technology. Once committed to these ideas, he was to remain an impassioned defender of Linnaeus and to become the author of numerous memoirs on such subjects as paper-making and the use of kaolin and petuntse in the manufacture of porcelain. At the same time, he showed a distinct aversion to scientific system-building, traceable in part to his fear of materialism and in part to his belief that not enough evidence existed to support any system; he therefore attacked with equal vigor the biological theories of Charles Bonnet and J. T. Needham and the cosmological ideas of Buffon. The building of systems he believed to be premature, doomed to failure, and a waste of valuable time and energy that might better be spent in the gathering of data; but systems were dangerous, too, because wrong ideas tend to pollute both the heart and the soul.[6]

The results of Guettard's beliefs and interests are evident in virtually all his writings. He was a fact-gatherer of inexhaustible energy, extracting information from scientific publications, travel accounts, encyclopedias, almanacs, newspapers, and personal correspondence. His published memoirs are long, erudite, and tedious, often consisting of minute descriptions of a single stone quarry or a particular fossil, and often accompanied by historical introductions summarizing the opinions of experts from Aristotle onward. The talent he most conspicuously lacked was that of generalization, of seeing the implications of his own observations.

There were times, of course, when Guettard did try to bring coherence to his diverse observations, and one such occasion arose in connection with his first mineralogical map. He had noticed during his earliest field trips that there was "a certain regularity" in the distribution of minerals and rocks over the earth's surface, and he decided to plot his observations on a map to see what patterns emerged. By comparing the different quantities of key substances found in each region he traversed or read about, he concluded that France could be divided into three concentric zones or *Bandes*: the Sandy, the Marly,

---

[6] These remarks are based on statements scattered through Guettard's published works and his unpublished papers at the Muséum d'histoire naturelle. The standard biography is still Condorcet's "Eloge de M. Guettard," *Histoire de l'Académie royale des sciences, 1786* (1788), 47–62. See, also, Roger in *Les Epoques de la nature*, p. cxxxix, note 7.

and the Schistose or Metalliferous (see Figure 1). The *Bande sableuse* consisted primarily of sandstones and limestones, the *Bande marneuse* of marl, and the *Bande schisteuse* of metallic ores, schists, slates, granites, marbles, and other less common substances; some clayey and calcareous materials were actually present everywhere, but were more prominent in one zone than another, and the same could be said of fossiliferous deposits. Although these *Bandes* have since been said to follow the general contours of the Paris Basin, it should be emphasized here that stratigraphy was of no concern to Guettard; he nowhere indicates the relationships among the three *Bandes*, and he placed his symbols without reference to the stratigraphic level of each formation or deposit.[7]

What, then, was Guettard's purpose in constructing his map? At one point he remarks that observations such as his will hopefully lead one day to the formulation of "une théorie physique & générale de la Terre," a comment that suggests he was thinking of the structure of the earth's crust.[8] The same interpretation is supported by the fact that Guettard went on in subsequent years to construct mineralogical maps of Switzerland, Poland, the Middle East, and much of North America; each of these maps is marked off in *Bandes*, and the impression is inescapable that Guettard was gradually applying his system to as much of the globe as his observations and reading would permit.[9] In the memoir of 1746, however, Guettard states his more immediate purpose as the practical one of locating important natural resources and learning where to search for valuable mineral deposits, good building stone, and useful agricultural and industrial soils. In 1752, announcing his plans to continue his one-man survey, he described his map as being of especial value to civil engineers, who had sometimes built roads of less durable materials when they were ignorant of the resources of particular French provinces.[10] Admittedly, Guettard was here trying to interest the government in his survey, and utility was eventually to prove a strong selling point. But a strikingly similar point of view is apparent in a later memoir that contains Guettard's only known comment on the vertical type sections constructed by Lavoisier: showing the stratigraphic "composition" of a given region, says Guettard, will enable one "to judge the cost of obtaining one or another of these substances." For the scientist, he continues, these sections are an important

---

[7] Guettard, "Mémoire et carte minéralogique sur la nature & la situation des terreins qui traversent la France & l'Angleterre," *Mém. Acad. roy. sci.*, *1746* (1751), 363–392.
[8] *Ibid.*, p. 363.
[9] *Mém. Acad. roy. sci.*, *1763* (1766), 228.
[10] *Journal œconomique*, June 1752, pp. 113–138.

V

illustration of "the varieties that nature can employ in the arrangement of the [strata] of the earth's mountains."[11]

A well-read naturalist, Guettard was not unaware of the newer currents in geology after 1750. He clearly knew about contemporary studies of sedimentation rates, the evidence for extinction of species, and many other ideas and observations that suggested a long history for the earth's crust, a history that could be charted stratigraphically. Indeed, he was himself the author of three lengthy memoirs on sedimentation and erosion, he came to believe that the earth's crust consisted of superimposed primitive and sedimentary formations, and he frequently emphasized the need for comparative anatomical work to shed light on the problem of the possible extinction of species. In attacking all hypotheses about the formation of the earth, he also stressed his own interest in the *trans*formations of the earth's crust. Furthermore, Guettard understood how to assemble a type section and occasionally gave verbal descriptions of "coupes générales." In connection with one such description, he even remarked that variations in particular outcrops are perhaps only "an accidental disruption in the general structure common" to a given region.[12]

Despite all the evidence that Guettard knew, understood, and contributed to the development of new geological ideas, most of his work reveals, nonetheless, that he tried hard to avoid thinking of the earth as having much of a history. Thus, when his observations presented him with a sizable sedimentary formation, he could only suggest that the whole mass had been deposited at once and that the effect of bedding had then been produced by a drying and cracking "en quelque sorte suivant une direction horizontale."[13] Similarly, he refrained from discussing the age of the earth, equivocated on the subject of the geological effects of the biblical flood, and constantly denied the likelihood of species becoming extinct. In one of his rare perceptive moments, Antoine Monnet summarized Guettard's problem thus: "he resisted [the implications of] his own observations and Moses was always for him a divine and infallible oracle, and his six thousand years of antiquity assigned to the earth was the barrier which halted his [Guettard's] observations in many respects."[14]

[11] *Observations sur la physique, sur l'histoire naturelle et sur les arts*, 5 (1775), 361. For an explanation of the meaning of "type section" as used throughout this paper, see later text and note 21.
[12] In manuscript, "Lavoisier et Guettard: Géographie minéralogique de la France," Vol. II, "Valois" (Lavoisier Collection, Cornell University).
[13] *Mém. Acad. roy. sci.*, 1756 (1762), 237.
[14] Monnet Papers, Ecole des mines, MS 8286, pp. 120f.

The tortuous paths of Guettard's reasoning do much to explain the peculiar nature of his contributions to the *Atlas*. Although still committed to his own system of *Bandes*, he decided to eliminate all theoretical controversy in the construction of the quadrangles of the *Atlas*.

We have not followed any *systèmes physiques* in the construction of these maps, that is to say, we have not classified terrains as primitive or secondary, as some scientists do, nor as schistose or metalliferous, marly, or sandy.[15]

As a result, the maps of the *Atlas* resemble in their general design Guettard's map of 1746: conventional symbols have again been used to locate the major deposits, and stratigraphy has been ignored (Fig. 2).

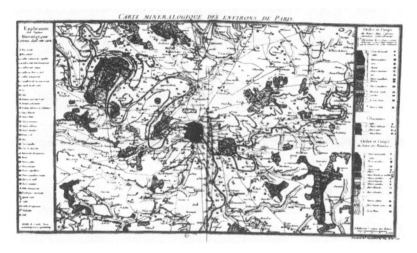

FIG. 2. *Guettard and Lavoisier, "Carte minéralogique des environs de Paris," 1766 (Map no. 40bis).*

The radical differences between Guettard and Lavoisier as geologists are symbolized in Figure 2 by the neat line separating the right-hand margin from the rest of the map. Lavoisier's type sections would undoubtedly have posed a technical problem had he attempted to show this third dimension on the map itself. The problem, however, was not insurmountable, but Lavoisier never had the opportunity to experiment with the technique he had devised.

Lavoisier's training in geology was of a fairly common sort for a young Parisian educated in the 1760's. As he later remarked, his three teachers were Rouelle, Buffon, and Guettard,[16] and the lectures of Rouelle and the writings of Buffon were easily accessible to many of

---

[15] *Observations sur la physique,* 5 (1775), 360.
[16] *Œuvres de Lavoisier* (6 vols.; Paris: Imprimerie impériale et Imprimerie nationale, 1862–1893), Vol. V, p. 226. See Henry Guerlac, "A Note on Lavoisier's Scientific Education," *Isis,* 47 (1956), 211.

his contemporaries. Rouelle's lectures included some discussion of the then-current theory that the earth's crust consisted of a primitive core, formed by a process of aqueous crystallization, and a series of more recent sedimentary formations. Here Lavoisier learned, too, about sedimentation, erosion, and the nature of marine fossils. In the same period, beginning in 1763, he gained considerable experience in the field during excursions often made in Guettard's company. Guettard also taught the younger man his own system of *Bandes*, and the two geologists are known to have discussed the merits of Buffon's *Histoire naturelle*. Lavoisier was impressed by Buffon's work, despite Guettard's criticisms of it and despite eventual doubts of his own about the accuracy of some of Buffon's geological observations.[17]

During his earliest field trips, in the years 1763 through 1766, Lavoisier devoted much of his time to applying the lessons of his teachers and developing certain practical skills. His field notes show his concern with surveying techniques and with the chemical analysis of minerals and rocks, the latter activity resulting in two memoirs on gypsum (1765 and 1766). In the same period, possibly following the example of Rouelle, he began to use the barometer to measure the altitude and thickness of geological formations.

Lavoisier's insistence on precise measurement is one of the striking differences between his work and that of his teachers. How he intended to use these data is not always as clear as one could wish, although an early note offers some hint: "I am thinking of surveying the bed of the Seine in order to produce a map."[18] His next statement of this kind, probably written in the summer of 1766, offers startling clarification of his intentions; here he announces that he has begun and expects to continue a series of barometric measurements that will yield

precise information about the former limits of the sea, the bed which it occupied, the former arrangement of land masses; in short, a system . . . [describing] the changes which the earth has undergone.[19]

This tantalizingly brief statement certainly suggests that Lavoisier was working toward a sophisticated view of the evolution of the earth's crust, and later in the same year, 1766, he was to explain his ideas in some detail.

[17] Lavoisier Papers, Académie des sciences, dossiers 419, 420 (notes dated 1763 and 1765).
[18] *Ibid.*, dossier 417 (May 1765). Rouelle or one of his friends had earlier made a series of similar measurements. "Cours de chymie de M. Rouelle," Bibliothèque municipale de Bordeaux, MSS 564–565, pp. 571–572, and "Cours de Chimie de m^r. Rouëlle Pour l'année 1761—30 octobre," p. 404, manuscript in collection of the late Dr. Pierre Lemay.
[19] *Œuvres de Lavoisier*, Vol. III, p. 109.

Whether the ideas implicit in Lavoisier's statement were based to any great extent upon those of his teachers or other contemporaries is at present only a matter for conjecture. Lavoisier himself implies that his thinking was stimulated by his teachers and then modified considerably by his observations in the field. Thus, for example, he remarks that Rouelle had taught him the differences between pelagic and littoral deposits — he could easily have learned this from Guettard as well — but, until 1766, he was under the impression that sedimentary formations in any one locality should be *either* pelagic *or* littoral. During his field trips he was therefore surprised to notice how frequently limestone was to be found superimposed on coarse sandstone; his reaction, as he describes it, was to begin to rethink his geological principles and to play with the notion that alternating pelagic and littoral beds were evidence of the "flux and reflux" of the sea.[20]

Once Lavoisier had arrived at this key idea, his barometric measurements took on added significance: they could be used to trace the lateral extent of formations, to estimate the time needed for deposition, and, eventually, to determine the changing "limits of the sea . . . [and] the former arrangement of land masses." But Lavoisier also recognized that comparing two outcrops of the same formation revealed some variations in the number and nature of strata and that to reconstruct the geological past required that he assemble vertical type sections. His own explanation of what such a section represents is worth quoting, since it shows not only his grasp of the problem but also his difficulty in expressing what was then a new concept.

These are the only two mountains on this part [i.e., quadrangle] of the map which wholly correspond to the [general] section just described; but the remarkable thing is that most of them or, more accurately, all the mountains of this region are almost precise fractions of the same section; so that, to explain myself more clearly, if one were to remove the top quarter, third, or any other fraction of the general section, what remained would describe [in turn] the [structure] of all of these mountains.[21]

Significantly for the *Atlas*, Lavoisier developed these stratigraphic ideas in 1766, the year in which his most active collaboration with

[20] Lavoisier Papers, Académie des sciences, dossier 421 (11–17 October 1766).
[21] *Œuvres de Lavoisier*, Vol. V, p. 75. Although undated, this document concerns a geological map completed and engraved in 1766. Lavoisier always used the phrase "coupe générale," and this is more precise than my own use of "type section." It was pointed out to me by several members of the New Hampshire Conference that, in its modern meaning, a type section is a section observed in the field and used as a standard for comparison with other observed sections. By contrast, Lavoisier's type sections were ideal, composite constructions based on a series of observations; but he used these constructions for a purpose similar to the modern one, i.e., as a standard of comparison.

V

Guettard began. To see how Lavoisier's ideas were put into practice, it is only necessary to look at the margin of one of the early quadrangles (see Figure 2). This map of the environs of Paris dates from 1766, and similar "coupes générales" are to be found on many of the maps to which Lavoisier contributed; in the more geologically disturbed regions of Alsace, Lorraine, and Franche-comté, he apparently found it impossible to construct type sections, and the margins of those maps show the strata in particular mines or quarries.

Lavoisier's long-range plan soon became to write not a traditional "theory of the earth," but rather a geological history of France in which he would examine the state of the crust during successive epochs. How fully developed this project was in 1766 is not known, since his only detailed explanations of it were written some 20 years later. But it can be said with certainty that the stratigraphic ideas on which he meant to base a geological history of France were worked out in 1766. It is equally clear that Lavoisier later considered his contributions to the *Atlas* as the beginnings of that treatise. Unfortunately, Lavoisier's treatise was never written, although it remained on his list of scientific projects as late as 1792.[22]

It need hardly be said at this juncture that Guettard and Lavoisier inhabited different geological worlds. To Lavoisier, at least, their points of view were potentially reconcilable, and he apparently suggested to Guettard at an early date that they work out a compromise in preparing the quadrangles of the *Atlas*. Exactly what Lavoisier proposed is no longer known, although he later explained that Guettard's system of symbols might simply have been modified slightly in order to reveal the stratigraphic arrangement of the earth's crust.

In order to indicate the size and extent of each deposit, it is only necessary to link all similar symbols by lines which would show not only the size and extent [of each deposit] but also their various points of intersection. . . .[23]

Whether this or some other method was suggested by Lavoisier in the 1760's, his proposals were rejected and he was permitted only the use of the margins of each quadrangle. It is tempting to attribute Lavoisier's partial failure to his scientific differences with Guettard, and this interpretation is, in fact, a reasonable one when it is recalled that Guettard had only a limited appreciation of the purpose served by type sections

[22] Lavoisier Papers, Académie des sciences, dossier 335C, fols. 17r–17v, and *Œuvres de Lavoisier*, Vol. V, pp. 205–213. See, also, Rappaport, "Lavoisier's Geologic Activities, 1763–1792," *Isis, 58* (1967), 375.
[23] Quoted in Henri Pigeonneau and Alfred de Foville, *L'Administration de l'agriculture au Contrôle général des finances (1785–1787): Procès-verbaux et rapports* (Paris: Guillaumin, 1882), p. 87.

and that he also preferred to eliminate from the *Atlas* any kind of geological "system." But Lavoisier's failure must also be attributed in part to the fact that the survey was a government-sponsored enterprise.

Too little is known about the relationship of Bertin to the *Atlas*. One can only assume that, in view of Guettard's seniority, reputation, and pioneer cartographic work, Bertin would have given greater weight to his opinions than to Lavoisier's. There is indisputable evidence that Bertin did share Guettard's practical views about the purpose of geological maps, and he referred to the survey as a project to be completed quickly for the benefit of "ceux qui s'occupent de la recherche des mines."[24] Furthermore, Bertin was an official in a chronically impoverished regime; much as he might desire economic reform based on scientific knowledge—and his other activities show that this was, indeed, the case—he had neither the funds nor the time to support a complex undertaking that might bear fruit in some 20 or more years. As Lavoisier put it:

. . . the work had been begun; the minister . . . had only very limited funds at his disposal and he could not even be sure of having them for long; it would perhaps have been unwise to undertake a project so disproportionate to the means available to carry it out.[25]

From Bertin's point of view, then, speed was essential, and Guettard's cartographic method, already in use on Guettard's earlier maps, would serve well enough.

Bertin's financial problems and his desire for a speedy conclusion to the project were also responsible for the appointment, in 1777, of Antoine Monnet as director of the survey. Progress had been slow in the preceding four or five years, and Monnet had suddenly offered to complete the *Atlas* quickly and economically. Since Monnet was already an Inspector General of Mines, making periodic tours at government expense, it is hardly surprising that Bertin should have accepted this proposal.[26]

Although Monnet took up his task so late in the history of the survey, he was still able to impress upon it some of his own geological ideas. Indeed, he had hoped to alter the maps even more radically but was constrained to follow the general pattern established by his predecessors.[27] His principal contribution to the maps was the elimination of the type sections in those quadrangles either begun by Lavoisier and

---

[24] Bertin to Guettard, 14 December 1767, Bibliothèque municipale de Clermont-Ferrand, MS 473, fol. 67.
[25] *Œuvres de Lavoisier*, Vol. V, p. 205.
[26] Monnet Papers, Ecole des mines, MS 5527, pp. 226–227.
[27] *Atlas*, pp. v–vi. The alterations Monnet had in mind are not specified.

V

completed later or prepared in their entirety by Monnet. This procedure can be fully documented in only one case, in which the Guettard-Lavoisier map and Monnet's revised version both exist (see Figures 2 and 3).[28]

FIG. 3.   Monnet, "Carte Minéralogique des environs de Paris" (Map no. 40).

Monnet's own geological views were strikingly different from those of Guettard and Lavoisier, but the net result of his beliefs, as far as the *Atlas* was concerned, was a certain sympathy with the aims and methods of Guettard. Monnet's training in chemistry and his early association with influential patrons had, by 1767, obtained for him a post as a traveling student and inspector of mines on behalf of the Bureau du Commerce. A few years later, Bertin began to avail himself of Monnet's talents, finally, in 1776, awarding Monnet the title, "Inspecteur général des mines et minières du royaume." By both training and profession, then, Monnet's interests were largely practical and technological. Thus, although lacking the broader education of Guettard, Monnet did share with Guettard a utilitarian outlook as well as considerable experience in the field.

In his personal and scientific career, Monnet was undoubtedly a maverick. Historians of chemistry know him as one of the last and most violent supporters of the phlogiston theory; historians of mineralogy describe him as a place-seeker and an egotist. He seems to have enjoyed

_____

28 The only known copy of the *Atlas* to contain both of these maps is now at the Ecole des mines. For a brief description, see Duveen, *Supplement*, p. 131.

attacking his contemporaries both personally and professionally, setting himself up as one of the few progressive, clear-sighted, and unappreciated thinkers and observers of his day. Accordingly, his philosophy of science was sweepingly negative: Nature's law is to have no law.

Nature never adopts any general rule without also creating many exceptions; it would in fact be contrary to her power and splendor if, once she had traced a path, she could never depart from it.[29]

This remarkable statement was not intended to have theological overtones. Indeed, Monnet prided himself on his emancipation from all religious dogmas and scientific "systems." His aim was pure and accurate observation, untainted by preconceptions.

Monnet's negative philosophy and his practical concerns combined to determine his attitude toward geological type sections. Although he understood Lavoisier's methods, he explained more than once that he preferred to eliminate type sections and to represent the order and nature of the strata *as he observed them* rather than as they might be reconstructed. In his verbal descriptions, but not on the maps themselves, he often pointed out that had he selected another outcrop for his diagram he would have had to include certain observed variations in the strata; and he did recognize that many more such variations would be taken into account in a general section. But, he concluded, a type section cannot represent all the infinite variety in nature, and the whole enterprise is pointless if one's aim is to "indicate precisely the place where one has found a [particular] mineral."[30]

Monnet's other major contribution to the *Atlas* was the text he published to accompany the maps. It was Guettard's habit to issue memoirs along with his early maps, and he probably would have continued the practice had he remained associated with the survey after 1777; in fact, Guettard's unpublished papers include drafts of memoirs on the geology of virtually all of France. Lavoisier, too, may have had some sort of text in mind, perhaps envisioning his treatise on the geological history of France as a commentary on a completed atlas. The text published in 1780, however, was entirely the work of Monnet, despite the appearance of Guettard's name on the title page. As one

[29] Monnet Papers, Ecole des mines, MS 4685, I, 451. Also, Hélène Metzger, *La Genèse de la science des cristaux* (Paris: F. Alcan, 1918), p. 69. For Monnet's career and personality, see Louis Aguillon, "L'Ecole des mines de Paris: notice historique," *Annales des mines*, 8e série, *15* (1889), 433–686.
[30] *Atlas*, pp. 29, 181, 198, and *Collection complette de toutes les parties de l'Atlas minéralogique de la France, qui ont été faites jusqu'aujourd'hui* (Paris, n.d.), p. 2. The latter work is a re-edition of the plates of the *Atlas* with a few additional quadrangles and a prefatory notice by Monnet; the only known copy is at the Ecole des mines (see note 28).

V

would expect, this work is in part idiosyncratic and in part filled with precise observations that were to draw admiring comments from nineteenth-century geologists.[31]

In conclusion, this analysis of the *Atlas* and its makers has hopefully shed some light — if perhaps a flickering and uncertain one — on the problems of geological cartography in the eighteenth century. It is at least clear that the *Atlas* itself is a hybrid creature that bears the impress of three distinctive approaches to geology as well as three clashing personalities. In a general sense, therefore, the story of this survey supports the original contention of this paper that geology in the eighteenth century was in enough of a state of confusion and change to permit and indeed to invite the individual geologist to treat the science, within broad limits, much as he saw fit. It is equally evident that the *Atlas* is not "typical' of early geological maps since it embodies only three points of view, and one could go on to find additional variations in the ideas and methods of other cartographers of the period.

Within the limits of this discussion, it has also been implied that the authors of the *Atlas* were in many ways representative of extreme points of view. In the decades immediately after 1750, one of the more important links among geologists was a common interest in the general classification of geological formations as Primary and Secondary, or Primary-Secondary-Tertiary, or some variation of these schemes. This simple stratigraphic arrangement, with the addition of more detailed specifications, is to be found in many memoirs and maps of this period. In the case of the *Atlas*, however, Guettard's contributions reveal a quickly outdated nonstratigraphic viewpoint, Lavoisier's type sections go far beyond the conceptions of his contemporaries, and Monnet's observations embody a stubborn refusal to be guided by any theoretical considerations. These peculiarities of the *Atlas* help to explain the fact that this pioneer survey had no imitators and serve also to illuminate the problems faced by all the early geological cartographers, namely, how to construct maps that combine observation and theory, and how to represent three dimensions on a two-dimensional surface. In a period when various solutions to these problems were being suggested, the *Atlas* provides a striking example of unresolved conflict among three geologists.

[31] For example, J. J. d'Omalius d'Halloy, *Mémoires pour servir à la description géologique des Pays-Bas, de la France et de quelques contrées voisines* (Namur: D. Gérard, 1828), pp. 236–237, and Brochant de Villiers, in A. Dufrénoy and L. Elie de Beaumont, *Explication de la carte géologique de la France* (2 vols.; Paris: Imprimerie royale, 1841–1848), Vol. I, pp. iv–v.

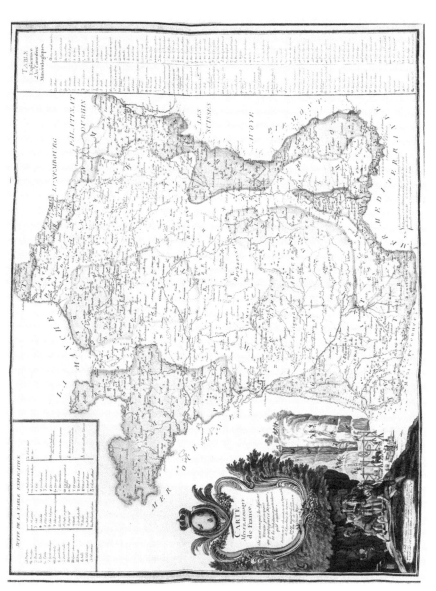

Fig. 1. Carte Minéralogique de France, Où sont marqués les différens Terreins principaux qui partagent ce Royaume, Et les Substances particulières qu'il renferme. Dressée sur les observations de M.ʳ Guettard de l'Accadémie des Sciences, par M.ʳ Dupain-Triel, pere, Géog.ʰᵉ du Roy et de Monsieur[.] Censeur Royal. 1784. Courtesy of the Academy of Natural Sciences of Philadelphia, Ewell Sale Stewart Library and the Albert M. Greenfield Digital Imaging Center for Collections.

V

Fig. 2 Carte minéralogique des environs de Paris. Exécuté [sic] par le S^r. Dupain-Triel Ing^r Géog. du Roy 1766. Courtesy of the Academy of Natural Sciences of Philadelphia, Ewell Sale Stewart Library and the Albert M. Greenfield Digital Imaging Center for Collections.

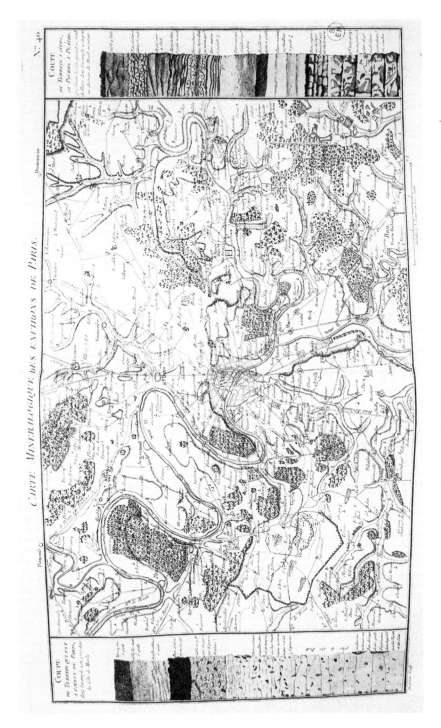

Fig. 3. Carte Minéralogique des environs de Paris. No. 40. (Engraved by Philibert Boutrois.) Courtesy of the Bibliothèque MINES Paris Tech (École nationale des Mines de Paris).

THE EARLY DISPUTES BETWEEN LAVOISIER
AND MONNET, 1777–1781[*]

THE list of Antoine-Laurent Lavoisier's opponents is a long and distinguished one, ranging from Joseph Priestley and Henry Cavendish to Jean-Paul Marat. Among the less distinguished members of this company is Antoine Monnet, a minor chemist and mineralogist whose fame rests in large part on the very fact that he and Lavoisier became enemies. Unlike his better-known contemporaries, Monnet remains almost wholly neglected, and no attempt has yet been made to sort out the issues in his controversy with Lavoisier; instead, Lavoisier's biographers have been content to accept Lavoisier's own version of at least a part of the affair, with the result that Monnet has emerged as the villain of the piece and Lavoisier as the injured victim. The two men probably first came into close contact in 1776 or 1777, the nature of their relationship has remained obscure for the period 1777–1780, and the situation after 1780 has been summarized by biographers in a manner resembling that of Edouard Grimaux:

"ayant en main tous les documents réunis par Guettard et Lavoisier, il [Monnet] ajouta de nouvelles cartes à celles qui étaient déjà gravées, et publia, en 1780, un altas minéralogique incomplet, qu'il signa avec Guettard, tout en s'attribuant la plus grande part du travail. Il cita, il est vrai, Lavoisier comme l'auteur des seize premières cartes, mais il utilisa sans son aveu et sans le nommer les matériaux préparés pour le reste du travail, et négligea d'indiquer que les coupes placées en marge de chaque carte étaient le résultat des nivellements faits au baromètre par Lavoisier. Celui-ci en fut vivement froissé . . . Il trouva toujours en Monnet un adversaire obstiné qui, en 1798, attaquait encore les doctrines nouvelles en publiant une soi-disant *Démonstration de la fausseté des principes des nouveaux chimistes.*"[1]

This summary is based on three well-known facts and one denunciatory statement written by Lavoisier himself. The facts are indisputable, since the title-page of the *Atlas et Description minéralogiques de la France* does contain the names of only Guettard and Monnet, Monnet nowhere refers to Lavoisier's barometric measurements and marginal sections, and Monnet's *Démonstration* is certainly a bitter attack on the new chemistry and its supporters. Equally well known is Lavoisier's statement

[*] Research for this article was supported in part by a fellowship from the American Association of University Women.

[1] Grimaux, *Lavoisier 1743–1794* (2nd edn., Paris, 1896), 25–26. Also, Douglas McKie, *Antoine Lavoisier: Scientist, Economist, Social Reformer* (New York, 1952), 70–71, and Denis I. Duveen and Herbert S. Klickstein, *A Bibliography of the Works of Antoine Laurent Lavoisier 1743–1794* (London, 1954), 238. For examples of other controversies, see Maurice Daumas, "Polémiques au sujet des priorités de Lavoisier", *Revue d'histoire des sciences*, iii (1950), 133–155. A striking case of Lavoisier's concern to rewrite his own scientific biography is presented by Henry Guerlac, "A Curious Lavoisier Episode", *Chymia*, vii (1961), 103–108.

234

that Monnet had used, without authorization and proper acknowledgment, geological data collected and transmitted to Monnet by Lavoisier.[2]
These charges suggest not only that Monnet was wholly lacking in integrity
but that he was, in effect, carrying on a personal campaign against
Lavoisier; and because Monnet's actions appear to have been so un-
justifiable, this interpretation has seemed a reasonable one.

The evidence against Monnet deserves reconsideration, however,
for it is of a limited and, in the case of Lavoisier's statement, a possibly
questionable nature. Although his abuse of the new chemists seems a
gratuitous insult when several of them were no longer alive, it should
be pointed out that his greatest scorn was reserved not for Lavoisier but
for a prominent phlogistonist, Carl Wilhelm Scheele.[3] Furthermore,
these personal attacks date from the later, embittered years of Monnet's
career; while he always had a few derogatory remarks to make about his
contemporaries, Monnet's earlier tendency had been to concentrate
on the interpretation of evidence rather than on personalities. To conclude,
therefore, from his attitude in 1798 that he hated Lavoisier in 1780 is at
best a dubious procedure. In fact, the two episodes—the dispute about the
Atlas and that about the new chemistry—are separated by a long period
in which neither scientist is known to have discussed the other, even in
contexts where such reference would have been appropriate; it would
thus seem that considerable bitterness remained after 1780, but there is
no evidence of a continuous personal vendetta.[3a]

Concentrating therefore on the earlier period, Lavoisier's charges
against Monnet require examination in their proper context and with
some attention to Monnet's point of view. Relevant documents have until
now been in short supply, but recently published information about the
construction of the *Atlas* provides a valuable supplement to the familiar
evidence; in addition, manuscript material found among the papers of
Guettard, Lavoisier, and Monnet sheds considerable light on the period
1777–1781, a period which began with Monnet's assuming control of the
geological survey and ended with Lavoisier's violent reaction to the publication of the *Atlas*. Although there remain important gaps in the
story, it is now possible to ask, and hopefully to answer, such questions
as: Was the early antagonism between Lavoisier and Monnet primarily

[2] *Oeuvres de Lavoisier*, 6 vols. (Paris, 1862–1893), v, 221.
[3] Monnet, *Démonstration de la fausseté des principes des nouveaux chymistes* (Paris, An VI),
passim. Cf. his "Dissertation et expériences relatives aux principes de la chimie pneumatique",
Mémoires de l'Académie Royale des Sciences, Années [1788–1789], Turin, 1790, pp. 123–205.
[3a] In later references to the geological survey, both Monnet and Lavoisier paid tribute to
Guettard, Monnet including Lavoisier's name once in passing and Lavoisier omitting Monnet
entirely. Cf. Monnet's prefatory remarks in the second edition of the *Atlas*, published after 1794
and described by Rappaport in Duveen, *Supplement to a Bibliography of the Works of Antoine
Laurent Lavoisier 1743–1794* (London, 1965), 131–132; and Lavoisier's remarks in Henri
Pigeonneau and Alfred de Foville, *L'Administration de l'agriculture au Contrôle général des finances
(1785–1787): Procès-verbaux et rapports* (Paris, 1882), 68–69, and *Oeuvres de Lavoisier* (2), v,
205–206.

a matter of personal rivalry? How did Lavoisier himself contribute to the quarrel? And, if one can legitimately talk about degrees of dishonesty, just how dishonest, in fact, was Antoine Monnet? This article will, in short, be devoted to the examination of a hitherto obscure period in the lives of Lavoisier and Monnet and to some discussion of the effects of their conflict upon the construction of the *Atlas*. Although the period is brief, many of the details petty, and one of the protagonists a minor figure in the history of science, it is hoped that this paper will illuminate such larger matters as the geological activities of Lavoisier, the questions dividing early geological cartographers, and the influence of government upon a scientific enterprise.

*   *   *   *

By 1777 Monnet had long since made his professional début by presenting two well-received papers to the *Académie royale des sciences* and by gaining entrance into several learned societies. The author of works on chemistry and mineralogy, he had already spent approximately ten years as a travelling student and then inspector of mines employed by the Bureau du Commerce. In 1776 he was appointed by Bertin, then director of the Department of Mines, to the position of "Inspecteur général des mines et minières du royaume", and, in the same year, began to supply geological data, noted during his travels, to Guettard and Lavoisier who were at work on the geological atlas of France. In this promising career there was only one source of frustration, but that a major one: Monnet's failure to become a member of the *Académie royale des sciences*. This was an important source of his conviction, which was to grow steadily throughout his life, that he was superior to and persecuted by members of the scientific establishment. Belief in his own ability and in the fallibility of other scientists is, in fact, apparent in most of Monnet's published works; towards the end of his career, and especially in his unpublished journals, it is clear that his delusions amounted to paranoia.[4] Nevertheless, Monnet's peculiarities were only dimly visible by 1777, and at that date he was undeniably considered a talented, if eccentric, scientist who had won the confidence of such men as Bertin, the Trudaines, and Malesherbes.[5]

[4] Cf. *Journal de médecine, chirurgie, pharmacie, &c.*, liv (1780), 559–561. Monnet's journals, now at the *École des mines*, Paris, must be used with care because, for example, while one series of copies of letters (MS. 5527) is accurate, another (MS. 4672) is full of gross inaccuracies; some of the originals are available for comparison (Archives nationales, F¹⁴ 1313–1314, Muséum d'histoire naturelle, Paris, MS. 283, and Bibliothèque municipale de Clermont-Ferrand, MS. 1339).

[5] The best summaries of Monnet's career are the anonymous eulogy in *Annales des mines*, ii (1817), 483–485, and the remarks scattered through Louis Aguillon, "L'École des mines de Paris: notice historique", *Annales des mines*, 8ᵉ série, xv (1889), 433–686. For the role of Bertin *et al.* as patrons of science, see Harold T. Parker, "French Administrators and French Scientists during the Old Regime and the Early Years of the Revolution", in *Ideas in History*, ed. Richard Herr and Harold T. Parker (Durham, N.C., 1965), and H. Guerlac, "Some French Antecedents of the Chemical Revolution", *Chymia*, v (1959), 73–112.

236

In 1777 the geological atlas, originally conceived by Guettard, had been in preparation for at least eleven years. Although actually begun before 1766, the work was only commissioned by Bertin in that year, and intensive field work and mapping took place in the period 1766–1770. By the end of those four years, Guettard and Lavoisier had completed 16 quadrangles describing the entire Ile de France, most of Brie and Champagne, and portions of Alsace, Lorraine, and Franche-comté.

The survey was planned to consist of some 200 quadrangles, to be issued in separate sheets and then, at the buyer's pleasure, assembled or bound as an atlas. Each quadrangle showed a geographical area marked by symbols to indicate the location of superficial mineral deposits and rock formations; such observations were gathered by both Guettard and Lavoisier. In the margin of each map was a vertical section showing either the stratigraphic arrangement of a particular mine or quarry or, as a "general section", a composite picture of the stratigraphy of an entire region; barometric measurements were used in the preparation of these sections, and the technique and ideas were wholly Lavoisier's.

In 1770 or shortly thereafter, the survey encountered financial difficulties which made impossible the continuation of field work, and the two collaborators had to resort to gathering data by corresponding with provincial naturalists and amateurs of science. Lavoisier attempted, in 1773, to persuade Bertin that the original project was too ambitious and that it would be desirable to reduce drastically the scale of the maps and to rely heavily upon the use of geological questionnaires. For reasons now unknown, Bertin rejected these proposals, and work on the atlas progressed slowly following the scheme approved in 1766. During the period 1770–1777 only 14 or 15 new quadrangles were partially completed, but Guettard and Lavoisier announced in March of 1777 that they hoped to be "en état de suivre avec autant de confiance que d'activité un Ouvrage aussi nouveau dans son genre qu'intéressant dans son objet".[6] On 30 June 1777, Monnet was named Guettard's successor as director of the survey.

The reasons behind Monnet's sudden appointment and a description of his relationships with Guettard and Lavoisier are summarized in a hitherto unpublished letter from Monnet to Lavoisier, dated 30 July 1777. (The full text is given at the end of this article.) On the whole, the letter is strikingly cordial, since Monnet apparently wanted and certainly needed both the geological data and the co-operation of Guettard and Lavoisier. Indeed he had already been in correspondence with Lavoisier— the earlier letters have not been located—and had learned of Lavoisier's dissatisfaction with the way in which the survey had been proceeding. Monnet therefore tried to reassure Lavoisier that the project would move ahead with vigour and efficiency.

---

[6] *Journal de Paris*, 14 March 1777. The construction of the *Atlas* is treated by Duveen and Klickstein, *op. cit.* (1), 236–244, and Rappaport, *loc. cit.* (3a), 129–132.

It is more than likely that both Guettard and Lavoisier were somewhat chagrined at being superseded by Monnet; Guettard had, after all, begun his pioneer cartographic work as early as the 1740's, Lavoisier had devoted at least eleven years to the survey, and Monnet was a newcomer. In writing to Lavoisier, Monnet was thus careful to emphasize his own reluctance to undertake so onerous a task and instead laid the blame squarely on his patron Bertin who had thought it only natural that a travelling Inspector of Mines should be responsible for the gathering of geological data. What Monnet neglected to mention was the fact that he had earlier written to Bertin promising that, given his own way with the atlas, he would finish the survey rapidly and at little additional cost; Bertin had approved his proposals, adding: "j'ai pensé ne pouvoir en charger personne plus en etat que vous de mener cette besogne à sa perfection."[7] From the beginning Bertin had considered the atlas a purely practical venture, useful for discovering new natural resources to revitalize the French mining industry; he was therefore concerned to have the survey completed with all possible speed, and Monnet's offer, coming as it did from a subordinate of proved ability, could not but have appealed to him.

To enlist the co-operation of his predecessors, Monnet outlined to Lavoisier three proposals. In the first place, he would welcome enthusiastically any geological observations contributed by Guettard and Lavoisier. Furthermore, he invited Lavoisier to send him the details of his own contributions to the survey so that these could be properly and publicly acknowledged. Finally, he asked that Guettard and Lavoisier send him any explanatory memoirs they might possess which would make a suitable accompaniment to the geological maps already completed; such memoirs, with the addition of some by Monnet, could be published as a companion volume to the maps.

Thus far, Monnet's letter to Lavoisier suggests a certain harmony between the two and goodwill on his own part. But the text also contains two significantly discordant notes. The first is the repeated reminder that it is he, Monnet, who is in control of the survey and is the final judge of the acceptability of all contributions. The second, related to the first, is his refusal to consider Lavoisier's proposal that data be collected by the use of questionnaires to be sent to provincial naturalists. Just such a proposal had been rejected by Bertin in 1773, and Lavoisier had informed

---

[7] Copy of a letter dated 30 June 1777, *École des mines*, MS. 5527, pp. 226–227; the letter includes a summary of Monnet's proposals to Bertin. Monnet was later to repeat the explanation offered Lavoisier, in *Observations sur la physique, sur l'histoire naturelle et sur les arts*, xi (1778), 461, and *Atlas et Description minéralogiques de la France* (Paris, 1780), v. Bertin's aims in sponsoring the survey are discussed by Rappaport, "Guettard, Lavoisier, and Monnet: Conflicting Views of the Nature of Geology", to appear in a volume of papers delivered at the New Hampshire Inter-Disciplinary Conference on the History of Geology, 7–12 September 1967; the editor of the volume is Cecil J. Schneer.

Monnet of his intention to submit a similar plan to Bertin once more; Monnet, in reply, refused to accept the inaccurate data which would inevitably be reported by untrained observers and amateurs. From Lavoisier's point of view, Monnet's veto meant that the atlas would not be completed for many decades; that his own plan seemed the more practicable one is evident in his memorandum to Bertin, in a further memorandum of 1781, and in his later attempt, in 1785, to find governmental support for his twice-rejected scheme.[8]

Lavoisier's immediate reaction to Monnet's letter remains unknown. While it is clear from Lavoisier's later statement, in 1781, that he did make at least some data available to Monnet, there is at present no way of discovering whether Lavoisier did transmit details about his own contributions to the early quadrangles of the atlas; nor, in fact, is there evidence that the two men continued to exchange correspondence. One can say with certainty only that each pursued his own scientific activities in the three succeeding years and that these activities were to bring them into conflict in 1780–1781.

Some of Lavoisier's time after 1777 was devoted to the beginnings of an independent geological survey. He probably presented no further proposals to Bertin during this period, but instead, ignoring Monnet's veto, simply began to put his plans into effect, hoping to achieve results which would impress Bertin and induce him to grant financial aid. The precise date at which Lavoisier made this decision cannot be determined, although in 1781 he claimed to have been working on the project for two years.[9] By February 1781, approximately two months after Monnet's publication of the *Atlas*, Lavoisier possessed and wanted to have engraved a map of France divided into some 25 quadrangles; after filling in the geological information already gathered by himself and Guettard, he intended to send copies of this map to the provinces so that local naturalists might furnish more details. In addition, the questionnaire devised years earlier called for some information useful in constructing vertical sections, and Lavoisier probably intended to send out similar queries once more. How far these plans and intentions had actually progressed by 1781 it would be difficult to say, since Lavoisier refers to them only in documents written to impress government officials with the practicability of his scheme. In general, however, there is no reason to

---

[8] *Oeuvres de Lavoisier* (2), v, 216–221, and Pigeonneau and de Foville, *op. cit.* (3a), 72–73, 116–117. Lavoisier expected that errors of observation would be corrected in future editions of his maps.

[9] Lavoisier Papers, Academy of Sciences, dossier 636. This dossier contains three documents, two of them partial drafts of the memorandum published in *Oeuvres de Lavoisier* (2), v, 220–221; one draft refers to a two-year period, a detail not found in the published version. Neither draft is dated, but both were written after the appearance of the *Atlas* (announced in *Journal de Paris*, 16 December 1780) and probably before 26 February 1781, the date of the published version.

doubt that, at some time before the appearance of the *Atlas*, he had decided to go his own way, independently of Monnet.[10]

Monnet meanwhile was unable to proceed with the collaborative volume of geological memoirs, and the obstacle here seems to have been Guettard's refusal to co-operate. As Monnet reported in his letter to Lavoisier, he had had a most unsatisfactory interview with Guettard who, although pleasant enough at the outset, had finally decided to have nothing more to do with the atlas. Unhappily, Guettard has left no comment on this interview, but Monnet's account is plausible and is supported in indirect fashion by other evidence. Monnet was not the only protagonist of uncertain temper; Guettard, too, was known to his contemporaries as an irascible soul—one observer called him an "animal disputeux"—and he may simply have been aroused by Monnet's manner or by his appointment to the post of director of the survey. It is possible, too, that Guettard had already formed a low opinion of Monnet's abilities, although he was to say so explicitly only in 1779.[11] Whatever the reasons behind Guettard's animosity, it is clear that Guettard never did hand over to Monnet his own data and the series of geological memoirs he had prepared to accompany the maps of the atlas.[12] Under these circumstances, Monnet's hope for a collaborative text gradually vanished.

On the other hand, Monnet did attempt to carry out one promise earlier made to Lavoisier. In 1778 he published a prospectus listing those maps prepared by his predecessors; here he commented briefly that Guettard and Lavoisier had also supplied "beaucoup d'observations pour la composition d'autres Cartes".[13] Although such general acknowledgment would seem less than satisfactory, no protest by Lavoisier has been discovered. In the period 1778–1780 he also published 17 new quadrangles, all but three of them based upon the work of Guettard and Lavoisier; although the authorship of these maps was not clearly explained until the *Atlas* itself appeared, there is again no evidence of a protest by Lavoisier.[14] This apparent lack of reaction on Lavoisier's

[10] Preceding note (9). The earlier questionnaire is reproduced in *Oeuvres de Lavoisier* (2), v, 214–216, and can be dated 1771 from the reply in *Oeuvres de Lavoisier—Correspondance recueillie et annotée par René Fric*, Paris, fasc. 2, 1957, 323–326.

[11] Guettard, *Mémoires sur la minéralogie du Dauphiné* (Paris, 1779), vol. i, pp. cxv–cxxiv. For Guettard's personality, see Condorcet, "Eloge de M. Guettard", *Histoire de l'Académie royale des sciences*, 1786 (1788), 47–62, and A. Birembaut, "L'Académie royale des Sciences en 1780 vue par l'astronome suédois Lexell (1740–1784)", *Revue d'histoire des sciences*, x (1957), 161.

[12] *Atlas* (7), x. Monnet was to claim that Guettard had retired from the survey before Monnet took control (*ibid.*, vi, and *École des mines*, MS. 5527, p. 272); that this was not true is implied in Monnet's account of their interview. Among the unclassified Guettard papers at the *Muséum d'histoire naturelle* are unpublished memoirs on the geology of virtually every French province; Guettard's own attempt to have these and other writings published by the Imprimerie Royale failed (Archives nationales, o¹610, pièce 230). Lavoisier was to act as Guettard's scientific executor after the latter's death in 1786 (cf. Muséum, MS. 2188, fols. 30–31, 167, 179), but he, too, failed to publish any of the memoirs. Whether Lavoisier had memoirs of his own to contribute to the survey is uncertain; see Rappaport, "Lavoisier's Geologic Activities, 1763–1792", *Isis*, lviii (1967), 375–384.

[13] *Observations sur la physique* (7), 460.

[14] *Ibid.*, 459–462, and xv (1780), 423; *Journal de Paris*, 26 May 1778; *Atlas* (7), xii.

240

part suggests that, with his own project in hand, he had no great interest in the progressive stages of the original survey. Furthermore, as he was to state in 1781, he was convinced that the original plan would require 60 to 80 more years for completion, and again the implication here is that his interest had waned. But he could not maintain his silence when in 1780 Monnet issued an *Atlas* consisting of only 31 sheets, omitted from the collection 10 of the 16 plates completed by Guettard and Lavoisier, published a text made up of accounts of his own travels, and decorated the whole with a title-page listing Guettard and Monnet as the authors. Despite prefatory comments about both of his predecessors, the inescapable impression conveyed by the *Atlas* is that it was primarily the work of Antoine Monnet. As Lavoisier put it, Monnet had had nothing to do with planning the survey, had contributed relatively little to the maps thus far published, and yet "il a lair de Sapproprier louvrage . . .".[15]

Examined in detail, certain of Monnet's actions do admit of reasonable explanation. Suppression of the 10 Guettard-Lavoisier maps was, in fact, a temporary measure; Monnet intended to revise them, did later include them in a second edition of the *Atlas*, and was aware that they had long been available for separate purchase.[16] As noted earlier, he could not have published a collaborative text without the co-operation of Guettard. Even his failure to mention Lavoisier's barometric measurements might be attributed to ignorance of the exact nature of Lavoisier's contributions or, as will appear below, to Monnet's decision to alter or omit most such data. In general, however, Monnet's letter of 1777 reveals that he was sufficiently aware, if only in some vague fashion, of the magnitude of Lavoisier's contributions to warrant including the latter's name on the title-page. That he did not do so is in itself inexcusable and lends to his other actions an air of deliberate malice.

It is in the light of the preceding discussion that Lavoisier's charges of dishonesty, cited by biographers, must be examined. Lavoisier did not draw up a detailed indictment, but instead selected two points, accusing Monnet of having used

"matériaux qu'il [Lavoisier] avait déposés de confiance entre les mains de M. Monnet, mais dont, au moins, il aurait dû n'être fait usage que de son aveu. Il [Monnet] aurait dû ajouter que toutes les coupes placées en marge de chaque carte . . . sont le résultat des nivellements faits par le baromètre, également par M. Lavoisier. On ne rappelle ces détails que pour faire sentir avec combien d'imprudence s'est conduit M. Monnet en s'emparant de planches qui appartiennent au roi, et sur lesquelles MM. Guettard et Lavoisier ont des droits avant lui ou pour mieux dire sur lesquelles il n'en a aucun".[17]

[15] Lavoisier Papers (9).
[16] Rappaport, *loc. cit.* (3a), 130–132, and *École des mines*, MS. 5050, p. 376 (note). Only one of the 10 was in fact revised by Monnet; see below (18).
[17] *Oeuvres de Lavoisier* (2), v, 221. Both Grimaux and McKie (1) use the word "impudence" rather than Lavoisier's "imprudence".

Despite Lavoisier's assertion, there is little reason to think that Monnet did in fact *use* the available barometric data when he completed the quadrangles begun by his predecessors or when he constructed additional maps for the survey. Since Lavoisier's original data, if extant, have not been located, totally convincing proof cannot be presented. In general, however, the vertical sections in the margins of Monnet's maps lack the precise measurements characteristic of Lavoisier's work. In addition, there is available Lavoisier's geological map of the environs of Paris and Monnet's revised version of the same quadrangle; comparison of these two reveals so many differences that one might reasonably conclude that Monnet's vertical sections were the result of independent field work. Indeed Monnet's philosophy and methods *required* that he *not* use Lavoisier's data, since Monnet firmly believed that the composite "general sections" prepared by Lavoisier were useless theoretical constructions which did not represent accurately the observations of the field geologist.[18] If Monnet did select, modify, or discard Lavoisier's data on most or all of his own quadrangles, then it would seem unthinkable that Lavoisier could have failed to notice this procedure when he examined the maps of the *Atlas*.

The second point in Lavoisier's accusation, namely, that Monnet had no right to use the data deposited with him, is even more questionable than the first. As Lavoisier himself remarked, these data "appartiennent au roi" and not to Monnet; nor, then, did the data belong to Guettard and Lavoisier. In practice, Monnet could not force his predecessors to transmit to him their geological observations, but, as Bertin's appointee, he could surely make use of any data deposited with him or with the engraver; morally, but not legally, Monnet was obliged only to make due acknowledgment.[19]

Why did Lavoisier base his accusation of dishonesty on grounds which were at best shaky? The answer, in all probability, lies in the full text of Lavoisier's memorandum, in which the denunciation of Monnet is only a fraction of a paragraph. When the published document is compared with the unpublished drafts, it is clear that Lavoisier was addressing someone in authority—possibly Bertin or Jacques Necker who had taken over many of Bertin's functions—and someone whom he wanted to impress with *his own right* to use materials collected for the geological survey. (Whether any version of this memorandum was subsequently sent to Bertin or Necker remains unknown.) Lavoisier was,

---

[18] *Atlas* (7), 181, 198. The two maps of the environs of Paris were published in the second edition of the *Atlas*; the relevant geological issues are discussed and the maps reproduced in Rappaport, *op. cit.* (7).

[19] In one draft of his memorandum, Lavoisier Papers (9), Lavoisier refers to having deposited data with engraver Dupain-Triel; the phrase was then crossed out and the reference to Monnet substituted.

242

in fact, outlining and defending the independent survey he had under-taken approximately two years earlier; in addition, he was defending his priority rights by enclosing a copy of the proposal submitted to Bertin in 1773. The burden of the memorandum might thus be described as an assertion of his own rights by an attack on those of Monnet.

Monnet's actions, too, become more understandable, if still not justifiable, if one assumes that he was aware of Lavoisier's activities, and this assumption is not unreasonable when it is recalled that he knew of Lavoisier's general intentions some years earlier, in 1777. Monnet was faced with a rival whose independent geological survey, which could be completed more quickly than his own, would offer dangerous competition in two respects: the new survey would anticipate his own maps, rendering them less original, less valuable, and less liable to sell, and it would eventually divert governmental patronage from Monnet's project to one which had shown more immediate results. Furthermore, Monnet felt keenly the need to convince Necker of his own worth to the Department of Mines, since he believed his job to be endangered by Necker's concern for administrative economy.[20] With Bertin's power declining and Monnet's career in jeopardy, Monnet may well have feared that Lavoisier's survey would find favour with the new administration. Threatened by Lavoisier's competition, insecure in his job, sensitive and vain about his scientific abilities—these factors do much to explain Monnet's reluctance to give Lavoisier more than minimal credit for his work on the *Atlas*. The same factors probably explain, too, why Monnet rushed into print with a relatively small collection of maps, rather than wait until he could complete a more substantial proportion of the projected 200 quadrangles.

It is apparent, in conclusion, that the quarrel between Lavoisier and Monnet cannot be reduced to a personal matter in which Monnet shoulders all responsibility. Instead, the affair seems to have been, fundamentally, one of conflicting ambitions. Both men were concerned to protect their own interests, and the differences in personality and scientific ideas precluded their co-operation. Both had been placed in an awkward position by Bertin's action in 1777 and by Guettard's intransi-gence at that time and later. Monnet's abnormal vanity and feelings of insecurity, coupled with a certain disregard for scholarly scruples, led him to misrepresent the facts both to the public and to his govern-ment patrons; Lavoisier, at the same time, engaged in a more subtle distortion of the facts, addressing himself only to a potential patron.

[20] Whether the danger was real or imagined hardly matters, but Monnet's fears are apparent in his correspondence with Necker during the years 1778–1780, *École des mines*, MS. 5527, pp. 227–229, 232–235, 253–254, 269–270. Bertin formally submitted his resignation only in May 1780, but, as these letters reveal, Necker was earlier assuming a controlling influence, and Monnet's attempts to ingratiate himself began in 1778. Furthermore, despite the actual date of Bertin's resignation, he continued to exert some influence at Court in the 1780's and it is thus possible that Lavoisier's memorandum of 1781 was addressed to him.

In all of this, no personal animosity appeared on either side until the publication of the *Atlas*, and the mutual antipathy, personal and professional, which arose at that time was mutually provoked.

APPENDIX

Monnet to [Lavoisier], 30 July 1777
(Bibliotheque municipale de Clermont-Ferrand, MS. 719, fols. 39–42)

J'ai reçu, Monsieur, avec plaisir la lettre que vous avez bien voulu m'ecrir en reponse à la mienne; les objets que vous m'y presentez, exigent de ma part de nouvelles observations, et je ne perds pas un instant pour vous les faire.

tout ceque je savais que vous aviez beaucoup contribué à l'execution de la carte mineralogique de la france. J'ignorais absolument que vous eussiez été si mal payé de votre travail,[11] et cette circonstance doit beaucoup augmenter mes regrets de mêtre chargé de cette besogne. mais il paraissait naturel à monsieur Bertin, que, voyageant pour le fait des mines, je m'occupasse en même tems de l'execution de cet ouvrage. quelque repugnance que j'aye dabord montré pour cela, il ma fallut ceder; mais cela na été qu'avec cette condition, que je serais le maitre absolument de cette besogne; c'est à dire qu'aucun maitre n'y ferait rien que de mon avis, et j'ai apporté pour preuve de la necessité de cet arrangement, l'injustice qu'on avait fait à vous et à m. guettard de ne pas parler de vos travaux suffisamment et dignement, de sêtre en un mot approprié vos travaux sans presque en dire mot.

Cet arrangement pris, j'ai courus pour ainsi dire chez monsieur guettard, pour partager avec lui ce pouvoir et lui faire offre de partager avec lui le peu de fond qui venait de mêtre assigné à cet effet. J'ai éte dabord enchanté de la réception qu'à [*sic*] faite m^r guettard à mes propositions et de l'acqu'iessement qu'il a donné dabord à l'arrengement que je voulais prendre avec lui et avec vous, car il s'etait chargé de vous en parler. Mais la réflexion a tout détruit chez lui sans doute, car il ma renvoyé avec dédain les projets de carte, que sa bonne volonté passagère lui avait subjeré de me demender, pour y travailler sur le champ; de sorte que me voilà privé de la satisfaction que je m'etais promise vis à vis de lui. il a sans doute des raisons pour se plaindre de la conduite qu'on a tenue à son égard, mais ce n'etait pas à moi à qui il fallait s'en prendre qui y allais de si bonne foi.

tout ceque je puis faire dans cette circonstance, est de vous faire rendre la justice qui vous est due à l'un et à l'autre. je pars à l'instant pour aller faire ma tournée, et à mon retour, je ferais graver un nouveau *prospectus*, ou il sera fait mention de vos travaux comme il convient. je vous serais fort obligé si vous vouliez bien m'indiquer les cartes auxquelles vous avez travaillé seul, je les citerai dans le *prospectus* et j'aurais soin de mettre sous les yeux de m. Bertin les raisons qui me porteront à en agir ainsi. Jespère aussi l'engager à faire imprimer à l'imprimerie Royale les memoires de m^r guettard sur la mineralogie de la france, si toutefois m. guettard veut bien se preter à cela; car d'après ce qui vient de se passer, je n'oserais encore me l'assurer; vous pourriez y joindre les votre; et cela pourrait faire un corp complet d'ouvrage sur les cartes que vous avez faites. il en sera de même de celles que je ferai exécuter d'apres mes

[11] This case of injustice, referred to again later in the paragraph, is not explained in any of the known documents.

244

observations particulieres; et chaqu'un sera garant de son propre travail vis à vis du public.

je sens combien vous devez être dégouté d'un travail où vous n'avez éprouvé que du désagrément, cependant connaissant votre honneteté naturelle, j'en ——— [illegible] assez bien pour penser, que je ne dirai rien d'inutile, quand je vous proposerais de rassembler les memoires qui ont rapport aux seize cartes qui paraissent d'en composer un ouvrage particulier et comme faisant suite à ces cartes. je ne doute pas que m. Bertin n'approuve fort la proposition que j'ai l'honneur de vous faire. ce travail ne vous oblige pas de vous distraire beaucoup de vos travaux ordinaires, et d'ici à l'hivers prochain vous pourriez l'avoir executé sans beaucoup de peine. mais il faut supposer le concours de m. guettard pour cela, et suposer en même tems qu'il vous livrera les memoires, et je ne sais en quel terme vous êtes ensemble. car il est bon de vous dire que je suis tellement dégouté de la conduite qu'il à [sic] tenu à mon egard, que je ne veux point absolument avoir à faire desormais avec lui. cela ne mempéchera pas de chérir toujours son savoir et ses lumieres.

à l'egard de votre projet pour l'execution de la carte minéralogique, et que vous avez intention de mettre sous les yeux de m Bertin, je crois devoir vous dire que si c'est celui là dont j'ai eu occasion de voir un énnoncé, il ne peut remplir le but que je me propose, qui est de l'exactitude et de la sureté, et comment en avoir de la part de gens non instruits dans la connaissance des minereaux?, qui se tromperaient, et qui diraient surement même de bonne fois des choses qui ne seraient pas; et qui n'etant animés d'aucun principe de gloire, citeraient au hazard et sans interêt les objets les plus inportants à connaitre. en vain enveraient ils les échantillons de ces objets, il y aurait surement des erreurs et des méprises, et l'on se trouverait avoir fait des fraix inutiles et sans fruit pour le public.

au Reste, Monsieur, le plan actuel de cet ouvrage, n'est pas de l'exécuter très promptement, mais avec sureté et avec exatitude j'ai eu tant d'occasion de voir qu'on avait enduit en erreur le public, et qu'on m'en avait ——— [illegible], que j'aime bien mieux qu'on fasse peu de carte, mais qu'on les fasse bien. c'est cette raison qui ma porte à déffendre à m^r Le graveur de recevoir des nottes indicatives qui n'auraient pas été faites par des personnes sures comme vous et m. guettard.

Je Suis, Monsieur, de
tout mon coeur
Votre tres humble
et tres obeissant serviteur
Monnet

ce mercredi 30 juillet 1777

Hooke on Earthquakes: Lectures, Strategy and Audience

Much has been written about Robert Hooke's so-called 'Discourse of Earthquakes', the series of lectures he delivered before the Royal Society of London over the years 1667–1700. The chief points of the lectures are thus well known: fossils (the word is used here in its modern meaning) are the remains of once-living organisms, and their burial in rather odd places within the earth's crust can be explained by the dislocations of land and sea resulting from earthquakes.

Historians analysing Hooke have seized upon these main arguments, often without realizing that the lectures form a long series in which Hooke introduced curious and seemingly non-geological material. On the whole, admiration has centered on Hooke's trenchant, cogent discussions of fossils, while one detects a degree of embarrassment that so 'modern' a thinker could have failed to find a more plausible mechanism for uplift. It has been noticed, on occasion, that Hooke's lectures, once they were published in 1705, seem to have been ignored by other naturalists. Indeed, few historians have asked why Hooke's contemporaries accorded the theory of earthquakes little attention, while many have deplored the same absence of response to his treatment of fossils.[1]

Why Hooke's geological work made virtually no impression on the writings of his British contemporaries (continental naturalists may have remained ignorant of his works because they were never translated into Latin or French) is a problem that defies exact solution.[2] Hooke himself, be it said, did offer a solution, when he suggested that Fellows

I am especially grateful to Dr Michael Hunter (Birkbeck College, London) for careful and provocative comments on the first version of this article; although I doubt that I have met his high standards, it is thanks to Dr Hunter that I have found more in both Hooke and his audience than I thought was there. A crucial matter, insisted upon by Dr Hunter (and with important results for this paper), was the consultation of the Journal Books of the Royal Society, and I would like to thank the Librarian of the Society, Mr N. H. Robinson, for the efficiency and courtesy with which he met my request for microfilms of these manuscripts.

1. Among the more valuable studies of Hooke's thought is D.R. Oldroyd, 'Robert Hooke's Methodology of Science as Exemplified in his Discourse of Earthquakes'. *Br. J. Hist. Sc.* (1972), 6, p. 109–130, which includes a bibliography of other pertinent works on Hooke's lectures. The paucity of published comments by Hooke's contemporaries is signalled by Victor A. Eyles, 'The influence of Nicolaus Steno on the Development of Geological Science in Britain', in Gustav Scherz, ed., *Nicolaus Steno and his Indice.* Copenhagen, 1958, p. 186–187.

2. On the lack of knowledge of English among French scientists, see Harcourt Brown: *Scientiflc Organizations in Seventeenth Century France (1620–1680)*, Baltimore, 1934, p. 198. 203–204, 225, 282. Even Buffon, who knew English well, cited John Ray, not Hooke, on the geological role of earthquakes. See *Oeuvres philosophiques de Buffon*. Jean Piveteau, ed. Paris, 1954, p. 88B and n. 16; the text published here is Buffon, *Histoire naturelle*. i, 1749, 'Preuves de la théorie de la terre, Article V'.

of the Royal Society had rejected the theory of earthquakes and thus could not accept his explanation of the nature of fossils. That these two issues, fossils and earthquakes, were intimately related in Hooke's mind cannot be doubted; whether the same was true of the minds of his audience is a proposition worth testing. Furthermore, the very fact that Hooke's lectures aroused debate within the Society suggests that it is here, to the living audience rather than to contemporary publications, that historians ought to turn to investigate the reception of Hooke's ideas.

To deal with Hooke's audience poses peculiar problems. First, any reader of Hooke's lectures will notice that the texts are repetitive and diffuse and that they include what appear to be odd digressions into matters usually thought to be irrelevant to geology, notably such matters as the explication of Ovid's *Metamorphoses*. These impressions, however, are the product of Richard Waller's rather primitive editing of the *Posthumous works* (1705), where the lectures are not always presented in chronological order, many dates are missing, and two or more lectures are printed as if they were single texts. If Waller's text is disentangled, so to speak, the lectures rearranged, and the missing dates supplied, a degree of coherence does emerge. Indeed, one can detect four distinct series of lectures, dated as follows: 1667 to 5 January 1687, 19 January 1687 to 9 March 1687, 2 November 1687 to 2 August 1693, and 25 July 1694 to 10 January 1700. (See the Appendix for a dated list of the lectures.) Examination of this sequence in chronological fashion not only eliminates some repetitiveness and explains Hooke's attention to Ovid, but it also suggests that Hooke engaged in tactical manoeuvres designed, at least in part, to persuade his audience.

The second problem is the nature of that audience. Most of the available information comes from Hooke himself, when he announces that he will reply to critics. Who were the critics, and, given Hooke's notorious sensitivity, how seriously should one treat his impression that his colleagues were arrayed against his ideas? Information about Hooke's audience is minimal. At times, we are told that there occurred 'much discourse' on some topic, with no reference to the names of the speakers. Those who attended meetings of the Royal Society are identified by name only if they made some substantive comment, with the result that, on occasion, we are told only that the President (or some other officer) was in the chair and that Hooke himself spoke. Furthermore, in an age of gentlemen-virtuosi, many Fellows doubtless felt themselves competent to speak on a variety of issues in areas where they had never done, and would never do, scholarly or scientific research. In short, information about Hooke's critics—who they were, and how numerous—is meagre indeed. To supply this deficiency, a combination of fragmentary evidence about certain Fellows, collateral information about some non-Fellows, and a degree of conjecture will be used.

In the lectures of 1667–1668, Hooke first presented to the Royal Society his fully developed views on fossils and earthquakes. Fossils had already received attention in the *Micrographia*, where Hooke, summarizing his own earlier communications to the Society, examined under a microscope a specimen of petrified wood and announced that it possessed the structural characteristics of living wood and thus could not be a mere sport of nature. Elaborating on the ancient maxim, 'nature does nothing in vain,' he insisted that fossil wood and shells had been produced by nature in the very same manner as their living counterparts. How marine shells in particular had come to be buried on land might be explained as the result of 'some Deluge, Inundation, Earthquake, or some such other means'.[3]

These statements were expanded, added to, and modified in 1667–1668, when lengthy comparisons of fossil and living forms were used to combat the concept of *lusus naturae*. The fact that fossils are rocky in substance should not lead one to imagine that they are the 'sportings of Nature . . . or the effects of Nature idely [sic] mocking herself, which seems contrary to her Gravity.' Indeed, why should nature have troubled to produce broken or fragmentary fossil shells? And why did nature avoid producing petrified roses or grass? Nature, Hooke concluded, had simply preserved in the rocks those materials least liable to decay, and their subsequent petrifaction might then have been the result of a 'great degree of Cold and Compression,' subterranean heat, hardening in the course of time, or the agency of 'petrifying water'.[4] As for fossils apparently without living counterparts, Hooke proposed that these had either been destroyed by a local earthquake or transformed by changes in 'Climate, Soil and Nourishment'.[5]

As a subsidiary argument, Hooke introduced an analogy that would presumably have held some appeal for an audience better versed in classical than in natural history.

> There is no Coin can so well inform an Antiquary that there has been such or such a place subject to such a Prince, as these [fossil shells] will certify a Natural Antiquary, that such and such places have been under the Water, that there have been such kind of Animals, that there have been such and such preceding Alterations and Changes of the superficial Parts of the Earth.[6]

Just as coins and other 'lasting Monuments' testified to events in civil history, so did fossils reveal episodes in the earth's history. A 'plastic virtue' could conceivably have formed coins buried in the earth, but no antiquary would think so; and no natural antiquary should invoke this explanation of fossils.

In an important departure from the *Micrographia*, Hooke undertook in 1667–1668 to examine in detail those causes responsible for the burial of fossil shells deep below the earth's surface, high on mountains, and far from modern seas. He quickly dismissed the agency of the biblical Flood since it had 'lasted but a little while'. Among other available causes, running water and 'the motions of the air' could also be discounted: these had certainly altered the earth's surface features, but could not explain what seemed to be

---

3. Hooke: *Micrographia*. London, 1665; facsimile reprint with preface by R.T. Gunther, New York, 1961, p. 107–112. Hooke's earlier comments on fossils, at meetings of 17 June 1663 and 24 August 1664, are in Thomas Birch, *History of the Royal Society of London*, 4 vols., London, 1756–1757, i, 260ff, 463. Hereafter: Birch. For the dating of Hooke's 1667–1668 lectures, see remarks in Appendix, lecture 1.

4. *The Posthumous Works of Robert Hooke*. Richard Waller, ed. London, 1705; facsimile reprint with introduction by Theodore M. Brown, London, 1971, p. 318. 290–295. Hereafter: *PW*.

5. Ibid., pp. 327–328. Hooke had in mind variability within species and used such common examples as the various breeds of sheep and dogs.

6. Ibid., p. 321; also, p. 319. For antiquarian interests in the Royal Society, see Cecil J. Schneer: 'The rise of Historical Geology in the Seventeenth Century'. *Isis*, (1954), 45, p. 256–268. For comparisons of coins and fossils, see the discussion of 'monuments' in Rhoda Rappaport, 'Borrowed Words: Problems of Vocabulary in Eighteenth-Century Geology', *Br. J. Hist. Sci.* (1982), 15, p. 27–44.

major dislocations of seas and land masses. Such large changes might have resulted from past shifts in the earth's centre of gravity, but Hooke knew that he had no evidence for this suggestion.[7] There remained one mechanism, earthquakes, whose effects were widespread and well documented. Hooke amassed examples from texts both ancient and modern, to show that earthquakes were recorded in many areas of the world and that observers had noted such effects as the elevation and depression of tracts of land. Since written records of such events could not be found for all periods and places, Hooke allowed that his theory was thus 'very hard positively to prove'. But the evidence nonetheless seemed sufficient to render his ideas 'more than probable'.[8]

After 1668, Hooke now and then found occasion to refer to his earlier views, notably in the famous Cutler lecture of 1678, *Of Spring*, where a long digression deals with the role of earthquakes in the history of the globe.[9] His intensive lecturing on topics geological did not resume until the end of 1686, however, and then the texts are almost wholly reminiscent of his 1667–1668 lectures. Here again we find the main topics Hooke would insist upon until 1700: the evidence for the organic nature of fossils, the role of earthquakes in the burial of organisms, the insistence that fossils are as trustworthy a record of the natural past as are coins and medals for civil history.

These renewed lectures of 1686 also contain Hooke's first comments on discord in his audience, discord so serious that he felt impelled to reflect on the basic problem of induction. He actually began with the disarming statement that he would defend the Society against critics who charged the Society with devotion to the mere collecting of ' a rude heap of unpolish'd and unshap'd Materials'. On the contrary, Hooke declared, such collections were essential to the process of induction. In 'Physical Inquiries', he explained, it was inappropriate to use the deductive method of geometers; instead, one had to frame hypotheses and then test or modify these by amassing experiments and observations.

So that tho' possibly we may not be able to produce a *Positive* Proof, yet we may attain to that of a *Negative*, which in many cases is as cogent and undeniable, and none but a willful or senseless Person will refuse his assent unto it[10]

Despite the ostensible purpose of this preamble, Hooke in fact echoed here sentiments he had earlier expressed when he described his theory of earthquakes as 'more than probable'. The same reasoning he now applied to the fossil question, examined as an instance of seeking positive and negative proofs. Towards the end of his discussion, Hooke's exasperation with critics of his own views at last became explicit. To those who doubted that fossil shells were true shells, he could only say: 'I would willingly know what kind of Proof will satisfie such his doubt'.

7. *PW*: p. 291, 312–313, 316, 320, 321–322.

8. Ibid., p. 324; also, p. 312. Accounts of earthquakes occupy p. 299–310. In addition to Oldroyd, op. cit. (1), see Barbara J. Shapiro: *Probability and Certainty in Seventeenth-Century England*. Princeton, 1983, ch. 2, *passim*.

9. 'Lectures De Potentia Restitutiva, or of Spring'. In R.T. Gunther: *Early Science in Oxford*, viii: *The Cutler Lectures of Robert Hooke*. Oxford, 1931, p. 380–384. As in 1667–1668, Hooke again refers briefly, p. 382, to possible changes in the earth's axis.

10. *PW*, p. 329, 331.

With such now as shall not think all, or any of these convincing Arguments to prove them Shells, I cannot, I confess, conceive what kind of Arguments will prevail, since these sensible Marks are, in all other things, the Characteriicksts [sic] and Proofs by which to determine of their Nature and Relation, and why they should not be allow'd to be so in this particular Case I cannot well conceive.[11]

This *cri de coeur* about the nature of fossils was accompanied in Hooke's lectures by two critical problems: why are some fossil shells much larger than any known living analogues, and how did fossil shells come to be buried on land?

These questions had long been anticipated by Hooke and answered, so he thought, in his early lectures of 1667–1668. Why the anguish of 1686? One answer may be that in the period between 1668 and 1686 the Society had acquired two Fellows more knowledgeable than most in natural history and distinctly opposed to Hooke: Martin Lister (elected 1671) and Robert Plot (1677).[12]

When Lister became a Fellow in 1671, he sent a letter to the Society disputing Nicolaus Steno's arguments for the organic origins of fossil shells. The challenge to Steno not only had general relevance to Hooke's arguments, but Lister also explicitly used the example of giant ammonites found in English strata: these were sufficiently different from living forms to suggest to Lister that the fossils had never been living organisms.[13] Like Lister, Robert Plot also had doubts about the organic nature of certain fossils, and he subscribed to the notion that a 'plastic force' within the earth had produced certain organic-like stones.[14] One cannot judge how large a role the presence of Lister or Plot may have played within the Society, but one of the rare instances when we have evidence of disagreement at a particular meeting of the Society was on 2 November 1671, when Lister's letter on fossils was read to the membership, 'some applauding' its contents and 'Mr Hooke endeavouring to maintain his own opinion, that all those shells are the *exuviae* of animals'.[15] Years later, in 1683, Hooke and Lister exchanged views at a meeting of the Society, when they continued to disagree about whether specimens of 'petrified oisters' sufficiently resembled living forms to be ranked as true organisms.[16]

Whether Lister and Plot were prime movers in the opposition to Hooke remains unknown, but there clearly was some confusion in the Society about petrified objects in general. One finds, for example, a reference to something 'said to be petrified snow', as well as numerous comments on 'petrifactions' like gallstones, kidney-stones, and tumours, not to mention various concretions and crystallizations.[17] Hooke did try to distinguish between fossils and other regular forms like crystals which he termed 'very

---

11. Ibid., p. 338, 342.
12. See the 'Catalogue of Fellows, 1660–1700'. In Michael Hunter: *The Royal Society and its Fellows, 1660–1700*. BSHS Monographs, no. 4, 1982. Useful articles on Lister and Plot are in the *Dictionary of Scientific Biography*.
13. Birch, ii, 487, entry dated 2 November 1671. Lister's letter was published in *Philosophical Transactions*, (1671), 6, p. 2281–2284.
14. Plot: *The Natural History of Oxford-shire*. Oxford, 1677, p. 111–123.
15. Birch: ii, 487.
16. Ibid., iv, 238, entry dated 12 December 1683.
17. Ibid., ii,13, entry dated 8 February 1665 (N. S.), for 'petrified snow' and ii, 487 (2 November 1671) for remarkable tumour. An admirable analysis of the confusion attending various kinds of 'petrifactions' is in Martin Rudwick: *The Meaning of Fossils*. London, 1972, ch. 1–2.

easily explicable Mechanically.'[18] Other Fellows clearly did not find the matter so easy, and this kind of incomprehension may account for the tone of despair detectable in Hooke's lectures in 1686.

Hooke's theory of earthquakes may have received a more cordial reception, at least initially, within the Society. So many recent seismic disturbances had come to the attention of the Fellows that they were at times willing to add to the examples Hooke himself offered in his lectures. One element of dissent did occur at an early date, when Seth Ward, bishop of Exeter, suggested that 'subterraneous canals,' rather than earthquakes, might have allowed the transport of marine shells to burial places in the earth.[19] This possibility, also mentioned by other of Hooke's contemporaries,[20] seems not to have stimulated Hooke to respond, but he took more seriously the kind of argument raised by Robert Plot in 1677. Examining the question of whether the earth's surface had been repeatedly altered by floods or earthquakes, Plot denied any grounds for such ideas because the historical evidence, 'the Records of time', contained little support for such great and frequent 'concussions'.[21] Hooke had, in fact, tried to anticipate such arguments in 1667–1668 by acknowledging that written records of earthquakes were rather sparse before the sixteenth and seventeenth centuries, when *curiosi* were becoming more numerous. But Hooke did not, at this early stage in his thought, deny the relevance of written records, and he actually invited members of the Society to test his own views by delving into ancient documents. The occasion of this invitation, late in 1686, was not a discussion of earthquakes, but rather of the presence in English strata of great ammonites without living analogues. If, said Hooke, the largest living organisms were typical of tropical climates, then England might once have been tropical. Such climatic change might be verified or rendered improbable if Fellows 'better versed in ancient Historians than I ever have been or hope to be' did the requisite searching in ancient documents.[22]

Despite his doubts that written history would supply enough evidence to confirm his ideas, Hooke never lost interest in this matter, and he would revive the subject late in 1687. Early in that year, however, he launched a new series of lectures that would, in effect, circumvent the problem of a paucity of written records. In what may be called a second series, Hooke tackled the possibility, earlier mentioned and dismissed, that there had been alterations in the earth's centre of gravity and thus in the distribution of seas and land masses on its surface. Why Hooke resurrected this subject at this time is uncertain, but his lectures include an explicit attack on Thomas Burnet's recently published views about the shape of the earth.[23] In addition, Hooke was aware of the fact that Isaac Newton had sent to press a book dealing with such topics as had long

18. *PW*, p. 281.
19. Birch, ii, 183, entry dated 27 June 1667.
20. See, for example, *Histoire de l'Académie royale des sciences*, Paris, 1703 (1705), p. 22–24.
21. Plot, op. cit. (14), p. 113.
22. *PW*, p. 343.
23. Reference to Burnet occurs rather late (ibid., p. 371), when Hooke had just decided to shelve astronomy and begin his analysis of fables. Still later, Hooke read accounts of Burnet's theory of the earth to the Society; see Journal Books of the Royal Society (hereafter: *JB*), entries dated 12 December 1688, 19 December 1688, 9 January 1689 (N. S.). Among the many accounts of Burnet's most famous work, the article in the *Dictionary of Scientific Biography* is convenient and accompanied by an excellent bibliography.

interested Hooke himself; it is thus conceivable that the Hooke-Newton rivalry played some role in stimulating Hooke's return to an old topic.[24] It is possible, too, that Hooke had recognized a flaw in his theory of earthquakes. Having assumed that earthquakes are always associated with subterranean fires, Hooke had found no way for the earth to renew its supply of combustibles. Hooke did not find repugnant this idea of the earth's decay, and he more than once suggested that a diminution of fuel was but one of several kinds of evidence in behalf of the idea that the earth is growing old; but early in 1687 he devised an explanation for earthquakes that would eliminate the fuel problem.[25]

Hooke launched his second series with a number of questions: what is the present shape of the earth? Is the earth's centre of gravity (and thus its axis of rotation) changing? If such changes occurred in the past, how had they progressed and what was their cause?[26] Should proof of these alterations be forthcoming, then one would have, so to speak, a built-in mechanism insuring the repeated displacements of land masses and sea basins. Not only would crustal movements be freed of dependence on a fuel supply, but the naturalist would no longer have to worry about the failures of historical texts to record great natural events. Taking precedence over mere texts, the laws of physics and astronomy would allow one to argue that such events had taken place 'in all Ages . . . tho' we have no Histories or Records that have preserved the Memory of them, but only such Signs and Monuments as they have left by the unequal rugged and torn Face of the Surface of the Land and the Bodies [fossils] that are discovered'.[27]

To answer the questions posed, Hooke began by comparing ancient and modern determinations of latitudes for some notable sites, to see if these had changed in the course of time. But ancient measurements like Ptolemy's proved to be too inaccurate by seventeenth century standards. Nor could Hooke discover whether any ancient building still in existence had originally been aligned with the meridian. Abandoning ancient evidence as unreliable or uninformative, he then proposed using medieval cathedrals or even modern structures whose alignments might have altered during a few centuries or decades. One might also, he suggested, compare early seventeenth century and more recent determinations of latitude for famous places like Uraniborg, London or Paris. If Hooke tried any of these expedients, the results must have been unsatisfactory, for he finally suggested that measurements carried out in his own day ought to be repeated by future observers with comparably modern equipment.[28]

When Hooke abandoned this line of investigation in 1687, he clearly did not abandon hope in its ultimate importance, and years later he would rejoice that French astronomers seemed to be pursuing those topics he himself had thought promising.[29] At

---

24. *PW*, p. 330. For knowledge in the Royal Society of some of the contents of the *Principia*, see Richard S. Westfall, *Never at Rest: a Biography of Isaac Newton*. Cambridge, 1980, p. 444–445; Newton was still putting Book Two into final form during the winter of 1686–1687 (ibid., p. 465).

25. On renewing the fuel supply, *PW*, p. 327. For early and later discussions of the earth's growing old, ibid., p. 325–326, 379, 422, 427. For subterranean fires always associated with earthquakes, see Birch, iii, 435, entry dated 7 November 1678.

26. *PW*, p. 345.

27. Ibid., p. 347.

28. In the Appendix, lectures 4–7. Analysis in Oldroyd, op. cit. (1), p. 127–129.

29. Text dated 1695, in *PW*, p. 536–540.

the same time, proceedings within the Royal Society may well have added to his decision to look about for some different way to support his theory of earthquakes. One letter, read to the Society in October of 1687, indicated that the latitude of Nuremberg had not changed in 200 years, a discouraging report for Hooke.[30] Probably of more significance to Hooke was the fact that one of the premier astronomers in the Society, Edmond Halley, thought Hooke to be mistaken in his evaluation of the accuracy of Ptolemaic measurements. According to Halley, Ptolemaic determinations of latitude contained errors so small that their correction suggested latitudes had in fact not changed in the course of some fifteen centuries.[31]

Although Halley continued to find worthwhile the study of changes in latitudes, Hooke abandoned the topic and turned elsewhere for support of his theory of earthquakes.[32] Halley's critique of Hooke's arguments had, in fact, been part of an exchange of letters between Halley and Dr John Wallis in Oxford, and Wallis's critique was read to the Royal Society by Halley. Although we do not know how the Fellows responded to the Halley–Wallis–Hooke controversy, it is very clear indeed that Hooke himself found Wallis to be a formidable opponent.

Halley had sent to Wallis an account of 'the Hypothesis of Mr Hook, concerning the changes which seem to have hapned in the Earth's surface, from the shells in bedds' found high in the Alps and far below sea level; Hooke had tried to show 'how the superficies of the earth may have been frequently covered with water, and again dry, so as to answer to all the appearances; if the change of the Earth's axis may be allowed'.[33] In reply, Wallis agreed with Halley that Hooke had no evidence for changes in the earth's axis, and he went on to report the consensus of the Oxford Philosophical Society to which he had read Halley's letter. The Oxford group 'seemed not forward, to turn ye world upside down (for so 'twas phrased) to serve an hypothesis, without cogent reason for it; not only, that possibly it might be so; but that indeed it hath been so'.[34] Furthermore, Wallis continued, the great eruptions and earthquakes supposed by Hooke had left no trace in historical records, and no historian could have failed to notice such large dislocations of land and sea. Since Genesis describes a postdiluvian world with much the same geography as today, when would Hooke's geological events have taken place? Unless one said 'before Adam', then there were no periods of upheaval such as Hooke imagined. Hooke's ideas thus could not be accepted 'without over throwing the credit of all History, sacred & profane'.

30. Birch, iv, 550, entry dated 26 October 1687.

31. Letters from Halley to John Wallis, London, 15 February 1687 and 9 April 1687. In *Correspondence and Papers of Edmond Halley*. E. F. MacPike, ed., London, 1932, p. 77–80, 80–82.

32. Halley presented to the Society (*JB*, entry dated 1 February 1688) a paper on changes in the earth's axis. He did in fact think such changes might have occurred, but so slowly that they would not be detected by any comparison of ancient and modern measurements.

33. In MacPike, op. cit., (31), p. 77–78.

34. Wallis to Halley, Oxford, 4 March 1687. In A. J. Turner: 'Hooke's Theory of the Earth's Axial displacement: Some Contemporary Opinion', *Br. J. Hist. Sci.* (1974), 7, p.167. The texts of this letter and a second one to Halley are here given in full. The meetings of the Oxford group, during which changing latitudes were discussed, are described in R. T. Gunther. *Early Science in Oxford*, iv: *The Philosophical Society*, Oxford, 1925, p. 199, 200, 201–202, entries dated 22 February and 1 March 1687.

When this devastating letter was read to the Society on 9 March 1687, Hooke had already abandoned the subject of changing latitudes. Ironically, however, the Fellows had not been fully persuaded by the arguments of Hooke, Halley and Wallis, and later in the same month they cautiously approved the proposition that 'some alteration of the poles of the earth' had taken place.[35] But Hooke had decided that future research, not past observation, would resolve this issue, and when he resumed his lectures late in 1687 he undertook to reply to Wallis's second argument: the gaps in historical records. As we have seen, Hooke had himself acknowledged the existence of such lacunae in his early lectures, but the Wallis letter apparently impressed him sufficiently to induce him, a year later, to echo its phrases, when he declared that 'some may say, I have turned the World upside down for the sake of a Shell'.[36]

Hooke's third series began with a summary of his views on the earth's axis of rotation, but he quickly reminded his audience that he really preferred to explain 'the ruggedness and inequalities of Hills and Dales, Mountains and Lakes, and also the alterations of these superficial Parts of the Earth' as 'most probably ascribable to another Cause, which was Earthquakes and Subterraneous Eruptions of Fire'.[37] In fact, Hooke had never abandoned emphasis on earthquakes, but had sought physical laws to explain their regular recurrence. Having failed to do so, he turned now to written history, candidly admitting: 'One of the most considerable Objections I have yet heard, is, that History has not furnish'd us with Relations of any such considerable changes as I suppos'd to have happen'd in former Ages of the World'.[38]

In earlier lectures, Hooke had suggested that ancient written records might be defective because the older strata of the earth's crust had perhaps been laid down before the invention of writing. Only a confused, unreliable memory of such events could have been transmitted to later generations, and the scientist would do better to rely on the fossil record itself to provide a 'natural Chronology' of the earth's past.[39] Like his contemporaries, however, Hooke also thought it possible to detect in ancient writings some allusions to those real events that time, oral transmission and human frailty had distorted and disguised. The Euhemerist historians of Hooke's day were, indeed, engaged in this task of discovering the traces of historical truth concealed in pagan tales of gods and heroes. Among the more famous exponents of Euhemerism, Athanasius Kircher and Pierre-Daniel Huet argued that pagan beliefs and myths were distorted memories of Hebrew history or of Christian beliefs prefigured in the Old Testament.[40] Hooke did not succumb to the common temptation to identify pagan flood stories with

35. Birch, iv, 527, 529: entry dated 9 March 1687, when Halley read the first of Wallis's letters, and entry for 23 March 1687, when the Society allowed that the earth's poles had shifted.

36. *PW*, p. 411.

37. Ibid., p. 372.

38. Ibid.

39. Ibid., p. 303, 308, 324, 337 (gaps in written records), 334, 335 (the invention of writing), 335, 338 (fossils as natural chronometers).

40. See Frank E. Manuel: *The Eighteenth Century Confronts the Gods*, Cambridge, Mass., 1959, ch. 1–3, and D. P. Walker: *The Ancient Theology*. Ithaca, N. Y., 1972. Hooke's knowledge of the works of Kircher and other antiquarians is evident in *The Diary of Robert Hooke, M. A., M. D., F. R. S. 1672–1680*. H. W. Robinson and W. Adams eds. London, 1935, p. 70, 163, 254, 266, 390.

the biblical Flood, but his grasp of the principles of Euhemerism is evident in his justification for using Ovid's rather late text to reconstruct the history of earlier, 'fabulous' ages. His lectures on Ovid would show

> That this Mythologick History was a History of the Production, Ages, States and Changes that have formerly happened to the Earth, partly from the Theory of the best Philosophy; partly from Tradition, whether Oral or Written, and partly from undoubted History.[41]

Ovid, in short, had gathered up older strands of idea, tradition, and fact and had used these, somewhat disguised, to weave what appeared to be myths. Like Ovid, Hooke declared, other ancient writers—indeed 'a Cloud of Witnesses'—would be found who had either observed or displayed knowledge of geological upheavals.[42]

Hooke began his new series not with Ovid but with the *Periplus* of Hanno the Carthaginian. The celebrated navigator had observed flaming islands located, so Hooke thought, west of the Straits of Gibraltar and in a region where no islands now existed. What Hanno had witnessed Hooke identified as the last stages in the disappearance of Atlantis, swallowed up by a great earthquake and eruptions of subterranean fire. Plato's myth of Atlantis doubtless contained some poetic inventions, but, Hooke concluded, the substratum of fact in Plato could be confirmed by the testimony of the Carthaginian.[43]

After this relatively straightforward exposition, Hooke turned to Ovid's *Metamorphoses*. Not surprisingly , one fable after another received a naturalistic exegesis, so that the wars of the Titans became great earthquakes and Pluto's rape of Proserpine represented a notable eruption (with attendant earthquake) of Mt Etna. That this elastic, imaginative technique could tempt its users into extravagances had already been shown in Huet's *Demonstratio evangelica* (1679), where the pious author had reduced a great array of pagan gods and heroes to representations of Moses. Hooke tried to be more discriminating. He admitted that some ancient tales are genuine fables rather than distortions of historical events, and he hoped that these two varieties of fable would one day be accurately classified. He allowed, too, that poetic texts might give rise to diverse interpretations: 'In these Matters Geometrical Cogency has not yet been applied, and where that is wanting, Opinion, which is always various and unstable, prevails'.[44]

---

41. *PW*, p. 384. For pagan floods as different from Noah's, ibid., p 389, 408; for a comparable refusal to conflate pagan tales with the Tower of Babel, ibid., p. 395–396. On the one occasion when Hooke tried to weave the Flood into his geological theories, he did allude to accounts in authors other than Moses that might have some bearing on the Flood, ibid., p. 412.

42. Ibid., p. 374. Theodore M. Brown (ibid., p. 9) suggests that Hooke's use of fable may have been a reply to Thomas Burnet. But Hooke had just referred to Burnet in a different context, the earth's shape (above and n. 23), and his *return* to the subject of fable in 1693 is more plausibly a result of his reading of Burnet's later work (see below) rather than Burnet's theory of the earth.

43. In the Appendix, lecture 9 and the preceding lecture for which there is no text. Hooke had earlier toyed briefly with interpretations of the Atlantis myth and Ovid, *PW*, p. 308, 320, 323, but at that time he probably thought he had enough evidence for earthquakes without pursuing the exegesis of fables. Hanno's voyage probably took place in the fifth century B.C.

44. Ibid., p. 391. In the two examples of the Titans and Proserpine, I have actually used lectures of 1693, rather than 1687–1688, since Hooke's method of exegesis did not change. There was, in fact, an earlier lecture on Proserpine (ibid., p. 395), but the text seems not to have survived.

Hooke's lectures on myths began late in 1687, suffered a significant interruption a few months later, and were resumed for a short time in 1693. According to Hooke, even his treatment of Hanno and Plato raised doubts among some Fellows who wanted more confirmation of the Platonic myth than the problematical narration of Hanno in which, as classicist Thomas Gale noted, there could be scholarly uncertainty about the precise rendering of ancient place names when these were translated from one ancient language into another.[45] If more evidence seemed necessary here, then one may readily suppose that Hooke's audience found even less persuasive his treatment of Ovid. Fellows who knew their classical texts would probably have thought it reasonable to quote or allude to Ovid, as did John Evelyn, when discussing floods or even the biblical Flood, but Hooke did not engage in literary allusion or charming parallel. Nor, in fact, did he do what contemporaries like Huet or Dr John Woodward would have considered normal, namely, use pagan tales to support the undoubtedly reliable history found in Genesis.[46] Instead, Hooke seemed to be saying that a whole array of pagan texts (Plato, Hanno, Strabo and Pliny) was sufficient to establish the truth of a myth.

Actually, Hooke never relied on pagan texts alone, for he insisted that nature provided independent confirmation about past upheavals, as shown by the curious burial places of fossil shells. If his colleagues demanded histories, however, he would provide them, and he turned from Ovid to such unpoetic authors as Herodotus and Aristotle who described changes in shorelines and river beds and the shells visible in areas where the waters had subsided. In addition, Hooke for the first time tried to weave the Flood into his geological theory. Earlier allusions to the Flood had referred to that event as merely a brief submersion of land masses and thus inadequate to explain the formation of strata and their enclosed fossils. Now, with so few non-fabulous ancient texts in his arsenal, Hooke resorted to Genesis and therefore found it necessary to alter his own earlier interpretation of that text. After lengthy exegesis, Hooke reached the conclusion that, in modified form, would later be suggested by more than one geologist: the Flood had been essentially a great earthquake, during which continents had been submerged and new land masses raised from the bottom of the sea.[47] This new theory, be it said, did not dispense with the need for additional earthquakes at other periods, since Hooke did not believe that all existing continents and islands had been produced simultaneously.[48]

---

45. In the Appendix, lecture 12, and Birch, iv, p. 555–556, entry dated 1 December 1687. (According to *JB*, this should be 7 December.) Dr Thomas Gale, F. R. S., formerly regius professor of Greek at Cambridge, was a celebrated classicist admired by Huet and Jean Mabillon. His distinguished career is chronicled in the *DNB*, and his active participation in the Royal Society can be inferred from Hunter, op. cit. (12) p. 218–219.

46. When Evelyn saw how the sea had encroached on areas in Holland, he was reminded of the Flood and went on to quote Ovid; see *The Diary of John Evelyn*, E. S. de Beer. 6 vols., Oxford, 1955, ii, 32, entry for 23 July 1641. For an example of Woodward's use of Ovid and other poets to confirm Genesis, see Joseph M. Levine; *Dr Woodward's Shield*. Berkeley, 1977, p. 70. On Huet, in addition to Walker, op.cit. (40), there is A. Dupront: *Pierre-Daniel Huet et l'exégèse comparatiste au XVIIIe siècle*. Paris, 1930

47. In the Appendix, lectures 13, 14. Earlier references to the Flood are in *PW*, p. 320, 341. Interpretations resembling Hooke's views of 1688 were devised late in the eighteenth century by such geologists as John Whitehurst and Jean-André Deluc, and as early as 1700 by Abraham de la Pryme; for the latter, see Roy Porter: *The Making of Geology*. Cambridge, 1977, p. 81.

48. *PW*, p. 320, 422.

Hooke's efforts were in vain, and apparently for the same essential reason: the paucity of historical texts.[49] But his failure in 1687–1688 did not discourage him from making another attempt, and 1693 found him again lecturing on Ovid's *Metamorphoses*. The stimulus for this renewed activity may well have come from the publication of Thomas Burnet's *Archaelogiae philosophicae* (1692), of which Hooke read an account to the Society. In reaction to this controversial work, the members were moved to debate such topics as the antiquity of writing, of astronomical observations and, in general, of pagan cultures, In addition to these matters with clear relevance to any interpreter of fables, an earthquake in Sicily moved Hooke not only to report on that event but to return to the tale of Pluto and Proserpine.[50] This lecture and the few to succeed it contain no departures from Hooke's established method of interpretation, with the significant exception of one effort to bolster his reading of Ovid with a passage from Genesis. The same learned critic, Dr Thomas Gale, who had earlier raised linguistic problems in connection with Hanno's *Periplus*, now queried whether the Hebrew Old Testament, in contrast to the Greek Septuagint used by Hooke, would readily support Hooke's views.[51] In so scholarly a realm Hooke could offer little reply, and it may be no coincidence that Hooke offered only one more lecture on Ovid after the exchange with Dr Gale.

Hooke's lectures on ancient texts were far from his last in support of his geological theories. Indeed in the interval between his first lectures on such texts (1687–1688) and his brief resumption of that topic (1693), he once again altered his approach to the subject and, despite the brief return to Ovid, continued in this vein for the rest of his career. This fourth 'series' perhaps hardly deserves such an appellation, since the topics of the discourses range from the discovery of fossil bones, to reports on particular earthquakes, to examinations of the nature of amber. One may find here, too, a few remarks on Ovid, as well as a report on research by French astronomers into the matter of the earth's axis of rotation. In these last two instances, it is clear that Hooke had not abandoned the views expressed in his second and third series of lectures. With one exception: he gave up his effort to reinterpret the Flood which once more became simply a brief 'soaking of the Earth'.[52]

The case of the Flood is no accident. Although Hooke's last lectures appear to treat diverse topics, what unifies them is a return to three themes discussed in 1667–1668, mentioned thereafter, and at last insisted upon. First, Hooke again urged the incompleteness of the historical record. Much that had happened in the past—man's past as well as the earth's—simply had not been written down. Second, despite the absence of such human records, the 'natural antiquary' possessed trustworthy documents provided by nature: fossils. And nature, unlike man, did not falsify evidence, i.e., there are no *lusus*

49. This is a supposition based on the absence of any new critical issues raised in *PW*. There were, as Hooke admitted in 1667–1668, very few ancient texts relevant to his own geological concerns, even Strabo and Pliny generally repeating older tales and adding little fresh, first-hand observation. Hooke may have come to a halt in 1688 for lack of material.

50. *JB*, entries dated 14 December 1692 and 18 January 1693, and *PW*, p. 402–403. In addition to earlier work on Burnet, cited above (23), see Paolo Rossi: *I Segni del Tempo*. Milan, 1979, esp. p. 59–60.

51. *JB*, entry dated 13 July 1693, and *PW*, p. 384–385 (see comments in Appendix under lecture 21).

52. *PW*, p. 440.

naturae. Third, nature offers many modern examples of the effects of earthquakes, and it is legitimate to conclude that events similar in kind, albeit greater in magnitude, occurred in all ages of the past.[53] In this scheme, the transient episode of the Flood, albeit a well documented event, could hold no place.

A defence of the theory of earthquakes had, indeed, been Hooke's grand strategy throughout his lectures, and his forays into astronomy and Euhemerism may be viewed at least in part as tactical manoeuvres designed to provide his audience with additional persuasive evidence. From Hooke's point of view, a mechanism to account for the burial of fossils was absolutely essential if his hearers were to adopt his view of the nature of fossils themselves.

> I confess it seemed to me a little hard, because I could not give the Pedigree of the Fish [shellfish], therefore I should not be allowed to believe it a Fish when I saw all the sensible marks of a Fish; and that, because I could not tell who it was, or upon what occasion that caused the Stones of *Salisbury* Plain to be dispersed in that irregular Regularity, that therefore I must allow them to be a *Lusus Naturae*, or placed there by *Merlin* or, some such unknown way . . . .[54]

Hooke thus devoted some of his final lectures to attacking once again concepts of *lusus* and plastic forces and to examining some of the great fossil forms with no known living analogues. As late as 1697, in a series of lectures on amber, a subject Hooke had broached more than 20 years before, he referred to his own earlier 'conjectures' about present continents having once lain under the sea, a matter of which he hoped he would one day 'be able to give a more particular, convincing, and satisfactory account'.[55]

In statements such as these, Hooke conveys the impression that he addressed an unyielding audience. Elsewhere, in fact, he went so far as to inveigh against the Baconian Idols 'which pre-possess the Minds of some Men' and prevent them from abandoning 'any unsound, unaccountable and unwarrantable Doctrines formerly imbrac'd'.[56] On one occasion, he felt obliged to remind the Fellows that a preference for written histories, rather than the plain testimony of nature, was 'contrary to the *Nullius in verba* of this Society'.[57] In a despairing passage, perhaps referring to his own unpublished lectures, he

---

53. The issue of *lusus naturae* occupies much of lecture 16 (in the Appendix) and part of a lecture summarized in *JB*, entry dated 4 January 1699. The other topics mentioned are recurrent themes in lectures 15, 17, 18, 23–27, and in other lectures during this period, recorded in *JB* and listed in the Appendix.

54. *PW*, p. 404.

55. Text in R. T. Gunther: *Early Science in Oxford*, vii: *The Life and Work of Robert Hooke*. Oxford, 1930, p. 774. Hooke's earlier remarks on amber are in Birch, iii, 75 (5 February 1673) and 440 f (28 November 1678). Hooke's renewed interest was prompted by the reading to the Society of a letter on amber from J. P. Hartmann, author of an earlier work on the same subject (*JB*, entries for 14 January, 3 February, 24 February 1697).

56. *PW*, p. 433.

57. Ibid., p. 450.

remarked that it was 'a discouragement to any one to Publish that which he finds by Discourse is generally disapproved'.[58]

Did Hooke's lectures really meet with wholesale disapprobation? The surviving records do not permit detailed reconstruction of discussions within the Royal Society, and this is not the place to rehearse debates conducted beyond the confines of Gresham College, in books, articles and correspondence of the late seventeenth century. As we have seen, however, some Fellows did want more evidence before they would countenance the theory of earthquakes in particular, and Hooke himself acknowledged late in his career that, for at least certain fossils, most people preferred to attribute their burial to the Flood, 'where they can think of no other Cause'.[59] In short, the fossil question and the theory of earthquakes were not as mutually dependent as Hooke thought.

The latter assertion will not surprise especially those historians of science familiar with the reception accorded John Woodward's *Essay toward a natural history of the earth* (1695). In the midst of considerable approval of his treatment of fossils Woodward encountered strenuous opposition to his explication of the Flood, his chief geological mechanism; Hooke himself was one of the vocal opponents.[60] It may be said, then, that the fossil question was becoming less problematical in Hooke's last years, even while matters of geological dynamics remained puzzling. Such men as John Evelyn and John Flamsteed found the many reports of earthquakes sufficiently intriguing to warrant philosophical study, but they did not conclude that events of this kind had had a major role in the formation of the earth's crust. Among others, Robert Boyle found 'subterranean fires' to be a plausible reality, but Boyle would have been the last person to erect a plausibility into the basis for an explanation of the earth's structure.[61]

58. Ibid., p. 446. This remark comes near the beginning of a long discussion of priority disputes, including the lack of public recognition when discoveries remain unpublished. An autobiographical element can be detected in Hooke's statement not simply because he did not publish his lectures on earthquakes, but also because his treatment of fossils in the *Micrographia* had been ignored by Philippe de LaHire who in 1692 published what purported to be an original analysis of petrified wood; LaHire is explicitly discussed in Hooke's lecture. That LaHire did not know Hooke's work is evident if one compares the *Micrographia* with the summary of LaHire in *Memoires de mathematique* [sic] *et de physique, tirez des registres de l'Académie Royale des Sciences.* Paris, 1692, p. 122–125.

59. *PW*, p. 437. For the debates of this period, see Rudwick, op.cit. (17), ch. 2.

60. Ibid., esp. p. 83–84, and Levine, op.cit. (46), ch. 2. Hooke's attack on Woodward occurred in a lecture devoted to the causes of petrifactions and of their burial, in which he included an assault on the Woodwardian notion of the Flood's dissolution of the antediluvian world. *JB*, entry dated 4 January 1699. Woodward was present at this meeting, but no debate on these issues is recorded.

61. On earthquakes, see letter from Evelyn to Tenison, 15 October 1692, in *The Diary and Correspondence of John Evelyn, F. R. S.*, William Bray, ed. 4 vols., London 1883–1887, iii, p. 325–330, and J. E. Kennedy and W. A. S. Sarjeant: "Earthquakes in the Air': The Seismological Theory of John Flamsteed (1693)'. *J. Roy. Astronom. Soc. Can.* (1982), 76, p. 213–223. For subterranean heat and/or fires, see Robert Boyle, 'Of the Temperature Of the Subterraneal Regions, As to Heat and Cold' in: *Tracts Written by the Honourable Robert Boyle*, Oxford, 1671 (each tract in this collection is separately paginated), and Henshaw's discussion of hot springs in Birch, iii, p. 433–434, entry dated 31 October 1678. A later revival of Hooke's theory of earthquakes met with no better success, as indicated in Rudolf Erich Raspe: *An Introduction to the Natural History of the Terrestrial Sphere* (1763), transl. and ed. A. N. Iverson and Carozzi, A. V. New York, 1970, p.xxxvii, where the occasional compliment to Raspe is coupled with no indication that the theory itself was admired. Apart from Raspe himself, Hooke's only known convert is Aubrey; see Michael Hunter: *John Aubrey and the Realm of Learning.* New York, 1975, p. 58–59, 223.

If opposition to Hooke's views on fossils waned as time went on, the theory of earthquakes did not fare as well. Nor, indeed, did John Ray's presentation of a roughly comparable theory—apparently not derived from Hooke and not as well developed or supported—arouse any enthusiasm.[62] Ray came at the problem as a tentative way to repair the damage done the earth's landforms by weathering and erosion. Although an advocate of Design, Hooke based no theory on Design, but held fast to the 'more than probable' truth of a theory derived from present processes as applied to past history. Hooke's contemporaries clearly had a different notion of what constituted proof.

APPENDIX

Chronology of Hooke's lectures

All the numbered lectures listed below are in Hooke's *Posthumous Works* (PW), p. 279–450. Other published texts, marked with one asterisk (*), have been inserted into their proper chronological places, as have some entries in the Journal Books of the Royal Society (*JB*) for which no texts seem to have survived; the latter entries are marked with two asterisks (**).

Dates given in *PW* have been verified in Thomas Birch (B) for the period to December, 1687, and thereafter in *JB*. Undated texts have been assigned dates when their content corresponds to descriptions in B or *JB*. Clear stylistic breaks in the *PW* texts, whether or not signalled by a marginal note by Richard Waller, are taken to mean the start of a new lecture; final and initial paragraphs in these cases are marked by the letter P in the left column, and the resulting two lectures separately dated.

The designation 'chief topics' (middle column) should not be taken to exhaust the content of each lecture. Hooke was very repetitive and often liked to remind his audience, sometimes at length, of arguments in earlier lectures. In the earlier lectures in particular, he touched upon subjects that he would not develop until later; such preliminary hints are not included among 'chief topics' and are to be found in the footnotes to this article.

It is worth calling to the reader's attention Birch's efforts to provide footnotes identifying the relevant texts in *PW*, notes then reproduced by Gunther, op. cit. (55). On examination, most such notes are inaccurate.

62. In addition to Rudwick, op.cit. (17), esp. p. 63–66, see Gordon L. Davies: *The Earth in Decay: A History of British Geomorphology, 1578–1878.* London and New York, 1969, for the subject of 'degradation' without a mechanism for 'renewal' of the earth's landforms. For influential opposition to Ray, see Buffon, op. cit. (2). One would like to know more than Davies offers about the apparent incompatibility of the ideas of immutable laws of nature and the irreversible processes of decay (chiefly denudation). Hooke did encounter some equivocal opposition to the latter idea from John Evelyn (*JB*, 30 July 1690), but there is no evidence that Hooke's notion of the diminishing intensity of earthquakes played any role in the general rejection of his theory.

144

Lectures	Chief topics	Dates
1. 279–328	Fossils, earthquakes	Series ended 15 September 1668, according to *PW*, p. 328, but B, ii, 313 says the Society resumed its meetings on 22 October 1668, after a 10 week recess. It is assumed here that the first lecture in the series corresponds to the entry in B, ii, 183, dated 27 June 1667.
*	Earthquakes	1678. Part of *Lectures De Potentia Restitutiva, or of Spring*, one of the Cutler series. Text in Gunther, op. cit. (9), viii, p. 380–384.
2. 365–370	Barometer	28 May 1684. *PW*, p. 365; B, iv, 309. Waller explains that he thought it suitable to add this text to the subject of lecture 8.
3. 329–345	Methods of inquiry, fossils, last pp. on latitudes	Probably the six lectures delivered in seven weeks from 8 December 1686 to 19 January 1687 (N. S.), described in B, iv, p. 511–521.
4. 346–350	Earth's axis of rotation, changes in sea basins	26 January 1687, described in B, iv, p. 521–522. Confirmed by opening lines of next lecture.
5. 350–354	Shape of earth	2 February 1687. *PW*, p. 350; B, iv, 523.
6. 355–360	Same topic	9 February 1687. *PW*, p. 355; B, iv, 525.
7. 360–362	Proposals on same topic	23 February 1687. B, iv, 527. *PW*, p. 360 dates text 16 February 1687, a date for which there is no entry in B.
8. 363–364	Shape of earth's atmosphere	9 March 1687. *PW*, p. 363: B, iv, 527.
**	Hanno the Carthaginian	2 November 1687. B, iv, 551, verified in *JB*.
9. 371–376	Plato and Hanno	7 December 1687. *JB*. B, iv, 555 says the Society met on 1 December.
10. 394 P2– 402	Fables in general; Ovid	14 December 1687. B, iv, 557.
11. 377–384	Ovid's *Metamorphoses*, book 1	4 January 1688. *JB*.
12. 403–406	Critics of Hooke	15 February 1688. *PW*, p. 403; *JB*. Critics focus in part on the interpretation of fables.
13. 407–410	Herodotus	22 February 1688. *JB*.
14. 410–416	Aristotle, the Flood	29 February 1688. *PW*, p. 410; *JB*.
15. 428–433	Recent earthquakes	18 July 1688. *PW*, p. 428; *JB*.
**	Recapitulation of views on changes in earth's past	15 May 1689. *JB*.

16. 433–436	Extinction, plastic forces	29 May 1689. *PW*, p. 433; *JB*.	
17. 416–424	Earthquakes in the Leewards	23 July 1690. *PW*, p. 416; *JB*.	
**	Decay (old age) of the earth	30 July 1690. *JB*.	
18. 402–403	Earthquake in Sicily, confirmed by Ovid's fable of Proserpine	12 April 1693. *JB*. *PW*, p. 402 dates text 8 March 1693.	
19. 389 P2– 391 P3	Ovid: fable of Phaeton	31 May 1693. *JB*.	
20. 391 P4– 394 P1	Same topic	7 June 1693. *JB*,	
**	Ovid's 'gigantomachy'	13 July 1693. *JB*.	
21. 384–385 P1	Thomas Gale on Giants, Hooke's reply (not in the form of a letter)	Gale's undated letter presumably followed the lecture of 13 July 1693, in which Hooke asked if the Hebrew word, rendered 'gigantes' in the Septuagint, could signify earthquakes (Gen. 6:4). Hooke's reply begins, 'But to me . . . ', and may thus be a note to himself.	
22. 385 P2– 389 P1	Ovid: fable of Python	2 August 1693. *JB*. Opening lines refer to lecture of 13 July 1693.	
23. 446 P3– 450	LaHire on fossils	25 July 1694. *PW*, p. 450; *JB*.	
*	French research on earth's axis of rotation	3 July 1695. *PW*, p. 536–540, gives full text and date; date in *JB*.	
*	Amber	24 February 1697. Date in *JB*, date and text in Gunther, op. cit. (55), vii,p. 769–773.	
*	Amber	3 March 1697. Date in *JB*, text in Gunther, op. cit. (55), vii, p. 773–779.	
*	Amber	19 May 1697. Date in *JB*, date and text in Gunther, op. cit. (55), vii, p. 779–786.	
24. 438–441 P1	Burial of ships and bones of elephants etc.	26 May 1697. *PW*, p. 438; *JB*.	
**	Buried bones, extinction, past changes in climate	2 June 1697. *JB*.	

146

**	Buried bones, changes in climate, latitudes, landforms	3 November 1697. *JB.*
**	Possible chemical causes of earthquakes	22 June 1698. *JB.*
**	Causes of petrifactions and of their deposition	4 January 1699. *JB.*
25. 441 P2– 446 P2	Burial of ships, analysing texts reporting such discoveries	22 February 1699. *JB.*
26. 424–428	Causes of earthquakes, decay of the earth; some remarks on Ovid	30 July 1699, according to *PW*, p. 424. No relevant entry in *JB*, and the chemical examples differ from those of 22 June 1698 (above).
27. 436–438	Effects of earthquakes	10 January 1700. *JB.*

Fontenelle Interprets the Earth's History*

In 1729 Swiss naturalist Louis Bourguet listed what he called
the three main hypotheses explaining the present state of the earth's
crust: the two kinds of diluvialism associated with Thomas Burnet
and John Woodward, and the idea that fossil shells, wherever found,
signal the former presence of the sea. The latter theory, said Bourguet, was supported by at least two members of the Academy of
Sciences in Paris, Antoine de Jussieu and René-Antoine Ferchault
de Réaumur. A dozen years later, conchologist Dezallier d'Argenville would identify the same theory with "plusieurs membres"
of the Academy, adding that the Academy itself had wisely refrained

---

*An earlier version of this paper was delivered at the Linda Hall Library, Kansas
City, Missouri, in October, 1984.

from endorsing any hypothesis (1). Both Bourguet and Dezallier were uneasy about a theory seemingly impious, and a critic of Buffon would later go so far as to claim that the Academy had for decades countenanced in its publications theories which undermined the truth and authority of the Bible (2).

The anonymous critic took his clue from Buffon, who had quoted extensively not from the technical papers by academicians, but from the "résumés" written by Fontenelle and published in the Academy's *Histoire*. Fontenelle did have his own views about the earth's history, and he used the *Histoire* to inform the public that a proper theory of the earth could dispense with the hypothesis of Noah's Flood.

To understand how Fontenelle created and used his opportunities, we must return to the statutes of 1699 where the Academy's secretary is charged with informing the public of those academic proceedings judged "plus remarquable[s]" at each year's end. In practice, this meant that members could have their papers published in the annual volume of *Mémoires*, while the *Histoire* would include the secretary's summaries of those papers, of other papers left unpublished, and of communications sent to the Academy by correspondents of various sorts. Fontenelle thus had some discretion about what would be included in the *Histoire*, what space to give each item, and how to present the "remarquables" aspects of academic research. Fontenelle himself wanted to educate the reader to the importance of the sciences, and this required that the abstruse work of his colleagues be made comprehensible. As he put it in the first volume of the *Histoire* (1699):

"On a tâché de rendre cette Histoire convenable au plus grand nombre de personnes qu'il a été possible; on a même eu soin dans les occasions d'y semer des éclaircissemens propres à faciliter la lecture des Memoires, quelques-unes de ces Pieces pourront être plus intelligibles pour la plûpart des gens, si on les rejoint avec le morceau de l'Histoire qui leur répond (3)."

(1) Louis Bourguet, *Lettres philosophiques sur la formation des sels et des crystaux* (Amsterdam, 1729), 177-180. A.-J. Dezallier d'Argenville, *L'Histoire naturelle éclaircie dans deux de ses parties principales* (Paris, 1742), 159.

(2) Anonymous review in *Nouvelles ecclésiastiques,* 13 février 1750, in John Lyon and Phillip R. Sloan (eds.), *From Natural History to the History of Nature: Readings from Buffon and His Critics* (Notre Dame, 1981), 250.

(3) *Histoire de l'Académie royale des sciences,* 1699 (Paris, 1702), préface (not paginated) and page 9 for the duties of the secretary. All quotations reproduce the spelling, capitalization, and accents of the original texts.

Such "clarifications" at times add substantially to the content of the *Mémoires* or are so selective in reporting content that, as one historian has remarked, it is dangerous to use Fontenelle's texts unless they can be compared with the memoirs he was ostensibly summarizing (4). On the whole, one may say that Fontenelle tried to put specialized research into larger philosophical contexts, and sometimes this entailed that he explain "la conclusion où M. de Jussieu [or some other writer] veut venir (5)."

* *

In the first decades of the Academy's existence, members paid remarkably little attention to theories of the earth or to such related issues as the nature of fossils. When one recalls the English debates of this period, the relative silence in Paris is puzzling. French academicians were, to be sure, few in number, and their specialties did not include subjects like conchology or natural history in general. Nor did they have ready access to English works on these matters because few read English; J. J. Scheuchzer was thus quite right in thinking that the dissemination of John Woodward's theory of the earth required its rendering into Latin. But Latin editions of the writings of Thomas Burnet and Nicolaus Steno did exist, and academicians even had available in French an excellent exposition of Steno, published in Paris in 1671 by a visiting Sicilian botanist, Paolo Boccone (6). For reasons unknown, such works evidently attracted little attention.

Before 1699 only one member of the Academy is known to have expressed occasional interest in fossils, although not in theories

---

(4) Lesley Hanks, *Buffon avant l' « Histoire naturelle »* (Paris, 1966), 41n, 50. Among Fontenelle scholars who have used the *Histoire* are J.-R. Carré, *La Philosophie de Fontenelle* (Paris, 1932); A. Birembaut, Fontenelle et la géologie, *Revue d'histoire des sciences,* X (1957), 360-374; Jean Rostand, Fontenelle, « homme de vérité », *in* Suzanne Delorme *et al., Fontenelle, sa vie et son œuvre, 1657-1757* (Paris, 1961); and Jean Dagen, Pour une histoire de la pensée de Fontenelle, *Revue d'histoire littéraire de la France,* LXVI (1966), 619-641. None of these explicitly compares the memoirs Fontenelle was summarizing.

(5) *Histoire et Mémoires de l'Académie royale des sciences* pour 1722 (Paris, 1724), dans *Histoire,* 4. (In the notes that follow, the references to this series are listed as if separate volumes, *Histoire* and *Mémoires,* because each of these two parts, in any volume of the series, has its own pagination.)

(6) Paolo Boccone, *Recherches et observations naturelles* (Paris, 1671), 38-66. Scheuchzer's translation of Woodward's *Essay toward a natural history of the earth* (London, 1695) appeared in 1704; no French translation was published until 1735. For knowledge of English among academicians, see Harcourt Brown, *Scientific Organizations in Seventeenth Century France (1620-1680)* (Baltimore, 1934), 198, 203-204, 225, 282.

of the earth. Astronomer Philippe de La Hire in 1688 offered the Academy a petrified shark's tooth and then in 1692 a specimen of petrified wood. Although brief, the latter text contains points of interest. First, even a casual reading of the text shows that La Hire could not have known the more incisive treatment of the same subject in Robert Hooke's *Micrographia* (1665). On the other hand, La Hire does indicate his awareness of dispute surrounding the fossil question:

" Mais les Naturalistes ne conviennent pas de l'origine de ces pétrifications, ny de leur cause. Quelques-uns prétendent que les corps que l'on croit avoir été pétrifiez n'ont jamais été que des pierres & des cailloux, qui en se formant dans la terre ont pris par hazard la figure des choses qu'ils représentent : D'autres veulent qu'il y ait des eaux qui aient la vertu de changer effectivement en pierre certaines especes de corps, quand ils y ont trempé long-temps. Et il y a des raisons assez probables de part & d'autre (7). "

Whoever the unnamed naturalists, they presumably were not English, for this paragraph contains none of the locutions so familiar in English texts of this period — such phrases as "sports of nature," "plastic virtue," and *"lapides sui generis."* La Hire did conclude that the petrified and living specimens were sufficiently alike to rule out any chance resemblance, but the evidence that for him decided the issue beyond doubt was of quite another sort: the observation by a missionary in China of a petrifying spring that turned wood as hard as rock. In Fontenelle's hands, such a projection of present observations back into nature's past would become a methodological principle attached to one of the most famous experiments in early geology.

Apart from La Hire's memoir, one may cite a short piece by chemist Moïse Charas "sur les causes de la chaleur des sources chaudes" (1692), and this exhausts the early Academy's interest in matters geological. Nor did this situation change dramatically in the first years after Fontenelle became secretary. In the first

(7) *Memoires de mathematique et de physique, tirez des registres de l'Académie Royale des Sciences* (Paris, 1692), 122. This volume is one of the two (1692, 1693) published when the Academy made an abortive attempt to inaugurate an annual series; the texts may not be the author's words, but those of the editor, *l'abbé* Jean Galloys (or Gallois). A very brief summary by Fontenelle is in *Histoire de l'Académie royale des sciences. Tome II. Depuis 1686 jusqu'à son Renouvellement en 1699* (Paris, 1733), 140; this volume also refers, page 43, to the shark's tooth and to another member's remark on a petrified willow. Cf. Robert Hooke, *Micrographia,* facsimile reprint with preface by R. T. Gunther (New York, 1961), 107-112.

seventeen volumes of the Academy's *Histoire et Mémoires,* for the period 1699-1715, only three memoirs had any content that might be deemed geological, and Fontenelle summarized some half-dozen more such communications from resident members, foreign members, and non-members. These few occasions, however, allowed Fontenelle to express his own views and provided him with material he would eventually use in his own sketch of a theory of the earth.

Among these early texts, a famous one has already been alluded to: Nicolas Lemery's simulation of a volcanic eruption. The key experiment, involving the combustion of a mixture of sulfur and iron filings, had earlier been published in Lemery's textbook of chemistry, and his memoir of 1700 included additional experiments and an attempt at an "Explication physique & Chymique des Feux souterrains, des Tremblemens de terre, des Ouragans, des Éclairs & du Tonnerre." As this title suggests, the memoir is something of a hodgepodge, for which Fontenelle supplied the unifying idea:

"Le meilleur moyen d'expliquer la Nature, s'il pouvoit être employé souvent, ce seroit de la contrefaire, & d'en donner, pour ainsi dire, des representations, en faisant produire les mêmes effets à des causes que l'on connoîtroit, & que l'on auroit mises en action."

This method had allowed Lemery to recreate Etna and Vesuvius. Paying less attention to all the other phenomena Lemery treated, Fontenelle reduced a diffuse text to the essentials that would be discussed for the next decades, namely, the experimental simulation of past events, and the association of volcanic eruptions with deposits of bitumens (8).

What may have been Fontenelle's own introduction to the fossil question came two years later, in a memoir by botanist Joseph Pitton de Tournefort. Here Tournefort suggested that various hard bodies — including rocks and petrified shells — may, like plants, grow from seeds. This idea appealed to Fontenelle as a potentially powerful generalization.

"Et si," he speculated, "quelques Pierres viennent de semence, il est presque necessaire qu'elles en viennent toutes; tel est le Genie de la Nature. Les Cailloux qui ne paroissent que des masses informes, suivront la même loi que ces Pierres curieuses qui ont beaucoup plus l'air de corps organisés."

(8) *Histoire...,* 1700 (1703), 51-52. *Mémoires...,* 1700, 101-110. Also, Lemery, *Cours de chymie,* 7th ed. (Paris, 1690), 175-177 on preparing "saffran de Mars apéritif."

One would have to await proof of Tournefort's idea, Fontenelle added, and he concluded with one of his more striking epigrams: "On ne sçauroit guere attribuer à la Nature trop d'uniformité dans les Regles generales, & trop de diversité dans les applications particulieres (9)."

Neither Tournefort nor Fontenelle suggested that fossils might be mere sports of nature. Using plants as his model, the botanist speculated that many bodies, organic and inorganic, might grow in the same way. By contrast, Englishmen like Martin Lister and Robert Plot had argued that inorganic concretions could serve as the model for how seemingly organic petrifactions had been formed within the earth. One critical problem for Lister and Plot was that of transport: if shell-like fossils had once been living organisms, how had they come to be buried on land, often far from modern seas and high above sea level? The same problem arose in the Academy in 1703, the year after Tournefort's memoir, but on that occasion the astronomer Maraldi did not question the organic origins of such petrifactions; instead, he tackled the matter of transport in a way that would harmonize with Tournefort's conjectures: the eggs or seeds of organisms had grown to maturity in underground waterways which carried them to resting places in the earth or to fluid surroundings which had later hardened to rock (10).

Fontenelle probably found these and other memoirs difficult to deal with in his first years as secretary. As he admitted on one occasion, to prepare the *Histoire* required that he become familiar with a variety of subjects "qu'il faut que i'entende, [et qui] sont quelquefois un peu éleuées pour moi, et me demandent du temps (11)." But the years 1706 and 1707 brought him rapidly to a new level of sophistication in geology. In 1706 Gottfried Wilhelm von Leibniz sent to the Academy a memoir on fossils

(9) *Histoire...,* 1702 (1704), 51-52. Tournefort, Description du Labirinthe de Candie, Avec quelques Observations sur l'accroissement & sur la generation des pierres, *Mémoires...,* 1702, 217-234, on 222-223.

(10) *Histoire...,* 1703 (1705), 22-24; Maraldi's memoir was not published. Also, *Histoire...,* 1705 (1706), 35 for the remark that tiny fossil shells resemble embryos. Robert Plot, *The Natural History of Oxford-shire* (Oxford, 1677), chap. 5, and Martin Lister, in *Philosophical Transactions,* VI (1671), 2281-2284.

(11) Fontenelle to Leibniz, 30 April 1701, in A. Birembaut, P. Costabel, and S. Delorme, La correspondance Leibniz-Fontenelle et les relations de Leibniz avec l'Académie royale des sciences en 1700-1701, *Revue d'histoire des sciences,* XIX (1966), 129.

that must have given Fontenelle much to think about. More than a decade earlier, Leibniz had published a brief sketch of his own theory of the earth, he possessed considerable knowledge of the literature of natural history, and his own observations and ideas surpassed in breadth and depth any thus far produced by members of the Academy. Leibniz's expertise is also obvious in Fontenelle's summary. Here for the first time in the *Histoire* occurs the phrase then so familiar to the knowledgeable naturalist, "sports of nature." Some writers, Fontenelle reports, apply this phrase to fossils,

> "mais c'est là une pure idée Poëtique, dont un Philosophe tel que M. Leibnits ne s'accommode pas. Si la Nature se joüoit, elle joüeroit avec plus de liberté, elle ne s'assujettiroit pas à exprimer si exactement les plus petits traits des Originaux, &, ce qui est encore plus remarquable, à conserver si juste leurs dimensions (12)."

Fossils having once been organisms, how could one account for their burial? Leibniz suggested mechanisms for two classes of fossils. On the one hand, lakes might have become muddy, the mud enveloping both fish and plant life; the mud would have hardened to become slate, while the enclosed organisms decayed and eventually left only impressions in the rocks. On the other hand, marine shells found on mountains had been deposited when the sea, which had covered almost all land, retreated into caverns within the earth.

These ideas appealed to Fontenelle who, as we shall see, framed his theory of the earth with Leibniz rather than Maraldi's transported seeds and eggs in mind. But Fontenelle also did not think that Leibniz had explained how species of plants still living in the Indies had made their way into the slates of Germany. That such species could still be found alive certainly meant that the fossil forms were not sports of nature, but Leibniz's "system" left room for more investigation of the "grands changemens phisiques sur la surface de la Terre (13)."

Additional investigation was forthcoming the next year, the materials being provided by an obscure academician named Saulmon.

---

(12) *Histoire...*, 1706 (1707), 9-11, on 11.

(13) *Ibid.*, 11. Although Leibniz's memoir has not been discovered (see Birembaut, art. cit. n. 4 *supra*, 365, n. 1), Fontenelle's text makes it clear that he is expressing his own reservations; he often ended his summaries with suggestions for further research.

288

In the course of a voyage to Normandy and Picardy, Saulmon noted deposits of gravels along the seacoasts and in areas far inland, while none were to be found nearby at some of the higher elevations; furthermore, larger and smaller gravels were not mingled in any one place. Reminding the reader of Leibniz's evidence that the sea had once covered the earth, Fontenelle noted that Saulmon had added to this evidence. In fact, Saulmon's observations meant that the higher peaks had *not* been under the sea, while ocean currents in the valleys had distributed the larger and smaller gravels according to the laws governing the motion of bodies in fluids. (Details of that distribution, Fontenelle remarked, could be the subject of future topographic research.) In addition, Saulmon had observed at one place on the coast the sea's destruction of a 16-foot falaise over the course of thirty years:

"En supposant qu'elle [la mer] avance toûjours également, elle mineroit 1 000 toises ou une petite demi-lieuë de Moëlon en 12 000 ans. Il est constant par les Histoires qu'en une infinité d'endroits la Mer s'est avancée ou retirée, & qu'en general elle a un mouvement, mais fort lent, pour changer ses premieres bornes (14)."

In a few lines, the reader is offered a glimpse of lengthy processes and changing shorelines.

Precisely how much Fontenelle may have contributed to these summaries of Leibniz and Saulmon is unknown, but he clearly found both writers sufficiently persuasive to refer to them approvingly in the next years. Thus, for example, he could dismiss the whole dispute about the nature of fossils as already resolved: these petrified organisms were neither "des jeux de la Nature, [ni] des peintures fortuites (15)." In the case of the Flood, the importance of which was being emphasized in communications from the Scheuchzer brothers, Fontenelle could remark that Saulmon's observations "ne demande[nt] pas absolument" such a hypothesis (16).

If Leibniz and Saulmon proved useful, Fontenelle also had a habit of looking into the histories of topics treated by academi-

(14) *Histoire...*, 1707 (1708), 5-7. The little known about Saulmon does not include his first name; see *Index biographique des membres et correspondants de l'Académie des sciences* (Paris, 1954). He is one of the few academicians for whom there is no *éloge*.
(15) *Histoire...*, 1708 (1709), 34.
(16) *Histoire...*, 1710 (1712), 21. In 1708 the Academy received communications from both the famous Johann Jakob Scheuchzer (1672-1733) and his brother Johann, and thereafter only from the former.

cians, and the *Histoire* of 1708 shows that he had begun to do so for the geological issues he was encountering. He thus could put Johann Scheuchzer's ideas into the context of older theories of the earth:

> "Descartes, car il arrive souvent que l'histoire de quelque recherche, ou de quelque découverte, commence par lui, est le premier qui ait eu la pensée d'expliquer mechaniquement la formation de la Terre, ensuite Stenon, Burnet, Woodward, & enfin M. Scheuchzer, ont pris ou étendu ou rectifié ses idées, & ont ajoûté les uns aux autres (17)."

By turning Woodward and the Scheuchzers into Cartesians, Fontenelle presumably meant not to rewrite history, but only to show what differentiated a geological approach to the past from any other. All these writers, as Fontenelle explained, dealt with secular changes of the earth's crust, acknowledging that the concentric and parallel strata of that crust must have been formed by settling out of a fluid. A non-geological alternative did exist, also provided by Descartes: instead of a crust evolving according to the laws of motion, God might have created the earth as we see it now, in a state resembling the end-product of such an evolution. "Il est indifferent que Dieu ait creé d'abord l'œuf ou le Poulet." These writers, Fontenelle admitted, did not all agree about the mechanisms of change, most of them attributing much to the Flood, while Steno preferred combinations of floods, earthquakes, and volcanic eruptions. Indeed, emphasis on the Flood seemed quite understandable, that event having caused "un aussi grand renversement" in the past, but to attribute all fossils to the Flood, as Woodward and the Scheuchzers were doing, left unexplained the fact that fossil-bearing strata are not arranged in the order of their specific gravities. Thus, Fontenelle concluded, the Flood might be useful chiefly to explain those fossils "qui se trouvent dans des Lieux où nul autre accident ne peut les avoir portés, & où l'on ne peut croire qu'il y ait jamais eu d'eau depuis ce temps-là [*i.e.*, ce temps du Déluge] (18)." For all his apparent willingness to accept some geological role for the Flood, Fontenelle here has restricted

---

(17) *Histoire...*, 1708 (1709), 30. Although Scheuchzer's memoir has not been discovered, Fontenelle's historical and anti-diluvial preamble accords ill with the pronounced diluvialism of both Scheuchzers. It is impossible to say how well Fontenelle knew the works of the other authors he cites, except for Descartes.

(18) *Ibid.*, 32-34.

290

the effects of a universal event to some local deposits for which no other explanation has been found.

For the five-year period after 1710, Fontenelle evidently had no opportunity to expound upon the earth's history, until, in 1715, an unprepossessing memoir by Réaumur allowed him to rehearse briefly a point or two relevant to the "Sistême général de la formation de la Terre en l'état où elle est presentement. " After promising years in which he had been educating himself and the public, Fontenelle's remarks in 1715 are disappointing; all he can tell us is that the earth's crust, at least to a certain depth, is

"un tas de differentes matieres, de rüines, de débris, de *décombres,* qui ont été assemblés pêle-mêle par des tremblements de terre, par des Volcans, par des Déluges, par des inondations, & par une infinité d'autres accidents plus particuliers. Une longue suite de siécles a produit dans cet amas confus differents changements (19). "

One might call this passage a sign of confusion, of Fontenelle's inability to sort out the geological fragments being offered by his colleagues. More likely, however, is the possibility that Fontenelle, after a five-year hiatus, had forgotten all about the earth's history and produced this list of phrases — Réaumur's memoir of 1715 certainly required no comment of this kind — in haste or perhaps as a reminder that even the smallest observations could have relevance to larger issues. In any event, the very next year, and in connection with a memoir as unprepossessing as Réaumur's, Fontenelle turned his attention once more to the earth's history, attempting for the first time to sketch a sequential history of the earth's crust.

In 1716 chemist Étienne-François Geoffroy offered to the Academy a memoir "Sur l'origine des pierres, " in which he suggested that earthy particles, suspended in water, had consolidated as a result of two processes operating simultaneously: while the water evaporated, the particles were being cemented together by a "petrifying juice. " The degrees of hardness of different kinds of rock depended on the amount of this hypothetical juice that had gone into their formation.

Such was the burden of the memoir Fontenelle had to summa-

---

(19) *Histoire...,* 1715 (1718), 2. The occasion was Réaumur's Observations sur les mines de turquoises du Royaume, *Mémoires,* 1715, 174-202.

rize. The subject, to be sure, was not without interest for Fontenelle who had earlier commented more than once on the evidence that the earth's crust had once been dissolved in a fluid. How the hardening process had occurred entailed some speculation about whether the petrifying juice had always had the same composition, and at this point Fontenelle seems to have lost interest in Geoffroy's chemistry. With an almost visible shrug, an abrupt "quoiqu'il en soit," Fontenelle changed the subject from Geoffroy's chemistry of rocks to the history of the earth (20).

From Geoffroy's particles and fluids, Fontenelle moved to a discussion of the horizontal and parallel position of sedimentary strata; the ubiquity of marine fossils supported the idea that strata and fossils had been laid down in a formerly universal ocean. In this ocean lived aquatic forms, "les plus anciens habitans du Globe (21)." To the then familiar question, posed by diluvialists, of how the waters retreated, Fontenelle gave the not unusual answer that a soft crust might have collapsed here and there, allowing water to drain from areas which now became dry land, some of it mountainous. That these cave-ins must have been sudden seemed a plausible way to account for the great numbers of molluscs unable to escape abandonment, death, and petrifaction — the more mobile fish, of course, had a better opportunity to survive by following the course of the retreating waters.

What Fontenelle called a first "revolution," crustal collapses here and there, exposed land on which there appeared flora and fauna; their origins are not explained, but, thanks to Leibniz and the Scheuchzers, Fontenelle knew about terrestrial forms in the fossil record. Since such forms are also found in horizontal strata laid down in a fluid, Fontenelle tried to explain how water might again have covered the land. There now occurred a series of "révo-

---

(20) Fontenelle's theory occupies the last three pages of his summary of Geoffroy, in *Histoire...*, 1716 (1718), 8-16. Although Geoffroy's memoir has not been located, the abrupt change of subject on page 14 and Geoffroy's apparent lack of geological interests in his other writings suggest that the theory was Fontenelle's; in later modifications of the theory (discussed below), there is no reference to Geoffroy. Like so many of his contemporaries, Geoffroy did possess a collection of shells; see Edouard Lamy, Deux conchyliologistes français du xviiiᵉ siècle : Les Geoffroy oncle et neveu, *Journal de conchyliologie,* LXXIII (1929), 129-132. Geoffroy's brother, Claude-Joseph, was also a member of the Academy in 1716, but he was generally called *le cadet* or *le jeune*. I am grateful to David Corson, Olin Library, Cornell University, for help in sorting out the brothers.

(21) Fontenelle actually says fish were the oldest inhabitants, but later in the text includes molluscs.

lutions particulieres & moins considerables'' during which parts of the exposed land were sunk beneath the sea or to the bottom of lakes. As in the case of the first great revolution, these smaller crustal collapses might also have produced some mountains, while other elevations might have been the result of earthquakes and volcanic eruptions. With so much of the earth's crust now consolidated, it seemed obvious to Fontenelle that the great periods of creation of the earth's landforms had ended, and the earth could now be described as "assés tranquille."

Fontenelle never claimed any novelty for this synthesis, and much of his theory does seem familiar in a general way. If the universal ocean and crustal collapse are reminiscent of Descartes and Burnet, it should be noted that Fontenelle, unlike these predecessors, supplies no mechanism for the formation of this ocean — it is simply there, stocked with marine life. Nor, of course, can Fontenelle's ocean be identified with the Flood which occurred when the earth already possessed a terrestrial population. For all that certain elements are familiar, Fontenelle's brief sketch is, in fact, unusual in its attempt to outline a sequence of geological events and to correlate these with the sequential appearance of different forms of life.

Perhaps the most significant difference between Fontenelle and his contemporaries, however, lies in what he omitted from his theory: all reference to human testimony. He offered what he considered to be a natural series of events, based wholly on the natural evidence he had been encountering for some years. This is not to imply that other writers ignored physical laws and natural evidence; they obviously could not do so, and John Woodward's lapses in this respect supplied critics with a major grievance against his theory of the earth. But Woodward and others habitually used nature *in conjunction with* ancient written records, and this Fontenelle did not do.

How naturalists used ancient texts, both sacred and profane, is in itself a subject worthy of special study. In a rather general way, one may wonder why they did so, especially for the earliest period in the earth's history, when, as Genesis made plain, there had been no human witnesses to the formation of the earth's crust and its inhabitants. Indeed, Biblical apologists wondered how Moses could describe the Creation to which he had not been a witness. Had this information been revealed to Adam and the account transmitted orally — when had writing been invented? — to successive generations and ultimately to Moses (22)? If one managed to surmount (or ignore) these hurdles, the next considerable obstacles were the reliability of ancient texts and the proper ways to interpret them. For reliability, Moses was unimpeachable, even if Woodward and the Scheuchzers thought it important to use natural evidence to confirm the text of Genesis. But the veracity of Moses still did not clarify the precise meaning of the early chapters of Genesis, and ancient texts by profane authors suffered the double liability of difficult interpretation and no assurance of reliability (23).

Despite such difficulties, naturalists habitually did use ancient texts, and to say that they did so out of attachment to the Bible amounts to an assertion rather than an analysis. If one looks not at uses of the Bible, but at citations of Ovid and other pagan authors, one begins to suspect that naturalists were, so to speak, nervous men. Their task of reconstructing the earth's past was, in fact, a new kind of enterprise, and their conviction that laws of nature were immutable did not fully allow them to conclude that the present operations of nature had been the same in the *remote* past. Leibniz put the matter perfectly in his one geological treatise, the *Protogaea,* when he declared that to look at the earth's present condition would not tell us that a unique event had taken place: the earth had once been entirely submerged by the sea (24). In a more modest vein, without recourse to assumptions about laws of nature, Tournefort expressed the essential insecurity of naturalists when dealing with the remote past. During Tournefort's voyage to the Levant (1700-1702), he noted that the Black Sea, although fed by rivers, did not change in level or extent because it had an outlet into the Mediterranean by way of the Dardanelles. But what if these straits had not always existed? Logically, the Black Sea would have overflowed and burst its bounds, flooding the countryside and also carving for itself an outlet into the Sea

(22) For such subjects as the invention of writing and other problems where Genesis had to be accommodated to other historical sources, see Paolo Rossi, *I Segni del Tempo* (Milan, 1979).

(23) For the use of natural "monuments" to confirm the evidence of ancient texts, see Rhoda Rappaport, Borrowed Words: Problems of Vocabulary in Eighteenth-Century Geology, *British Journal for the History of Science,* XV (1982), 27-44.

(24) Leibniz, *Protogée ou de la formation et des révolutions du globe,* transl. B. de Saint-Germain (Paris, 1859), 65-66. The full text was published for the first time, in Latin and German, in 1749; a brief sketch appeared in the *Acta eruditorum,* January, 1693.

294

of Marmora. However logical these propositions, Tournefort was relieved to discover that an ancient historian did report a tradition of flooding in the very region in question; thus, "ce que nous venons de proposer comme une conjecture de physique, devient une verité historique (25)."

This search for human witnesses held no attractions for Fontenelle. Early in his career as a propagandist for science, his *Entretiens sur la pluralité des mondes* had lightheartedly alluded to the unreliability of ancient texts, when he declared the existence of Alexander the Great to be a little more probable than the existence of life on other planets, but the latter to be more probable than innumerable other "facts" of history (26). In a more serious vein, his *Origine des fables,* probably written in the 1690s, argued that the recorded beliefs of primitive peoples revealed their mentalities, but could not reliably be used as accounts of actual events. Furthermore, he declared, any historical events buried in such ancient myths were gone beyond hope of recovery, the myths having been repeatedly distorted in the course of their transmission. Although Fontenelle exempted the ancient Hebrews from his analysis, we need not take seriously this bow to convention, especially when in later years he could at times refer to Noah's Flood not as an event but as a "hypothesis" employed by some naturalists (27).

In the first volume of the Academy's *Histoire,* Fontenelle continued to reflect on human history, which he described as the study of the capricious, unpredictable behavior of men. By contrast, the scientist deals with regular, discoverable patterns in nature. We do not, he continued, know everything we could wish to know about how nature works; and the next year, in commenting on Lemery's experiments, he pointed out that one path to increasing our knowledge lay in imitating nature in the laboratory. By 1710 he could say that fossils were "sans comparaison plus anciennes,

(25) Tournefort, *Relation d'un voyage du Levant,* 3 vols. (Lyon, 1717), II, 403-409. A two-volume edition appeared in Paris in the same year. This was a posthumous work, Tournefort having died in 1708. For an examination of how Robert Hooke used pagan texts for geological purposes, see Rhoda Rappaport, Hooke on Earthquakes: Lectures, Strategy and Audience, *British Journal for the History of Science,* XIX (1986), 129-146.

(26) *Entretiens sur la pluralité des mondes,* critical ed. by Alexandre Calame (Paris, 1966), 160-161. This passage is in the sixth part of the work, added to the edition of 1687 (first edition, 1686).

(27) *De l'origine des fables* was first published in the 1724 edition of Fontenelle's *Œuvres.* For the problem of the date of its composition, see Dagen, art. cit. n. 4 *supra.*

& plus importantes, & plus sûres'' than the man-made monuments of classical Antiquity (28). In short, by the time Fontenelle produced his theory of the earth, his own intellectual commitments were such that any search for human witnesses to ancient events had become impossible.

*
* *

Fontenelle's comments on geology were not numerous after 1716, but a few examples will illustrate the ways in which he modified his theory of the earth as he incorporated into it the work of other academicians. The first such occasion arose in 1718 in connection with Antoine de Jussieu's description of fossil plants found in Lyonnais — a memoir which would give later writers the impression that there existed a "school" of theorists in the Academy.

That the plants Jussieu had discovered were deposited by the sea seemed evident from the fossil shells found nearby; these shells resembled no known freshwater forms, but similar marine specimens could be found on European coasts and in distant seas. More mysterious than the shells, the fossil plants differed so much from any living French forms that "il me sembloit herboriser dans un nouveau Monde." Differences between living and fossil forms had led some naturalists to doubt the organic nature of fossils, but Jussieu thought such recourse to the idea of sports of nature both "inutile" and evasive; unlike Fontenelle, who had earlier described extinction as an "idée un peu hardie," Jussieu did not hesitate to say that at least some fossil species no longer exist (29).

Because some fossil plants resembled, but were not identical to, living specimens recently sent to Europe from the East and West Indies, Jussieu discussed but could not resolve the problem of transporting exotic species from the Indies to France. His discoveries allowed him to draw only one conclusion: "La pluspart des terres qui semblent avoir été habitées de temps immemorial,

(28) *Histoire...*, 1699 (1702), préface (not paginated), and 1710 (1712), 22. Fontenelle's views of history are discussed by Dagen, art. cit. n. 4 *supra*, 634.

(29) Jussieu, Examen des causes des Impressions des Plantes marquées sur certaines Pierres des environs de Saint-Chaumont dans le Lionnois, *Mémoires...*, 1718 (1719), 287-297, on 288, 289, 296. For Fontenelle on extinction, see *Histoire...*, 1710 (1712), 20. Réaumur evidently had no qualms on this subject; see his letter to Crousaz, 3 October 1717, in Jacqueline E. de La Harpe, *Jean-Pierre de Crousaz (1663-1750) et le conflit des idées au siècle des lumières* (Berkeley and Los Angeles, 1955), 164.

ont été originairement couvertes de l'eau de la Mer qui les a depuis ou insensiblement, ou tout à coup abandonnées." He would not, he declared, postulate shocks that allow water to drain from the earth's surface. Nor would he identify the sea with the Flood, which had been too turbulent an event to account for fossil plants whose leaves were not curled or folded but laid out "de même que si on les y avoit colé (30)."

Jussieu apparently agreed with Fontenelle that life had existed on earth before the appearance of mankind, for he called his fossil-bearing rocks "autant de volumes de Botanique qui dans une même carriere composent, pour ainsi dire, la plus ancienne Bibliotheque du monde (31)." For his part, Fontenelle tried to integrate Jussieu's discoveries into their chronological place in the earth's history. After the first "revolution" which uncovered some land masses, terrestrial plants had begun to exist. This second period could have been a time when various floods occurred, thus transporting plants from the Indies to France. But, Fontenelle concluded, the problem of transport required more research. One would like to know, for example, about the geographical distribution of living species and about the species to which the fossil forms belonged; then one could begin to ask whether the fossils in any one stratum might have come from a single region of the earth (32).

Louis Bourguet's second member of the Academic "school," Réaumur, produced in 1720 a memoir that seems to have startled even Fontenelle, although Réaumur dealt with two topics already broached by Jussieu: an unusual deposit of fossils, and the question of transport. Réaumur's fossils, however, differed from Jussieu's in a significant way. The sheer quantity of marine shells, assembled in the massive faluns of Touraine, would, said Réaumur, prove beyond doubt what had already been known for some time, namely, that at least parts of the present continents had been "& pendant long-temps" at the bottom of the sea (33).

(30) *Mémoires...,* 1718 (1719), 289, 290, 292. Jussieu begins, page 287, with a critique of the way various writers (Woodward, Lhwyd, Leibniz, Scheuchzer, and others) have used the Flood.

(31) *Ibid.,* 289.

(32) *Histoire...,* 1718 (1719), 3-6.

(33) Réaumur, Remarques sur les Coquilles fossiles de quelques cantons de la Touraine, & sur les utilités qu'on en tire, *Mémoires...,* 1720 (1722), 400-416, on 400. A considerable part of the memoir describes how local people quarry the rock and use the shells for fertilizer; this did not interest Fontenelle at all.

Réaumur's description of these faluns includes such salient facts as the horizontal, undisturbed position of both whole shells and fragments, as well as the absence of many of the species from among the living forms found on French shores. Having concluded that the sea is responsible for deposits now 36 leagues from the nearest coast, he then compares with so great an assemblage of shells the much sparser populations found on modern sea floors. This comparison leads him to the further conclusion that a wholly tranquil body of water, however many centuries one may allow it, could not produce so large a population of organisms; the Touraine must therefore have been a gulf, into which marine currents might bring creatures not already living in the gulf itself. The sea might then have retreated suddenly, perhaps because of a crustal collapse; or it might have done so slowly, the rate observable on modern coasts suggesting that a period of 30 to 40 centuries would be sufficient for the sea to have moved 36 leagues (34).

Fontenelle's enthusiasm is evident, and he made sure the reader would see the significance of a discovery he called "une chose étonnante." After summarizing Réaumur's observations, he explained implications not spelled out by Réaumur. Chief among these was that the faluns of Touraine could not have been produced by the Flood, since the shells "n'ont pû être apportées que successivement." Although, he added, "il reste effectivement sur la Terre beaucoup de vestiges du Déluge universel rapporté par l'Écriture Sainte" (what these vestiges are he does not explain), the Touraine deposits give no evidence of the violent, turbulent behavior of flood waters. Putting these discoveries and ideas in the context of his theory of the earth, he concluded:

"Il faut donc ou qu'avant ou qu'aprés le Déluge la surface de la Terre ait été, du moins en quelques endroits, bien differemment disposée de ce qu'elle est aujourd'hui, que les Mers & les Continents y ayent eu un autre arrangement, & qu'enfin il y ait eu un grand Golfe au milieu de la Touraine. Les changements qui nous sont connus depuis le temps des Histoires, ou des Fables qui ont quelque chose d'historique, sont à la verité peu considerables, mais ils nous donnent lieu d'imaginer aisément ceux que des temps plus longs pourroient amener (35)."

(34) *Mémoires...,* 1720 (1722), esp. 411-415.
(35) *Histoire...,* 1720 (1722), 5-9. Réaumur's biographer, Jean Torlais, *Réaumur : Un esprit encyclopédique en dehors de "l'Encyclopédie"* (Paris, 1936), 74, mentions this memoir briefly and quotes a passage that is not to be found in the memoir; the closest approximation to this passage in Torlais is Fontenelle's remark on the need to construct maps of fossil deposits, a remark which ends Fontenelle's discussion of Réaumur's memoir.

298

As he had for Jussieu's memoir, Fontenelle again suggested how future research might chart the past condition of the earth, namely, by the construction of "des especes de Cartes Geographiques dressées selon toutes les Minieres de Coquillages enfoüis en terre."

Although Réaumur did not return to the subject of his memoir, Jussieu was to produce in rapid succession three more studies of fossils, in which the problem of transport continued to puzzle him. In fact, close attention to the language of Jussieu and Fontenelle suggests that a little dialogue was being carried on in public: had there been floods in the past? had the sea changed its basin? could changes of the latter sort be attributed to floods, to exceptional tides, or to "revolutions" of some kind (36)? ("Revolutions," of course, could include floods and tides, but left the door open for other possibilities.)

The several memoirs of Jussieu, Réaumur's study of the Touraine, and Fontenelle's "clarifications" of these and earlier works contain a common core of ideas shared by these three friends. One such idea was their willingness to imagine a long history of the earth, its surface markedly different from the present, and mankind not yet in existence. Jussieu treated these subjects allusively: French soil had an older history than its "immemorial" occupation by man, and fossils were an older "library" than man's books. Réaumur, never apt to make large statements, allowed 30 to 40 centuries for one minor geological event, and this surely implies his readiness to contemplate far longer spans of time. In his role as interpreter, Fontenelle permitted himself to go further than his restrained colleagues, perhaps because early in his career he had already contemplated the vastness of space, a plurality of worlds, and relative perceptions of time. Indeed, he probably took pleasure in saying that the earth's past was a subject for scientists, not for historians, and that scientists of this kind had as their materials "des espéces d'Histoires écrites de la main même de la Nature (37)."

(36) Jussieu, in *Mémoires...*, 1721 (1723), 69-75; 1722 (1724), 235-243; 1723 (1725), 205-210. Summaries in *Histoire...*, 1721, 1-4; 1722, 1-4; 1723, 15-17. The latter text summarizes two memoirs by Jussieu, one of them cited below, n. 38.

(37) The last sentence combines two texts by Fontenelle: *Histoire...*, 1721 (1723), 3-4, and 1722 (1724), 4. In Fontenelle's *Entretiens* the last pages of part five comment on perceptions of time.

The three men also shared an aversion to using the Bible in conjunction with the evidence of nature. Neither Réaumur nor Jussieu would assign to the Flood any causal role, and both deplored some of the uses to which that event had been put. Jussieu, furthermore, produced a study of primitive flint implements that was notably devoid of questions usually asked by his contemporaries — such questions as, how did primitive peoples lose that knowledge of iron which had existed in Noah's day (38)? Here, too, Fontenelle went further than his colleagues in pointing out the non-diluvial message of Réaumur's memoir. In fact, in 1727, when Hans Sloane sent to the Academy a memoir on the discovery in Siberia and elsewhere of the bones of mammoths, Fontenelle did not bother to report in the *Histoire* that Sloane attributed the transport of these animals to the Flood; to Fontenelle, Sloane's evidence pointed to those successive "revolutions" he had been discussing since 1716 (39).

In conclusion, Fontenelle's handling of geological topics allows us to glimpse his own growing familiarity with issues relatively new to French academic circles. More significantly, one may see how, in his discussions of one of the many sciences he had to deal with, he used the forum of the *Histoire* to communicate with the non-specialist and to place narrow, technical articles into larger contexts. For Fontenelle, with his so-called "libertine" background, these contexts were consistently non-Biblical and wholly naturalistic. It is no surprise, of course, to find Fontenelle interpreting scientific research *en philosophe,* but his ability and originality in this task emerge with especial clarity if one compares his writings with the articles in the Academy's *Mémoires.* As a larger result of such an examination, one may suggest that the Academy's *Histoire* deserves more attention than it has thus far received. Decades ago Hélène Metzger reported her impression that Fontenelle's résumés were cited "plus souvent que le travail de l'auteur, et cela par l'auteur lui-même (40)." Random examples might be men-

(38) Jussieu, De l'origine et des usages de la pierre de foudre, *Mémoires...,* 1723 (1725), 6-9. In Robert F. Heizer (ed.), *Man's Discovery of his Past: A Sourcebook of Original Articles* (Palo Alto, 1969), pages 14-21 discuss the problem of interpreting such implements in Jussieu's day, and pages 78-80 give Jussieu's text in English translation.

(39) *Histoire...,* 1727 (1729), 4. Sloane, Mémoire sur les dents et autres ossemens de l'éléphant, trouvés dans terre, *Mémoires...,* 1727, 305-334.

(40) Hélène Metzger, La littérature scientifique française au xviiie siècle, *Archeion,* XVI (1934), 12.

300

tioned to support at least the first part of Mme Metzger's statement (41), so that the *Histoire,* at least in Fontenelle's day, ought to be studied as a vehicle for the dissemination of ideas only obliquely visible in the pages of the *Mémoires.*

(41) In Buffon's theory of the earth (*Histoire naturelle,* vol. 1), for example, long quotations from the Academy's publications are generally from the *Histoire,* not the *Mémoires.* Unacknowledged echoes also turn up, as when Pierre Barrère, *Observations sur l'origine et la formation des pierres figurées* (Paris, 1746), 51, refers to "des especes d'Histoires écrites, pour ainsi dire, de la main même de la nature" (cf. Fontenelle's phrase, quoted above, at n. 37). With less excuse, modern historians have sometimes used the *Histoire* as if it were identical with the *Mémoires*; see the remark on Torlais, above, n. 35, and the comment on Pierre Brunet by Henry Guerlac, *Essays and Papers in the History of Modern Science* (Baltimore, 1977), 488, n. 1.

IX

Leibniz on Geology: A Newly Discovered Text

Although Leibniz never quite completed his most ambitious geological treatise, the *Protogaea* (published posthumously in 1749), he did during his career submit for publication three or perhaps four short papers on geological topics. In chronological order, the earliest, an outline entitled *Protogaea*, appeared in the *Acta Eruditorum* (Leipzig) in 1693. As a possible second, a letter to the Royal Society of London in 1697 may have been intended for the *Philosophical Transactions*, but it was not printed. The fourth, a commentary on C. M. Spener's fossil 'crocodile', was published by the new Berlin Academy in its first volume of *Miscellanea* (1710). Problematical for historians is the third text, dated 1706 and sent to the Academy of Sciences in Paris – the original has never been found, the only known version being the abridgment prepared by Fontenelle and published in the Academy's *Histoire* for 1706. Given Fontenelle's habit of commenting upon texts he was ostensibly summarizing, to distinguish his contributions from Leibniz's has long been a desirable goal, if a seemingly unattainable one[1].

---

1 Published texts as follows: *Acta Eruditorum*, January 1693, pp. 40-42; *Histoire de l'Académie royale des sciences*, 1706 (1707), pp. 9-11; *Miscellanea Berolinensia* I (1710), pp. 118-120. Also, MS Journal Books of the Royal Society of London, under date 17 November 1697. A useful analysis is by B. Sticker: *Leibniz' Beitrag zur Theorie der Erde*, in: *Sudhoffs Archiv* 51, 3 (1967), pp. 244-259. Geological materials in Hannover have been used by G. Scheel: *Leibniz historien*, in: *Leibniz. 1646-1716. Aspects de l'homme et de l'œuvre. Journées Leibniz, organisées au Centre International de Synthèse les 28, 29 et 30 mai 1966*, Paris 1968, pp. 45-60 and J. Roger: *Leibniz et la théorie de la terre*, in: ibid., pp. 137-144. Historians who have failed to find the 1706 text in Paris include A. Birembaut: *Fontenelle et la géologie*, in: *Revue d'histoire des sciences* X (1957), pp. 360-374 (at p. 365, n. 1), and R. Rappaport: *Fontenelle Interprets the Earth's History*, in: ibid. XLIV (1991), pp. 281-300 (at p. 287, n. 13).

7   Leibniz on Geology: A Newly Discovered Text

Leibniz's text of 1706 has been sought in both Paris and Hannover, and a transcription has now been found in the Archives of the Academy of Sciences in Paris. To understand how this text could have lain hidden requires some explanation of the organization of the Archives. Most members and correspondents of the Academy are represented by so-called 'biographical' dossiers. One would thus naturally hope to find the Leibniz text in his dossier. In addition, the Academy has 'pochettes' containing materials pertinent to single meetings and sometimes to the proceedings of an entire calendar year; such files are not very ample for the early decades of the Academy's existence. Finally, there are the minutes of meetings, the 'Registres des procès-verbaux des séances'. To historians familiar with these large tomes, turning the pages in search of the texts of articles would probably seem futile, since the common practice in the Academy was merely to record in the minutes that someone presented a paper on a particular subject; relatively few such presentations – perhaps no more than one-third of those papers read at the meetings – were transcribed into the minutes. As it happens, Leibniz's text is one of these few. We now have, therefore, not the original manuscript but a transcription of what Fontenelle read to the Academy on 4 September 1706. That text is appended here, published with the authorization of Mme Christiane Demeulenaere-Douyère, 'conservateur en chef' of the Archives.

Commentary in the next paragraphs will first address the content of the text and then Fontenelle's editing.

From beginning to end, despite apparent digressions, Leibniz's main concern was to undermine the concept of "jeux de la nature" ("lusus naturae", sports or tricks of nature). In this he was hardly alone, as other naturalists like Nicolaus Steno, Robert Hooke, and John Woodward had also argued that fossils, particularly marine shells, were the remains of organisms rather than the products of playful natural forces or powers[2]. Like his contemporaries, Leibniz knew that some rocks do show peculiar 'images' – dendritic markings, for example – and these he was willing to attribute to natural powers of a sort. His knowledge of chemistry here came to his aid, as he suggested that "l'arbre metallique" (or tree of Diana, i. e., crystallized silver salts) provided a ready analogue of processes used in nature. More easily explained were 'images' of identifiable organisms, as Leibniz devised an experiment to show how such impressions could be left in the rocks while the remains of the organisms themselves had otherwise completely vanished.

Less explicit in Leibniz's text is concern with the Biblical Flood. As he indicates, finding shells on high mountains, far from modern seas, suggests that

2   The best analysis of these issues is by M. J. S. Rudwick: *The Meaning of Fossils. Episodes in the History of Palaeontology*, London – New York 1972, pp. 1-100 (chs. 1-2). Discussion of "lusus" occurs in the *Protogaea* (see *G. W. Leibniz: Protogaea. De l'aspect primitif de la terre et des traces d'une histoire très ancienne que renferment les monuments mêmes de la nature*. Texte latin et traduction B. de Saint-Germain. Édition, introduction et notes J.-M. Barrande, Toulouse 1993, pp. 66-71, 74-101), sect. XVIII, XX-XXIX.

'almost the whole earth' was once under the sea[3]. What he does not explain is that the remains of large animals, such as the elephant, were then commonly being attributed to diluvial transport. His allusion here to a discovery in Thuringia signals a topic familiar to many contemporaries and of concern to Leibniz himself for a decade before 1706. In 1696 Wilhelm Ernst Tentzel had published a pamphlet describing his discovery, arguing that these bones, tusks, and grinders resembled those of the elephant, and concluding that their presence in Germany should be attributed not to transport by ancient Roman legions but to the Deluge. Leibniz had quickly written to Tentzel himself and to the Royal Society of London, proposing what in 1706 he would also communicate to Paris: these remains might belong to marine or amphibious animals once native to Europe. As he wrote to Fontenelle, it was possible that "la Longueur du tems" had produced changes in the species with which we are now familiar[4]. He would repeat this suggestion in 1710 in connection with Spener's fossil crocodile.

In abridging Leibniz's text for publication, Fontenelle remained faithful to the main point: the rejection of the concept of "lusus naturae". He recounted, too, Leibniz's comments on insects in amber and his 'curious' experiment to reproduce the impression of a spider. At the same time, he simplified the treatment of "lusus" – omitting, for example, the metaphysical remarks at the end of the original text – to emphasize one basic point: nature would 'play' more freely and would not confine itself to precise duplication of even the smallest details of organic structures. For one detail, however, Fontenelle changed the focus of the text, in connection with those fossil plants that Leibniz had described as resembling those 'brought from the Indies'. To Leibniz, this detail meant that the fossils had been genuine organisms; to Fontenelle, it was important to explain transport from the Indies to the slates of Germany[5].

Omitted by Fontenelle is much of the detail that testifies to Leibniz's experience and to his familiarity with the writings of earlier naturalists. Understandably, omissions include virtually all German place names, as well as personal names and technical terms that would have been unfamiliar to his readers. Less easy to explain, also missing is all reference to fossil bones, although Tentzel's discovery had earlier been given some attention in the *Journal des sçavans*; but members of the Academy had not yet shown any interest in this topic, and Fontenelle's own education in geology had barely begun[6]. Indeed, Leibniz's text, as presented by Fontenelle, now dealt wholly

---

3 See the *Protogaea* (see Barrande, ibid., pp. 26-33), sect. VI, on the Flood.

4 See W. E. Tentzel: *Epistola de sceleto elephantino tonnae nuper effosso* (Gothae 1696), text reprinted in *Philosophical Transactions* 19 (1697), pp. 757-776; letters to Tentzel, 1696, in: A I, 13, 204-205, 293-294, 345-347; letter to Royal Society (see note 1).

5 It has been argued that Fontenelle added the Indies to his account of Leibniz, largely because no such reference occurs in the *Protogaea*. See M. Carozzi: *Voltaire's Attitude Toward Geology*, in: *Archives des sciences* 36 (1983), pp. 3-145 (at pp. 26-27).

6 See review of Tentzel, *Journal des sçavans*, 20 August 1696, pp. 393-395. There was a

with the much-debated plants and animals common in marine and freshwater sediments. His initial sentence informs the reader that this topic had recently been broached in the Academy's *Histoire* for 1703, and Leibniz was confirming the analysis that had appeared at that time.

Text[7]

J'ay Lû L'Ecrit suivant que j'avois reçû de Mr. Leibnits
Sur les Pierres figurées.

Il y a plusieurs savans auteurs entêtez de ce qu'ils appellent j e u x d e l a n a t u r e . Ils y rapportent les figures des Plantes et des Animaux ou de leurs parties, qui se trouvent sous la Terre, et particulierement dans les pierres. Et ils ne sauroient comprendre comment les Coquilles de Mer et autres dépoüilles des animaux marins soyent montées sur les hautes montagnes dans les Lieux éloignez aujourdhuy de la Mer. Et comme on rencontre quelque fois des ossemens des animaux inconnus, ils y croyent trouver la confirmation de leur sentiment. Mais il se peut que la Longueur du tems ait fait un grand changement dans les especes que nous connoissons.

Il y a une Grotte fameuse (B a u m a n s h ö l e) dans Les Montagnes du Harz, d'où l'on tire assez souvent de ces ossemens incorporez dans une espece de Talc, ou dans d'autres pierres formées à l'entour des os. Il y en a aussi, qu'on trouve simplement dans la terre, ou dans les carrieres ordinaires comme ce qu'on a déterré à Tonna en Thuringue, où toutes les apparences ont fait juger qu'il y avoit eû là un Elephant, ce qui s'est trouvé aussi en plusieurs autres endroits de l'Europe; sans qu'on le puisse bien rapporter à quelques Elephants amenez de l'Affrique ou de l'Asie; et on pourra plûtôt soupçonner, qu'ils ont habité autres fois dans nos païs, à moins qu'on ne croye, que ce sont des reste [sic] des animaux marins approchans des terrestres, dont il y en devoit avoir bien plus autres fois (quand presque tout étoit sousmergé) qu'il n'y en a presentement. Et on a trouvé prés de Wolfenbutel des machoirs d'une grandeur pareille, mais d'une autre forme que de celle de l'Elephant. Feu Mr. Guerike parle aussi de l'esquelette d'un animal inconnu trouvé dans le païs d'Halberstat. Il y a une Montagne tout au prés de Lunebourg, nommée la montagne de Chaux (K a l c k b e r g) où l'on déterre des glossopetres comme à Malthe, qu'on sait être les dents des Lamies ou du Canis carchariae; quoy que feu Mr. Reiskins, qui a fait un Livre de nos Glossopetres, ait encore voulu soûtenir que

---

little more interest in geology in the early Academy than I realized in Rappaport (see note 1), pp. 283-286, but the additional fragments do not substantially alter the argument presented there.

7   From the Archives of the Academy of Sciences, Procès-verbaux des séances, t. 25, fols. 336r°-338v°, under date 4 September 1706. Abbreviations are spelled out, so that, for example, -mt is rendered –ment. The "Je" in the first line is the secretary, Fontenelle.

ce sont de ces L u s u s n a t u r a e , dont le P. Kircher, Beckerus, et autres ont été prevenus[8].

Plus on considére ces choses, plus on est obligé de croire, que presque tout nôtre Globe, excepté peut être les plus hautes montagnes, a été un jour couvert de la Mer; et qu'il peut avoir été découvert depuis, lors qu'une grande partie de l'eau a trouvé un passage pour entrer dans quelques abymes creux, qui étoient au dedans du Globe. Je ne dis rien maintenant de l'origine de la Mer, qui s'explique aisément dans mon hypothese. Mais j'expliqueray icy la generation des figures des Plantes et des Animaux qui se trouvent sur l'ardoise, parfaitement bien tracées; en sorte qu'il est aisé de reconnoître les especes, et leur moindres parties; ce qui a fourni aux deffenseurs des jeux de la nature leur plus apparentes instances, et servira aussi le plus à les desabuser, lors qu'on en verra la vraye origine.

Il faut donc savoir, que dans le païs de Bronsvic aux environs d'Osteroda, dans la Comté de Mansfeld aux environs d'Eislebe, et en beaucoup d'autres endroits, on trouve des veines de la pierre d'ardoise, horizontales, ou à peu prés; où il y a des representations des Poissons de toute sorte. Et on en reconnoît non seulement les figures en gros, mais les écailles, les barbes, les nageoires, et jusqu'aux moindres traces de l'extérieur de l'animal. Ces tracent [sic] consistent souvent dans un mélange de Cuivre, qui contient même de l'argent. Il y a d'autres endroits, où l'on voit un grand nombre de plantes desseignées sur la même ardoise par de semblables traits metalliques. La plûpart des Plantes sont reconnoissables jusqu'aux moindres filamens; et il y en a parmy, qui ne se trouvent plus dans ce païs cy, mais dont en rencontre les representations parmy les figures des plantes apportées des Indes. Un de mes amis, nommé Mr. Heyne, en a formé un Cabinet, et on donnera bientôt des Essais fort curieux de cette B o t a n i q u e S o û t e r r a i n e , aussi bien que de cette L i t h i c h t h y o - l o g i e [9]. Et il est remarquable que la grandeur naturelle du Poisson ou de la Plante paroit sur la pierre, c'est à dire, la longueur et la largeur; car l'epaisseur n'y sauroit être, l'animal ayant été comprimé.

Voicy comme j'en conçois la génération; il paroît qu'une espece de terre a couvert des Lacs et des préz, et y a enseveli des poissons et des plantes, ou bien quelqu'eau bourbeuse et chargée de terre les a envelopez ou emportez. Cette terre a été endurcie depuis en ardoise, et la longueur du tems ou quelqu'autre cause a détruit la matiére peu durable du poisson ou de la plante; à peu prés comme l'on a trouvé que les corps des Mouches, des Araignées, et d'autres animaux, qui paroissent dans le Carabé ou ambre jaune ont été dissipez, et ne

---

8 The allusions to Becher, Guericke, and Kircher are identified in detail in Barrande's notes to the *Protogaea* (see note 2), pp. 239, 242-243. "Reiskins" was J. Reiske (Reiskius): *Commentatio physica aeque ac historica de glossopetris luneburgensibus*, Norimbergae 1687; I have not seen what appears to be an earlier edition, published in Leipzig in 1684.
9 Presumably Friedrich Heyne, but conceivably his father, Michael Martin. References to both men occur in Leibniz's letters. I have been unable to discover any evidence that either man published a catalogue of a natural history collection.

presentent plus rien de palpable, quand on casse le morceau, ne consistant qu'en simples délineations, qui ne sont perceptibles qu'aux yeux. Or la matiére du poisson ou de la plante étant consumée et dissipée pareillement, a laissé sa forme empreinte dans l'ardoise par le moyen du creux, qui en est resté. Et ce creux a été rempli enfin d'une matiére metallique; soit qu'un feu soûterrain effectif cuisant la terre en ardoise, ait forcé le Métail mêlé dans la terre d'en sortir, et d'entrer dans le creux; soit qu'une vapeur metallique pénétrant l'ardoise, se soit fixée dans ces mêmes creux. Et l'on peut éclaircir cet effet par une operation mechanique assez curieuse. L'on prend une araignée ou quelqu'autre animal convenable, et l'on ensevelit sous de l'argille, en sorte pourtant qu'on fasse rester une ouverture, qui entre du dehors dans le creux, ce qui se peut obtenir de plusieurs façons; et puis on met la masse au feu pour l'endurcir; la matiére de l'animal va en cendres, qu'on fait sortir par le moyen de quelque liquide, et puis de l'argent fondu y est versé par la même ouverture; lequel étant refroidi et la masse cassée avec quelque precaution, on trouve la figure de l'animal assez bien representée en argent. La disposition aussi de la veine, qui est horizontale à peu prés, S c h w e b e n d e r g a n g , comme nos gens des mines l'appellent (ce que le nom de veine pendante n'exprime pas assez bien) confirme cette origine, puisque les lacs et les prez approchent aussi du plan horizontal. Enfin si cette délineation n'étoit pas tirée sur le naturel, elle ne garderoit pas si exactement la grandeur même de l'original aussi bien que sa figure. Quand cette exactitude ne s'y trouve point, je consens plus aisément, que ce sont des jeux; comme ces petites figures d'arbres sur le Marbre, qui se trouve prés de Florence et ailleurs, dont on orne des Cabinets[10]. Et la production de ces figures par une Liqueur qui coule et se partage, n'est pas difficile à concevoir, c'est à peu prés comme fait l'huile sur une table de bois ou d'autre matiére, ne trouvant pas partout la même facilité de couler, ou le hazard produit quelques fois des figures assez curieuses. Sans parler de la génération de l'arbre metallique dans de l'eau forte, qui est encore une autre maniére de jeu, ou la nature se plait à imiter par une simple mechanique, mais fort imparfaitement la production des corps organiques; laquelle demande plus de mistére, puis qu'elle suppose une p r é f o r m a t i o n , sans laquelle point de mechanique ne sauroit produire une machine organique qui soit organisée dans les parties des parties à l'infini; ce qui fait une distance immense entre ces machines de la nature, et celles de l'art; aussi bien que entre les Animaux ou les Plantes, et entre ces jeux de le nature, dont je viens de parler, qui n'en sont qu'une imitation grossiere et au dehors.

---

10 See A. Balfour: *Letters Write* [sic] *to a Friend*, Edinburgh 1700, p. 107, who wants his friend to acquire in Florence those kinds of stones "whereof the one Naturally represents Townes and Landscapes, the other [...] Trees and Forrests". In British texts of the period, a common locution is 'landskips' in stone.

X

GEOLOGY AND ORTHODOXY: THE CASE OF NOAH'S FLOOD IN EIGHTEENTH-CENTURY THOUGHT *

THE view that religious orthodoxy stifled geological progress has had many distinguished exponents, one of the earliest being Georges Cuvier. To Cuvier, however, efforts to combine Genesis with geology ended before the middle of the eighteenth century, and opened the way not for progress but for wild speculation. We may admire the genius of Leibniz and Buffon, he declared, but this should not lead us to confuse system-building with geology as 'une science positive'.[1] While Cuvier's younger contemporary, Charles Lyell, agreed that 'extravagant systems' had retarded progress, he insisted that 'scriptural authority' had had a similar effect until late in the eighteenth century.[2]

In his interpretation of history, as in much else, Lyell has for a long time prevailed over Cuvier. Although modern scholars have rejected or modified many of Lyell's historical judgements, it is still possible to find the eighteenth century described in Lyellian terms:

> [By 1785] many geological observations had been made and recorded in the literature; but previous attempts to synthesize these observations into a general 'theory of the earth' were unscientific and had not proved acceptable. The issue had been confused and progress retarded by a literal belief in the biblical account of the creation and the universal flood.[3]

It has even been said that geologists—many of them clerics—were appalled by their own observations, and, fearing to be heterodox or fearing punishment for heterodoxy, 'scurried off in strange directions, or returned to earlier [i.e. more orthodox] positions'.[4]

To scholars outside the history of geology, the idea that the eighteenth century was a period of 'scurrying off in strange directions' to avoid heterodoxy will rightly seem bizarre. For the physical sciences, the condemnation of Galileo had proved to be a victory for the condemned, and Catholic apologists were reaching the Galilean conclusion that the Bible could not be used as a source of physics or astronomy. Scientists themselves had become self-consciously empirical and determined to seek explanations in terms of natural causes and uniform laws. Nor is much scurrying to be detected even among the genuinely heterodox philosophers, who found numerous ways to get their subversive views into print. Furthermore, to discuss heterodoxy at all assumes the existence

* An abstract of this paper was presented on 8 April 1976 at a meeting of the American Society for Eighteenth-Century Studies, at the University of Virginia, Charlottesville, Virginia.

of a well-defined orthodoxy, and to discuss religious scruples or timidity implies the existence of well-oiled machinery for intimidation and censorship. But historians who deal with these aspects of eighteenth-century life often stress the secularism of the clergy itself, have uncovered doctrinal conflicts even in state churches with hierarchies capable of defining dogma, and continue to argue about the efficiency of the machinery for repression.

Although our general picture of eighteenth-century intellectual life differs somewhat from that often suggested by historians of geology, Lyell's interpretation does have some basis in fact. One undeniable fact is that some geologists—those usually cited are British writers of the first and last decades of the century—explicitly sought to defend Genesis. It is undeniable, too, that virtually all geologists discussed the biblical flood and that many of them attributed to the flood some role in the history of the earth's crust. To say this much, however, should not necessarily imply that geologists felt constrained to demonstrate their orthodoxy; mere use of the flood tells us nothing about religious convictions or geological competence unless we ask such questions as: Why and how did geologists treat the flood? To what extent did geological observation—and Lyell did not query the observational skills of his predecessors—support the notion that such a catastrophe had occurred?

The examination of these questions requires some preliminary discussion of what might be called 'the status of Genesis' in the eighteenth century: was there an 'orthodox' interpretation? Rather than attempt an impossible survey of all religious denominations, my emphasis will be on Catholic France. Not only was France—more accurately, Paris—the acknowledged intellectual capital of Europe, but there were also to be found in France all those conditions needed for the suppression of philosophical speculation: a vigilant, conservative Faculty of Theology at the University of Paris, elaborate machinery for censorship, and an established church with hierarchically imposed dogma and discipline.

For the geologists, the two relevant biblical texts are the narratives of the creation and the flood. I have chosen to deal with the flood because, while many geologists avoided cosmogony, few could avoid discussing the deluge and its possible effects upon the earth's crust. To what extent did geologists accept the historicity of the biblical account? Why was the flood, sent to punish mankind, often considered a natural event with geological consequences? What geological effects, in fact, was the flood supposed to have had?

This article does not pretend to survey all the geological literature of the period. An effort has been made to select works ranging in date and in country of origin, and written by men with some international reputation. The latter criterion dictated the neglect of some writers

Geology and Orthodoxy   3

(Raspe, for instance) and the use of others who were widely read but not as widely admired for their scientific expertise (Buffon being the obvious example). Some of the writers discussed—Linnaeus is one— earned their reputations not as geologists but as botanists or chemists. Unlike these older disciplines, however, geology was not a well-defined speciality in the eighteenth century, and so it seems worthwhile here to include those writers whose ideas carried the stamp of authority even when they ventured from their specialties into the amorphous field of natural history, of which geology was one branch. Finally, no attempt has here been made to give pride of place to such figures as Werner and Hutton, since the object of this article is not to determine which general theory excited most controversy or convinced most thinkers; instead, because the flood was discussed in connexion with virtually every theory, it has seemed more relevant to focus on this issue as it was treated by an international community of scientists.

*   *   *

In *The legend of Noah*, Don Cameron Allen provided a classic account of sixteenth- and seventeenth-century attempts to fill in the gaps, explain away the anomalies, and supply scientific details to substantiate the story of the flood. To the modern reader, the questions asked by these rationalists are both delightful and startling: Could one design an ark sufficient to house all the refugees from the flood? Could fish survive in the flood, or did they, too, have to be passengers in the ark? Were peculiarly American species in the ark, or had the flood not reached the western hemisphere? More significant for this article is Allen's contention that rationalist analyses produced so much confusion and disagreement and sheer doubt about the literal accuracy of Genesis, that the flood had either to remain a miracle, or be reduced to the status of one of many ancient legends. Allen argued, too, that rational explication was primarily a Protestant concern, Protestants seeking to understand 'the plain words of Scripture' while Catholics more readily resorted to miracle or allegory.[5] Allen to the contrary notwithstanding, Catholic clergy in eighteenth-century France shared some of the concerns of the earlier rationalists and had not reached a miracle-or-legend consensus about the flood. Nor, in fact, had they arrived at a generally accepted mode of interpretation of Genesis.

The old tradition of literal interpretation is often said to have been dominant during this period in France, and its representatives included men as different as the Oratorian Richard Simon, the Benedictine Augustin Calmet, and the doctors of the Sorbonne. Generally considered most 'typical' is Dom Calmet, whose verse-by-verse commentary aims at a more precise understanding of the Bible on the basis of philology

4

and ancient history. When treating the flood in 1707, Calmet was with obvious reluctance drawn into a brief discussion of problems raised by earlier rationalists, but he preferred to consider the event a miracle. Too much analysis, he believed, would destroy the unity and integrity of the Pentateuch.[6]

Calmet's older contemporary, Simon, also considered himself a literal expositor, and he believed that the obscurities, repetitions, and contradictions in Genesis could be explained as the historical result of compilation from older sources. Simon's scholarly efforts in the 1680s to show that the Pentateuch is a compilation rather than wholly the work of Moses aroused the wrath of censors, Jansenists, Oratorians, and Bishop Bossuet.[7] If this kind of literalism provoked a charge of 'impiety and libertinism' in the late seventeenth century, the same cannot be said of the mid-eighteenth. One of Simon's few followers, the Montpellier physician Jean Astruc, created no such storm in 1753, and reviewers of his work expressed opinions ranging from praise, to reservation of judgement, to the denial that his arguments were conclusive.[8] That the Sorbonne ignored Astruc is worth noting, since it is often said that the theologians had only recently been roused to extreme vigilance by the scandalous thesis of the abbé de Prades. If the Faculty's silence about Astruc cannot be explained, one may at least suggest that the doctors were less vigilant and less united doctrinally than is usually alleged.[9]

Although the literal tradition was strong, it was neither united nor unrivalled in France, and the rationalist tradition continued to be an important rival. One member of this school, the abbé Pluche, was the author of the widely read *Le spectacle de la nature* (1732), a derivative work of natural theology.[10] More interesting than Pluche were other rationalists who, like Simon and Calmet, realized that variant texts of Genesis resulted in different systems of chronology; some were aware, too, that all existing chronologies might be undermined by historical and scientific evidence. The abbé LeMascrier, for example, declared that in some of these matters the Bible is not 'our sole and unique guide', and he was willing at least to countenance the theory that the earth might be millions of years old.[11] The abbé Grosier similarly found no fault in interpreting the six days of creation as six epochs should the scientific evidence so warrant—but he did not think the available evidence conclusive.[12] LeMascrier and Grosier, it should be added, were dealing with three of the most 'subversive' chronologies of their day: De Maillet's *Telliamed* (1748), Mirabaud's *Le monde* (1751), and Buffon's *Les époques de la nature* (1778).

These few examples clearly suggest that, in France at least, there was no single, established, approach to Genesis. If anything united the 'orthodox', this was the belief that God had revealed Himself to man and that the fact of revelation was the basis of the authority of the

Catholic Church.[13] The implication of such belief for the story of Noah was simply that the flood had been sent to punish mankind. Whether the flood was universal or confined to regions inhabited by man; whether it was a miracle or produced by natural causes; whether it had appreciably changed the surface of the earth; whether the earth had had a long antediluvian history—all these matters remained subject to debate.

Diversity of opinion among the clergy is all the more noteworthy in a period when believing christians might have been expected to close ranks against the radical French philosophers. The views of Voltaire *et al.* are too well known to require discussion here, although it seems appropriate to rehearse briefly some direct and indirect attacks upon the early chapters of Genesis. A good many philosophers were producing wholly nonbiblical accounts of the history of primitive man, the origin of language, and the evolution of civilizations, but few would have gone as far as Rousseau did in 1755 in explicitly pointing out their divergence from Genesis:

> Religion commands us to believe that since God Himself took men out of the state of nature immediately after the creation, they are unequal because He wanted them to be so; but it does not forbid us to form conjectures, drawn solely from the nature of man and the beings surrounding him, about what the human race might have become if it had remained abandoned to itself.[14]

Rousseau's version of what 'religion commands' is in itself controversial, since theologians disagreed about what constituted a 'state of nature' and whether such a state was descriptive of conditions before or after the fall. Philosophers, too, failed to agree about whether the state of nature had been brutish or blissful, but their reconstructions of origins had little in common with the Mosaic account of the creation of Adam, the fall, the special status of Hebrew as the language of revelation, and the story of the tower of Babel. Boulanger might use the flood in his *Antiquité dévoilée* (1766), but his account of the disaster and of its effects on mankind shows no resemblance to Genesis.[15]

However difficult it is to define orthodoxy among theologians, much of the work by French men of letters was clearly heterodox, by any standard. In a country possessing an elaborate system of censorship, heterodoxy developed an equally elaborate system of evasion of the law, and historians of the Enlightenment know much about such techniques as anonymity, false attribution of authorship, false imprints, the smuggling of tracts from Holland or England, the clandestine circulation of manuscripts, and the pious disclaimer attached to the subversive book. In addition, evasion was helped because the will to censor was subject to variation and because the system itself was as anarchic as it was arbitrary. The royal corps of censors, the Sorbonne, and the Parlement of Paris all behaved erratically for a variety of reasons, not least among them the fact that these bodies contained individuals who sympathized with some of the radical ideas propounded by the men of letters.

While historians agree that the system was cumbersome and inefficient, there remains some divergence of opinion about the effects of censorship. Should one emphasize the philosophers' fear of persecution or the fact that their writings did get into print? Certainly, it is the easier course to stress the famous cases of imprisonment, suppression, confiscation, and condemnation. The very nature of the problem should make us wary, however, since we obviously know more about the reasons behind condemnation than we do about the reasons behind the failures to molest. In the context of this article, the example of Buffon is instructive: although the Sorbonne condemned parts of his *Theory of the earth* (1749), and although we know that the learned doctors took issue with his *Epochs of nature* (1778), why Buffon was not censured after 1778 remains a matter for conjecture.[16]

The preceding pages are intended to cast doubt on the view that French naturalists had to contend with a well-defined orthodoxy, enforced by religious and civil authorities. A similar analysis for England or Prussia or certain Italian states would probably also yield varying tones of gray rather than sharp definitions. In England, for example, the established Church was so 'comprehensive', so latitudinarian, that orthodoxy would be hard to define in any but the broadest terms.[17] For Prussia, it is gross simplification to claim that heterodoxy was limited 'mainly' to 'anticlerical jokes', especially in a period when members of the Berlin Academy were being permitted to publish speculation as subversive as Rousseau's.[18] For all countries, including France, analysis is further complicated by changes in policy and practice during the course of the century. In France, royal censors became more erratic after 1750, while they became more vigilant in Prussia after the death of Frederick II in 1786. After 1789, the revolution in France affected writers and censors everywhere, arousing both admiration of France and fear of French materialism. Problems such as these suggest that it would be unwise to assume that there existed widespread, well-defined, long-lived constraints upon naturalists in particular. If most naturalists discussed the biblical flood, their reasons for doing so should be sought elsewhere than in coercion or timidity in the face of authority.

*   *   *

For the purposes of this article, the story of Noah, interpreted literally, has three salient features: the flood was universal, covering even the tops of the highest mountains; there were forty days during which the waters increased, pouring out upon the earth from the 'great deep' as well as from the heavens; and the whole duration of the flood was less than one year. On the universality of the flood, there is a large literature by scientists and laymen (i.e. non-scientists, whether clerical or lay). The turbulence and duration of the event were of special concern to scientists.

The universality of the flood had been the subject of ancient debate, revived with vigour in the eighteenth century. Scientists and laymen all were aware of the existence of flood legends in many civilizations; but were these legends all memories of the biblical deluge, or were all, including the story of Noah, memories of purely local disasters? With rare exceptions, the naturalists tended to use flood legends as evidence for a once-universal catastrophe. Among non-scientists, many influential writers used the same evidence to suggest that there had never been a universal flood. The former harmonized with ancient christian tradition (and, as I hope to show, with some geological evidence), while the latter view was shared by people ranging from an Anglican bishop to Voltaire.

Robert Clayton, Bishop of Clogher, was the author of more than one tract in defense of biblical chronology and related subjects. While much of his evidence was based on the conformities between sacred and pagan histories, he was also a spiritual descendant of those rationalists who had wondered whether uniquely American fauna had been passengers in the ark. He concluded that the flood need not have been universal, as long as it was sufficient to destroy all mankind but for the family of Noah.[19] No such pious intent can be detected in the work of several philosophers who also denied the universality of the flood. A good example of their reasoning comes from J.-B. Mirabaud, Secretary of the *Académie française*, in his explanation of the psychology underlying the Greek legend of the flood of Deucalion:

> In those rude days (*temps grossiers*), men . . . knew only those parts of the world surrounding them and judged the rest by their surroundings. Thus it is that the first inhabitants of Greece told themselves that a flood which had affected them had demolished the whole human race; and it is probably in the same way that Noah, protected in his ark, . . . borne by the waves to a normally uninhabited region, or to where the inhabitants had died during the same catastrophe, thought that everyone not with him had been swallowed up by the waters.[20]

Variants of Mirabaud's message were being spread by Voltaire and the encyclopaedists. In the *Philosophical dictionary*, Voltaire summed up Genesis as a collection of folk tales not unique to the Hebrews, and went on to state that a universal flood was a scientific impossibility. We should, he added, accept the deluge as a miracle, but some pages later shrugged off miracles which always, suspiciously, originate among illiterate peoples.[21] In 1754 in the *Encyclopédie*, Boulanger implied comparable conclusions, weighing both the legends of floods and the scientific evidence for and against a universal flood.[22] The most competent scientist among the philosophers, the Baron d'Holbach, repeatedly denied that geological phenomena could be explained by the flood; in his polemical *System of nature* (1770), he went on to question the flood's universality.[23]

The few naturalists to argue that the flood had been a local disaster included De Maillet, Buffon, J.-C. de Lamétherie, J. G. Sulzer, and J. F. Blumenbach. All considered a universal flood to be a scientific impossibility—their reasons show some variation in detail—and all but Blumenbach felt obliged to discuss and reinterpret ancient flood legends, much in the fashion of Mirabaud. This did not prevent the three Frenchmen from espousing a theory (to be discussed below) of a universal ocean different from the flood, or Blumenbach from suggesting that the earth had undergone two world-wide catastrophes, and such theories were held by other geologists who often continued to believe that there had been a universal flood. In other words, one may fairly say about four of these five naturalists that they were willing to grapple with problematical concepts of world-wide change, but they objected to the particular kind of world-wide event described in Genesis.[24] In this they were not alone, for, as we shall see, those naturalists who made use of the flood often departed considerably from the biblical text.

Most naturalists did accept the historicity of the universal flood and turned their attention to what was, for them, the more immediate issue: had the flood had any geological effects? In particular, since geological strata had once had to be 'soft' in order for fossils to become embedded in them, was the flood responsible for the dissolution of large parts of the earth's crust and the deposition of marine fossils? The problem of marine fossils divided naturalists into three camps: those who considered them to be relics of the flood, those who allowed some lesser role to the flood, and those who denied to the flood any such geological role.

Parenthetically, it should be noted that the fossil problem often highlighted but was not always an integral part of discussions of the flood. The two issues were closely linked early in the eighteenth century, especially in the writings of natural theologians, and again late in the century when explanations were sought for the burial of large mammals. Fossil specimens of the latter sort were, of course, known much earlier, but they were rare enough—and often so poorly identified—that they could be safely ignored or explained away as the victims of some purely local disaster. But additional specimens found in several parts of the world made the problem an urgent one after about 1770. As the following pages will show, debaters often did ignore fossils and concentrated instead on whether sediments of some thickness could have been deposited by the flood or whether significant nonconformities in the strata might be explained as the result of such a cataclysm. Nonetheless, since debate about the flood was at first linked with the fossil record, it is appropriate to begin with the issues then posed.

Among those who considered fossils to be relics of the deluge were the Swiss naturalist J. J. Scheuchzer and the French author of a widely read guide to fossils, Dezallier d'Argenville. Scheuchzer had been converted to diluvialism by the English geologist and natural theologian John Woodward, and he was to remark in 1731 that the aim of all his scientific work had been to find natural evidence for the truths of Scripture.[25] Writing later, Argenville was well aware that there were serious objections to the Woodwardian theory, but he believed alternative explanations to be even more problematical and 'hypothetical'. By mid-century, in fact, Woodward's views had been so severely undermined that they could no longer be accepted even by those naturalists who considered the flood to have played a major role in sedimentation.[26]

One anti-Woodwardian argument repeatedly raised was that the strata of the earth's crust are not arranged in order of their specific gravity, as they ought to be if all were deposited in a universal flood; and, as a later writer put it, 'the whole of Dr Woodwards [sic] theory of the Earth hangs on this affirmation'.[27] Just as fundamental was the objection that the flood must have been turbulent, and sedimentary strata show every sign of having been laid down under tranquil conditions. In 1720 a third argument joined these two, when Réaumur presented to the Paris Academy of Sciences a memoir with the innocuous title, 'Remarks on the fossil shells of some parts of Touraine'. The memoir records his observations of tertiary faluns of gigantic dimensions: more than 30,000 square metres in area and about 7 metres in thickness. No flood of short duration could have produced such fossiliferous deposits, he declared, and we must therefore conclude that this region was formerly at the bottom of the sea.[28]

These arguments convinced most naturalists that it was necessary to find a more acceptable explanation for the accumulation of fossiliferous sediments. Any such explanation would have to satisfy the corollaries of arguments used by the critics: deposition during a time span longer than a year, under tranquil conditions, and with some interruptions to account for the problem of specific gravity. Two theories soon became widespread. One group adopted the view that the present land masses had long been at the bottom of the sea, while another thought the surface of the earth had undergone a series of 'revolutions'. For the first group, the flood played a variety of roles or none at all; for the second, the flood could be described as simply one of several revolutions.

The theory of a 'long sojourn of the sea' was advocated by such men as Buffon, d'Holbach, Lamétherie, Réaumur, Soulavie, Valmont de Bomare, Bergman, Linnaeus, and Whitehurst. Of these, only the latter three clearly indicate their belief in the historicity of the universal flood, but they differed in assessing its geological effects. Linnaeus, indeed, remarked that he had looked long but without success for any geological traces of the deluge. Bergman had the same difficulty, but concluded that the violence and short duration of the flood probably meant that it had merely shifted some loose, superficial deposits. For Whitehurst, the flood followed a long period of sedimentation; the earth's crust ruptured, permitting uplift and the draining of water from the surface of the present continents.[29]

The views of Linnaeus and Whitehurst deserve closer examination, since the former's piety pervades much of his work while the latter's magnum opus—in its first chapters, at least—bears a strong resemblance to the cosmogonies of earlier natural theologians. Neither had any doubt that the Mosaic account of creation and flood was essentially true, but the ways in which they treat these two events suggest that in geology, if perhaps not in cosmogony, observational data might be used to reinterpret the Bible, but not vice versa.

Both men argue that the earth must have had a beginning— Linnaeus using, among other things, the traditional logic in favour of the existence of a first cause, and Whitehurst suggesting that only a once-fluid mass could have assumed the shape of an oblate spheroid— and thence they proceed to reconstruct the earth's early history on the basis of those sciences with which they were most familiar. Linnaeus' expertise led him to a discussion of how species of plants and animals, originating in one locality, could achieve wide geographical distribution. Whitehurst, in contrast, provides a discussion of the 'separation of chaos' by the operations of the laws of gravity and elective attraction. Both realized full well that their accounts were pious, scientifically plausible, and unproven, but based on the most up-to-date information as well as on what was commonly considered to be the oldest written document. Indeed, their technique of reconstruction was common to all cosmogonists, including those who did not rely on Genesis—one need only recall the systems of Descartes and Buffon.[30]

When Linnaeus and Whitehurst move from cosmogony to geology, their emphasis undergoes some change, since they no longer need to speculate about the most remote, primordial, past. They do not lose sight of Genesis, but, after all, the flood had not occurred at some 'chaotic' time, but when the earth was fully formed and inhabited; in other words, the flood interrupted some sort of recognizably normal history. One ought, therefore, to be able to find antediluvian, diluvian, and postdiluvian geological formations. Just what Linnaeus looked for in the strata seems to be unknown, but he admitted his inability to find traces of the deluge. What he did find was a succession of strata formed during 'a long and gradual lapse of ages', and, in addition, he had reliable evidence that the Baltic Sea was measurably lower than it once had been. The Baltic was thus a remnant of a universal sea which had progressively retreated from the continents, leaving behind it the observed accumulation of sediments. To equate this universal sea with the flood was impossible, the flood having been both brief and turbulent.[31]

Much the same considerations troubled Whitehurst who took pains to show that the earth's sedimentary crust had been formed in successive stages. What, then, had the flood done? His ingenious answer was that the flood was not really a universal inundation, but a figurative description of orogenesis and a resultant shift in sea basins. Subterranean forces had elevated the antediluvian ocean floor, producing the present continents and mountain ranges (with marine fossils at high elevations) and turning antediluvian continents into the present ocean floor. If Linnaeus perhaps sought literal confirmation of Genesis, Whitehurst found that a geological disturbance could be reconciled with Genesis, provided only that one did not interpret Moses literally.[32] One may conclude, then, that the Bible posed a historical problem to which these men offered different solutions, while they agreed in thinking that sedimentary formations must be the work of an ocean different from the flood.

The theory of the long sojourn of the sea was said in 1767 to be 'held by all philosophers, ancient and modern'.[33] Although this assessment is far from true, the theory did have much to recommend it. Not only did it provide sufficient tranquillity and time for deposition, but it also could explain the occurrence of thick, uninterrupted series of strata containing marine organisms. Furthermore, since this ocean was believed to have steadily retreated from the continents, the theory seemed to receive confirmation when Scandinavian scientists reported that the level of the Baltic was falling at a measurable rate.[34]

Serious objections to this theory were to be heard through much of the century from naturalists who themselves reached no agreement about a more satisfactory alternative. One common argument was based on the discovery by Robert Boyle that the bottom of the sea remains calm even when the surface is agitated by storms; such excessive tranquillity clearly meant that little sediment could be built up on the ocean floor.[35] Another argument resembled that used against the flood: in a universal ocean, however long its duration, deposits ought to have been laid down in order of their specific gravity. One naturalist also recalled the flood when he protested that one would have to account for the amount of water needed to produce such an ocean; if the flood could be called a miracle, the idea of a universal ocean required a natural explanation.[36] These objections received a substantial addition when Elie Bertrand asked how the sea could have deposited thick accumulations of fossils when modern seas are far less densely populated. Summarizing these and other arguments, Bertrand declared it evasive to say that no one has ever seen what lives on the ocean floor. Nor was it legitimate to invoke long periods of time to solve these dilemmas. As Bertrand put it, the oldest historical accounts show that there has been little change in the position of sea basins, and why should we suppose that major changes took place in prehistoric, 'fabulous' ages, and not within historical time?[37]

Seemingly insoluble problems in the theory of a universal ocean help to explain why many naturalists instead adopted the view that the earth had undergone a succession of 'revolutions'. The word 'revolution' was synonymous with 'change', and it was applied not only to violent events like earthquakes but also to marine incursions, whether universal or local, short or long in duration.[38] This type of explanation was common for more than a century, in 'theories of the earth', in monographs, and in scientific handbooks. In so long a line of thinkers from Leibniz to Cuvier, it is hardly to be expected that they should agree on the nature of these revolutions or on the effects of the flood.

A philosophical defence of the concept of revolutions was provided early in the century by Antonio Vallisneri. There is, he declared, no need to have recourse to extraordinary events when the world— repeatedly called a 'machine' by this good Cartesian—follows 'the ordinary laws of nature'. We ought in fact to be able to explain the phenomena of nature 'without violence, without fictions, without hypotheses, without miracles'. To Vallisneri, therefore, the explanation that was 'more likely, simpler, and more natural' than the flood was a series of 'many local floods', occurring over 'many centuries in succession'.[39] Like Vallisneri, other naturalists saw no philosophical inconsistency in emphasizing the uniformity of nature's laws and in utilizing a series of unpredictable geological upheavals. Indeed, it was commonly said that a normal part of nature, in the past as in the present, is the occurrence of such events as floods, earthquakes, landslips, and volcanic eruptions. Nor did it seem philosophically inconsistent to suggest that cataclysms in the past might have been more violent than at present; not only was this said to be plausible in the epochs when parts of the earth's crust had not yet consolidated, but it was also argued that such upheavals, however great and unpredictable, could be explained by using the known laws of physics and chemistry.[40]

Revolutions proved to be a most flexible, versatile, explanatory device. How else, if not by some revolution, could one account for the burial of elephant-like bones in Siberia and North America? In the same vein, some writers attributed to a revolution the transport to Europe of fossil genera now found only in the Indian Ocean or the Caribbean. And the occurrence of erratic boulders was commonly said to be the result of some upheaval. Late in the century, it was the theory of A. G. Werner that turned out to be the most flexible of this genre, since Werner combined revolutions with the theory of a universal ocean. By positing alternate periods of turbulence and tranquillity, Werner could account for evidence of disturbance in the strata as well as for the occurrence of horizontal sediments.[41]

Naturalists who employed revolutions disagreed sharply about the role of the flood: was the flood one of these upheavals, and, if so, what geological effects could be attributed to it? For Lehmann and Wallerius the flood was unique in its universality and in its having deposited much of the sedimentary crust; but earlier and later revolutions had also played some role in producing the present arrangement of the strata.[42] In the 1770s Arduino and Pallas, on the other hand, seem to have been uncertain that the flood had had identifiable effects; if Pallas was tempted to use the flood to explain the burial of large mammals, he and Arduino nevertheless preferred to say that the present state of the earth resulted from 'the successive effects of volcanoes and of other subterranean forces, plus the effects of a flood or of several incursions (*débordements*) of the sea'.[43] Vallisneri, Moro, and Desmarest were critics of all diluvial theories, preferred to consider the flood a miracle, and developed their own theories of local revolutions.[44]

In this range of views, one feature is in fact common to all descriptions of the flood: departures from a literal interpretation of Genesis. Even Wallerius and Lehmann—unusual in attributing to the flood a large role in producing sedimentary strata—felt obliged to point out that periods of violence and of calm during the deluge had doubtless had different effects, and Wallerius suggests that 'perhaps several centuries' were needed to achieve the retreat of the waters.[45] Their theories, indeed, are reminiscent of Werner's later version of a non-diluvial, gradually diminishing universal ocean. More common, however, was the use of the flood to explain the extinction of terrestrial forms and the transport of exotic genera, marine and terrestrial, to regions which they had not inhabited during their lifetimes. Particularly in the case of the great mammals, it was argued that they must have been catastrophically annihilated because only rapid burial could have preserved them from decay, and only sudden disaster could have prevented them from migrating to escape from local upheavals or uncongenial changes in climate. If the case for the extinction of marine forms remained unclear, there could be less doubt about terrestrial ones, even when the identification of species often lacked precision; when the latter problem was tackled by Cuvier, he, like his older contemporaries, recognized at once the crucial importance of this subject for geological theory.[46]

As noted earlier, Pallas was one of the several naturalists who believed that the flood might have been responsible for the transport and burial of (for example) the Siberian mammoth, since the flood possessed the requisite characteristics of having been sudden, violent,

X

and recent. Other writers were less certain, but, as one of them remarked, the suitable explanation was surely the flood or a very similar catastrophe.47

From this array of oceans, floods, and upheavals, one may safely draw several conclusions. First, the flood was being reduced in status, not to a legend but to one of a series of natural upheavals. Those writers who doubted or dismissed the possible effects of the flood usually did not deny its historicity, while those who stressed local upheavals usually did not reduce the flood to a local disaster. In short, the universal flood was generally accepted as historical fact, but the scientific evidence permitted no consensus about the geological significance of the event. Furthermore, descriptions of the flood and its effects—the possible centuries of Wallerius, Whitehurst's orogenesis, Bergman's shifting of loose soil, Kirwan's transport of exotics—indicate how far geologists had abandoned the original tale of the animals two-by-two.

At this point, one may well wonder why the flood was used at all. Why resort to the Bible in a period when the physical sciences no longer did so? Why not be content—as only few geologists were—with Vallisneri's argument in favour of using explanations 'more likely, simpler, and more natural'? If to use Genesis necessitated modifying the meaning of the text, would it not have been sensible to abandon the attempt?

A number of answers might be suggested to these questions, but one possible answer must be discarded, namely, the idea that geologists were constrained to be orthodox. Indeed, the variety of geological interpretations of the flood provides additional evidence that there was no standard biblical exegesis to which the scientists had to adhere. On the affirmative side of the problem, it is worth remembering the old common assumption that religious and scientific truths cannot be in conflict. As Galileo had long since argued, the study of nature furnishes us with 'hard' information, while the Bible is notoriously difficult to understand; the two sources will be found to harmonize, however, if we reinterpret the Bible on the basis of scientific fact.48 Except for those few naturalists who set the flood aside as a miracle, most writers seem to have shared this Galilean belief. When so many theories are all said to be in harmony with Scripture, it is more than likely that theories were based primarily on geological evidence, and the Bible then reinterpreted suitably.

Probably more important than this general desire to reconcile Genesis with geology was the methodological principle enunciated by Richard Kirwan:

> In effect, past geological facts being of an historical nature, all attempts to deduce a complete knowledge of them merely from their still subsisting consequences, to the exclusion of unexceptionable testimony, must be deemed as absurd as that of deducing the history of ancient Rome solely

from the medals or other monuments . . ., to the exclusion of a Livy, a Sallust, or a Tacitus.[49]

The testimony Kirwan had in mind was that of Moses, confirmed by other ancient traditions. However staunch a defender of Genesis in the face of the 'immorality and infidelity' rampant during the French Revolution, Kirwan depicts Moses as neither prophet nor scientist, but *historian*. Not only was the interpretation of ancient traditions a subject of lively interest in the eighteenth century, but Kirwan's insistence that 'monuments' alone provide incomplete knowledge was also a topic debated for some decades. Both points merit further discussion, which can only be undertaken here all too briefly.

Quite early in the century, professional historians had tackled the question of the credibility of ancient legends and oral traditions. They had concluded that tradition could transmit accurately—in outline, if not in detail—any event that was large, dramatic, and public; and when essentially the same event entered into apparently independent traditions among many peoples, a mere historical probability became a virtual certainty.[50] Such critical precepts applied in all their force to the flood. When geologists assiduously compared ancient flood legends, discarding some details and reinterpreting others, they were being very 'modern' indeed. Furthermore, they were in the enviable position of being able to combine with the ancient texts the testimony of nature.

Early in the century, too, professional historians had debated the degrees of confidence one could repose in ancient texts as compared with the 'medals or other monuments' referred to by Kirwan, and the faction that considered 'monuments' to be more reliable than texts won at least a temporary victory. To be sure, 'monuments' should ideally be correlated with other (written) evidence, but what if no texts existed? While Fontenelle could grandly announce that geologists had available 'histories written by the hand of nature itself', geologists often felt reassured when 'nature itself' could be supplemented by the evidence of human witnesses.[51] Lehmann, for example, hesitated to comment on the nature and extent of antediluvian geological changes for which 'we lack historical monuments'. Desmarest considered it quite understandable that in seeking to interpret the rocks, geologists should 'have recourse to the greatest, most universal catastrophe mentioned in history'. And as one writer put it, if we do not use the flood to account for the burial of mammals, we would be in the position of having to suppose another such catastrophe 'without being able to base ourselves on history'.[52]

For the geologist, then, there were three options: one could use historical texts when these existed, or one could construct those 'hypotheses' so roundly condemned by Newton, or one could, like Blumenbach or Cuvier, make an effort to distinguish between speculation and geology as 'une science positive'.

NOTES

1 Report by Cuvier *et al.*, 1806, in N. André, *Théorie de la surface actuelle de la terre*, Paris, 1806, pp. 320–1, 326. See also, Cuvier, 'Discours préliminaire', *Recherches sur les ossemens fossiles des quadrupèdes*, 4 vols., Paris, 1812, i, 26–35, and William Coleman, *Georges Cuvier, zoologist*, Cambridge, Mass., 1964, p. 113.

2 Charles Lyell, *Principles of geology*, 3 vols., London, 1830–3, i, 29–30. M. J. S. Rudwick, 'The strategy of Lyell's *Principles of geology*', *Isis*, 1970, *61*, especially 8–11.

3 V. A. Eyles, 'Hutton'. in *Dictionary of scientific biography*, New York, 1972, vi, 580. The same view is to be found in many discussions of Hutton and Lyell, primarily to explain the resistance encountered by their theories. For a recent sample of revisionist literature on Lyell, see the 'Lyell Centenary issue', *The British journal for the history of science*, 1976, *9*, especially the articles by Porter and Ospovat, pp. 91–103, 190–8.

4 Francis C. Haber, *The age of the world*, Baltimore, 1959, pp. 112–13.

5 Don Cameron Allen, *The legend of Noah*, Urbana, Ill., 1963, especially pp. 68, 84–5, 181.

6 Calmet, *Commentaire littéral sur tous les livres de l'Ancien et du Nouveau Testament*, 23 vols., Paris, 1707–16, i, 176–9, 186. Also, F. Dinago (éd.), *Publication des oeuvres inédites de Dom A. Calmet*, 2 vols., St-Dié, 1877–8, ii. 39–67.

7 Richard Simon, *Histoire critique du Vieux Testament*, nouv. éd., Rotterdam, 1685, especially p. 33 for the flood. Discussion with long quotations can be found in Edward M. Gray, *Old Testament criticism*, London and New York, 1923. chapter IX. Also H. Margival, *Essai sur Richard Simon*, Paris, 1900, chapter V, and J. Steinmann, *Richard Simon et les origines de l'exégèse biblique*, Paris, 1960, pp. 100–16, 124–30.

8 [Jean Astruc], *Conjectures sur les memoires originaux dont il paroit que Moyse s'est servi pour composer le Livre de la Genese*, Brussels, 1753, pp. 3–18. Also, Gray, op. cit. (7), chapter XII. Reviewers are cited at length in A. Lods, *Jean Astruc et la critique biblique au XVIIIᵉ siècle*, Strasbourg, 1924, pp. 62–71.

9 For changes in the Sorbonne, see R. R. Palmer, *Catholics and unbelievers in eighteenth-century France*, Princeton, 1939, pp. 40–1, 51, 123, 129. Palmer considers the Sorbonne to have become more rigid and vigilant after the Prades affair (1751–3), but his book shows a continuing variety of thought in orthodox circles. An example of the Sorbonne's naive literalism is in its response to Buffon's two great works, in 1751 and after 1778, respectively; see J. Piveteau (ed.), *Oeuvres philosophiques de Buffon*, Paris, 1954, pp. 106–9, and P. Flourens, *Des manuscrits de Buffon*, Paris, 1860, pp. 254–80.

10 Noël-Antoine Pluche, *Le spectacle de la nature*, Paris, (1756), iii. especially 515–36. For the popularity of Pluche, see D. Mornet, *Les sciences de la nature en France, au XVIIIᵉ siècle*, Paris, 1911, pp. 248–9. Also, abbé Mallet, 'Arche de Noé', in Diderot and d'Alembert, *Encyclopédie*, Paris, 1751, i. 606–9.

11 [J.-B. LeMascrier], 'Essai sur la chronologie', in [J.-B. Mirabaud], *Le monde, son origine, et son antiquité*, 2nd edn., London, 1778, ii. 163–5, and the same writer in Benoît de Maillet, *Telliamed* (ed. and tr. by A. V. Carozzi), Urbana, Ill., 1968, p. 381, nn. 52, 54. Carozzi insists that LeMascrier held to the 'orthodox' view of 6,000 years as the age of the earth; pp. 30 and 380, n. 50. LeMascrier was less flexible about the flood, in fact, than about the age of the earth; see below, note 24.

12 *Journal de littérature, des sciences et des arts*, 1779, *3*, 412–15.

13 Palmer, op. cit. (9), p. 221. The whole of Palmer's book shows a range of belief on such vital matters as sin, grace, and the nature of man.

14 Rousseau, 'Discourse on the origin and foundations of inequality among men', 1755, in Roger D. Masters (ed.), *The first and second discourses*, New York, 1964, p. 103.

15 For Boulanger, see Frank Manuel, *The eighteenth century confronts the gods*, New York, 1967, pp. 214–19. Manuel's book is rich in examples of nonbiblical searches for origins, but see especially pp. 132–4. Theological conflict about the state of nature is treated by Palmer, op. cit. (9), chapter II.

16 The harshness of censorship is emphasized by Peter Gay, *The Enlightenment, an interpretation*, 2 vols., New York, 1967–9, ii. 70–9; less extreme views are to be found in Palmer, op. cit. (9), pp. 16–17, and in Jacques Roger, 'Introduction', in Buffon, *Les époques de la nature* (ed. by J. Roger), *Mémoires du Muséum national d'histoire naturelle*, Série C, Tome X, Paris, 1962, p. cxii. The only searching discussion of the failure of the Sorbonne to condemn Buffon after 1778 is in Roger, pp. cxxxii-vi; his conclusions are tentative because he has had to rely on Paris gossip in the absence of 'harder' information.

17 One of the many useful treatments of liberal Anglicanism is Roland N. Stromberg, *Religious liberalism in eighteenth-century England*, London, 1954, especially chapter IV. See also Charles R. Gillett, *Burned books: neglected chapters in British history and literature*, 2 vols., New York, 1932, ii. chapters XXVII-VIII. Gillett has discovered virtually no 'heterodox' books condemned after 1720; 'virtual' is necessary because of unsubstantiated reports that there may have been two.

18 The quotation is from Gay, op. cit. (16), p. 71. The Berlin Academy was under far greater royal control than its Paris counterpart, but it awarded its prize in 1772 to Herder's non-biblical

account of the origin of language, and it boasted Maupertuis as its President for a time. Maupertuis published several heterodox works during this period; see Bentley Glass, 'Maupertuis', in *Dictionary of scientific biography*, New York, 1974, ix. 186–9.

19 Clayton's arguments are summarized by one of his critics, Alexander Catcott, *A treatise on the Deluge*, 2nd edn., London, 1768, pp. 11–12. Also, K. B. Collier, *Cosmogonies of our fathers*, New York, 1934, pp. 229, 234. At the time of his death, Clayton was in imminent danger of being charged with heresy for his Arian views; but this was so unusual that I suspect the problem stemmed from his being a bishop rather than a lesser cleric or a layman. See A. R. Winnett, in Derek Baker (ed.), *Schism, heresy and religious protest*, Cambridge, 1972, pp. 311–21.

20 Mirabaud, op. cit. (11), i. 95–6. See *Telliamed*, op. cit. (11), pp. 298–9, and its editor's comments, p. 300, n. *c.*

21 Voltaire, *Philosophical dictionary* (tr. by Peter Gay), New York, 1962, pp. 284–97, 327–8, 394: articles *Genèse*, *Inondation*, *Miracles*. For Buffon's objections to miracles, see Roger, op. cit. (16), pp. lxxxv, xlviii.

22 Boulanger, 'Déluge', in *Encyclopédie*, op. cit. (10), 1754, iv. 795–803.

23 D'Holbach's transparent compromises with Genesis can be found in his translations of J. F. Henckel, *Pyritologie*, 2 vols., Paris, 1760, i. 122n, 123n, 131n; and J. G. Lehmann, *Traités de physique, d'histoire naturelle, de mineralogie et de métallurgie*, 3 vols., Paris, 1759, iii. pp. v–x, 83n, 192n. See also his article, 'Terre, couches de la', in *Encyclopédie*, op. cit. (10), 1765, xvi. 170; Collier, op. cit. (19), p. 283; and Manuel, op. cit. (15), pp. 234, 238.

24 Buffon, *Epoques*, op. cit. (16), pp. 182–4. *Telliamed*, op. cit. (11), compare pp. 213–15 with pp. 297–300. Lamétherie, *Théorie de la terre*, 3 vols., Paris 1795, iii. 189–224, 258–84, and *Leçons de géologie*, 3 vols., Paris, 1816, especially ii. 325–34; also, K. L. Taylor, 'Lamétherie', in *Dictionary of scientific biography*, New York, 1973, vii. 602–4. J. F. Blumenbach, *Beyträge zur Naturgeschichte*, 2nd edn., 1806, i. in *The anthropological treatises of J. F. Blumenbach* (tr. by T. Bendyshe), London, 1865, pp. 285–6 and 286, n. 2. J. G. Sulzer, 'Conjecture physique sur quelques changemens arrivés dans la surface du globe terrestre', *Histoire de l'Académie royale des sciences et belles-lettres*, Berlin, 1762 (1769), pp. 90–8. Sulzer developed a theory of local eruptions of lakes, which, he believed, could have given rise to flood legends; this theory apparently attracted little attention in the eighteenth century.

25 Melvin E. Jahn, in C. J. Schneer (ed.), *Toward a history of geology*, Cambridge, Mass., 1969, pp. 198, 200. Also, Henckel, op. cit. (23), i. 110–11, 123, 131, and Louis Bourguet, *Traité des pétrifications*, 2 vols., Paris, 1742, i. 53–94. Henckel's work first appeared in 1725.

26 [Dezallier d'Argenville], *L'histoire naturelle éclaircie dans deux de ses parties principales, la lithologie et la conchyliologie*, Paris, 1742, pp. 156–60; and additional remarks on the deluge in the 1757 edition of this work, pp. xix, 58, 66–71. Early critiques of Woodward are mentioned by Rudwick, *The meaning of fossils*, London & New York, 1971, pp. 82–3, 93; others were produced before 1750 by such writers as Vallisneri, Moro, and Buffon. Later naturalists who gave the flood a sedimentary role were Lehmann and Wallerius, discussed below; see notes 42, 45.

27 John Walker, *Lectures on geology* (ed. by Harold W. Scott), Chicago & London, 1966, p. 181. The lectures seem to date from about 1780.

28 Réaumur, in *Mémoires de l'Académie royale des sciences*, 1720 (1722), pp. 400–16. References to Réaumur and the Touraine are frequent throughout the century.

29 A. G. Nathorst, 'Carl von Linné as a geologist'. *Annual report of the Board of Regents of the Smithsonian Institution*, 1908, pp. 713, 721; Haber, op. cit. (4), p. 160; and Desmarest, 'Linné', *Encyclopédie méthodique: géographie physique*, Paris, 1795, i. 304. For Bergman, see Hollis Hedberg, in Schneer, op. cit. (25), p. 189. John Whitehurst, *An inquiry into the original state and formation of the earth*, 2nd edn., London, 1786, pp. 58–9, 118–22.

30 In addition to works cited in the preceding note, see Linnaeus, 'On the increase of the habitable earth', in *Select dissertations from the Amoenitates Academicae*, tr. F. J. Brand, 2 vols., London, 1781, i. 71–127. The allusion is to Descartes' *Le monde*, available in many editions and translations, and to several features of Buffon's cosmogony, e.g. his experiments to measure the rate of cooling of an incandescent globe.

31 See notes 29, 30, 34. The phrase by Linnaeus is quoted in F. C. Haber, 'Fossils and the idea of a process of time in natural history', in B. Glass, O. Temkin, and W. L. Straus (eds.), *Forerunners of Darwin: 1745–1859*, Baltimore, 1968, p. 242.

32 Whitehurst, op. cit. (29), especially p. 131.

33 J.-C. Valmont de Bomare, *Dictionnaire raisonné universel d'histoire naturelle*, nouv. éd., 6 vols., Paris, 1767–8, ii. 708, article 'Fossiles'. The articles 'Déluge' in the 1764 and 1767 editions recount the views of those who think marine fossils relics of the flood; discussion of evidence is reserved for articles 'Falun', 'Fossiles', and 'Terre', which deal with the universal ocean and the concept of successive revolutions.

34 Details of this research and the different interpretations of results are in Desmarest, 'Ferner', *Encyclopédie méthodique*, op. cit. (29), i. 133–50. Cf. Wegmann, in Schneer, op. cit. (25), pp. 386–94, who believes that these issues had little impact outside Scandinavia during the eighteenth century.

X

18

35 Boyle's little tract, *De fundo maris* (*Relations about the bottom of the sea*), was first published in English and Latin in 1670. Allusions to this work are numerous, some perhaps based on the summary in the better known study by L. F. Marsigli, *Histoire physique de la mer*, Amsterdam, 1725, pp. 1, 48. Marsigli could reach no conclusion about bottom currents, later discussed in *Telliamed*, op. cit. (11), especially pp. 60–9.

36 A. L. Moro, *De' crostacei e degli altri marini corpi che si truovano su' monti*, Venice, 1740, pp. 15–23, 142–55. Also, J. G. Wallerius, *Mineralogie* (tr. by d'Holbach), 2 vols., Paris, 1753, i. 139.

37 Élie Bertrand, *Memoires sur la structure interieure de la terre*, Zurich, 1752, pp. 23–31, 50–1, 56–7, 64–6.

38 For an early use of 'revolutions' in this non-astronomical sense, see Fontenelle, in *Histoire de l'Académie royale des sciences*, 1718 (1719), p. 5. Also, Roger, op. cit. (16), p. 270, n. 10, and 'Révolutions de la terre', *Encyclopédie*, op. cit. (10), 1765, xiv. 237–8. I am indebted to Professor Henry Guerlac for calling the last to my attention. 'Revolution' was used primarily in English and French; the corresponding terms common in other languages have no violent connotations: *Veränderung, mutatio, mutazione.*

39 Vallisneri, *De' corpi marini, che su' monti si trovano*, 2nd edn., Venice, 1728, pp. 34, 35, 41, 47, 73.

40 R. Hooykaas, *Natural law and divine miracle: the principle of uniformity in geology, biology and theology*, Leiden, 1963, pp. 4–17. See also Desmarest, in *Encyclopédie méthodique*, op. cit. (29), i. 417, and iii. 197, where he distinguishes between disorderly *bouleversements* and orderly *révolutions* produced by known causes.

41 A. G. Werner, *Short classification and description of the various rocks* (tr. with introduction and notes by A. Ospovat), New York, 1971, pp. 17–24.

42 Lehmann, op. cit. (23), especially iii. 192–8, and John C. Greene, *The death of Adam*, New York, 1961, pp. 67–72. Wallerius, op. cit. (36), ii. 123.

43 The quotation is from P. S. Pallas, *Observations sur la formation des montagnes et les changemens arrivés au globe*, St Petersburg, 1777, pp. 35–6. See Greene, op. cit. (42), pp. 80–1. G. Arduino, 'Saggio fisico-mineralogico di lythogonia, e orognosia', *Atti dell' Accademia delle scienze di Siena*, 1774, *5*, 254. Reprinted in Arduino, *Raccolta di memorie*, Venice, 1775, which was translated into German in 1778.

44 Vallisneri, op. cit. (39), pp. 49, 76, 83–4; Moro, op. cit. (36), pp. 426–32; and Desmarest, in *Encyclopédie méthodique*, op. cit. (29), iii. 197–8, 606–15, 618–32.

45 Lehmann, op. cit. (23), iii. 284–92, 297ff, 314–15. Wallerius, *De l'origine du monde et de la terre en particulier*, Warsaw, 1780, pp. 354–7.

46 Cf. Cuvier's early statement in 'Notice sur le squelette d'une très-grande espèce de quadrupède inconnue jusqu'à présent, trouvé au Paraguay', *Magasin encylopédique*, 1796, *1*, 310; and Rudwick, op. cit. (26), chapter III. The case for marine invertebrates was complicated by problems of identification and classification, but it was also argued that apparently extinct forms might still be alive in unexplored ocean depths. One of the rare efforts to gather together evidence on this subject is F. X. Burtin, 'Reponse a la question physique, proposée par la Société de Teyler, sur les revolutions generales, qu'a subies la surface de la terre, et sur l'ancienneté de notre globe', *Verhandelingen, uitgegeeven door Teyler's tweede genootschap*, Haarlem, 1790, viii. He concludes that the great number of extinct invertebrates demonstrates that at least one catastrophe did occur. See also Blumenbach, op. cit. (24), pp. 283–6, his *Manuel d'histoire naturelle*, tr. Soulange Artaud, 2 vols., Metz, 1803, ii. 148–9, and the synopsis of one of his works by Héron de Villefosse, in *Journal des mines*, 1804, *16*, 5–36.

47 J. F. Esper, *Description des zoolithes nouvellement decouvertes d'animaux quadrupedes inconnus et des cavernes qui les renferment*, tr. J. G. Isenflamm, Nuremberg, 1774, especially p. 81. For Pallas, see above, note 43. An excellent discussion of the problem of extinction, with emphasis on the great quadrupeds, is in Greene, op. cit. (42), chapter IV. Further examples of scientists who questioned whether the recent catastrophe should be identified with the flood are given by Leroy Page, in Schneer, op. cit. (25), p. 267 and *passim*.

48 Galileo, 'Letter to the Grand Duchess Christina', in *Discoveries and opinions of Galileo* (tr. with introduction and notes by Stillman Drake), New York, 1957, pp. 175–216.

49 Kirwan, *Geological essays*, London, 1799, p. 5; also, pp. 5–6, 54–86.

50 See articles by Pouilly and Fréret, in *Memoires de litterature tirez des registres de l'Academie royale des inscriptions et belles lettres*, Paris, 1729, *6*, especially 153, 156, 71–114. Also, Manuel, op. cit. (15), *passim;* Lionel Gossman, *Medievalism and the ideologies of the Enlightenment: the world and work of La Curne de Saint-Palaye*, Baltimore, 1968, pp. 153–7; and Rénee Simon, 'Nicolas Fréret, académicien', *Studies on Voltaire and the eighteenth century*, 1961, *17*, 120–30.

51 Fontenelle, in *Histoire de l'Académie royale des sciences*, 1722 (1724), p. 4. On the question of 'monuments', see the crucial article by Arnaldo Momigliano, 'Ancient history and the antiquarian', *Journal of the Warburg and Courtauld Institutes*, 1950, *13*, 285–315; reprinted in two collections of Momigliano's articles: *Studies in historiography*, New York, 1966, pp. 1–39, and *Contributo alla storia degli studi classici*, Rome, 1955, pp. 67–106.

52 Esper, op. cit. (47), p. 81. Lehmann, op. cit. (23), iii. 192. Desmarest, in *Encyclopédie méthodique*, op. cit. (29), iii. 606.

XI

BORROWED WORDS:
PROBLEMS OF VOCABULARY IN
EIGHTEENTH-CENTURY GEOLOGY *

EVERY science has its technical vocabulary, consisting in part of terms coined for explicit purposes and in part of words borrowed from ordinary discourse and used with greater or lesser degrees of precision. Words of the latter sort pose curious problems, some of them familiar to those historians of science concerned with, for example, what Galileo meant by *forza* and Newton by *attraction*. Indeed, analogous problems face any historian seeking to understand the older meanings of terms still in use today.

The technical vocabulary of early geology included at least three terms—monuments, revolutions, and accidents—so commonplace in eighteenth-century writing that only one of the three, monuments, has been in part recognized as fulfilling a peculiar function at that time. By contrast, the large literature on 'revolutions' and the small one on 'accidents' do little to clarify geological usage in particular. Since all three terms were not only common but proved to be ephemeral, disappearing from geology sometime during the earlier part of the nineteenth century, to examine how they were once used can shed light on concepts characteristic of one stage in the development of what was then a new science.

The words to be discussed here are, of course, cognates in English and French, and these languages have served as my point of departure. Texts in other languages have been chosen with special attention to cognates or to equivalent words in Italian or German (e.g., *rivoluzioni* and *Veränderungen*). On some occasions, where the specific words or their equivalents are lacking, certain texts are used to supply additional illustrative material, but only after the main argument has been developed by use of cognates or their equivalents.

Monuments: the legacy of antiquarianism
Most historians of geology are aware that early naturalists often referred to fossils as 'medals' of the Flood, that the term 'monument' was frequently applied to both rocks and fossils, and that even such words as 'documents' and 'archives' had some currency in geological texts. Why

---

* Different sections of this article were presented at meetings of the Northeastern American Society for Eighteenth-Century Studies, Amherst, Mass., 5 October 1978, and the 26th International Geological Congress, Paris, 9 July 1980.

28

such terms found favour has been clarified to some extent by Cecil J.
Schneer in a classic article showing that individual members of the early
Royal Society of London were often interested in both man-made and
natural antiquities, and this is certainly evident in the titles of many
weighty tomes devoted to 'the natural history and antiquities of'
Oxfordshire or Staffordshire or Northamptonshire. More recently, Joseph
M. Levine has argued that for John Woodward natural history and civil
history were the same kinds of activity, dealing with, so to speak, two halves
of the same story: the history of the earth and of its human inhabitants.[1]

These admirable studies, however, do not wholly explain the function
in geology of words borrowed from the historians, and to do so one must
turn first not to the borrowers but to the lenders. An important clue is
provided by antiquarian Charles Patin, who declared: 'without medals,
history, *stripped of proofs*, would seem to many minds to be either the record
of the opinons of historians writing about their own times or a summary of
memoirs which could be false or prejudiced.'[2] Patin was writing in 1691,
when the study of history had long been under fire from a variety of critics
who argued that any discipline based essentially on human testimony
could only yield uncertain results. That history could never be demonstrative in the fashion of geometry was the view shared by such authorities as
Descartes, Malebranche, and Leibniz.[3] There were indeed ways to
evaluate the mere memoirs cited by Patin, and the medievalist Muratori
listed some of the questions one ought to ask: Had the authors witnessed the
events they described? Had they used older documents no longer extant?
Did they have some discernible personal or political bias? In these matters,
however, no rules could infallibly guide the scholar, and Muratori, like
Mabillon before him, could only conclude that historians should learn to
exercise 'il buon gusto.' Leibniz put the problem succinctly when he said he
wished there were a science of weighing evidence.[4]

In the decades around 1700, scholars like Mabillon, Montfaucon, and
Spanheim were providing the critical tools that allowed historians to claim
a considerable degree of certainty for one kind of historical evidence:

[1] Cecil J. Schneer, 'The rise of historical geology in the seventeenth century,' *Isis*, 1954, 45, 256–68.
Joseph M. Levine, *Dr. Woodward's shield: history, science and satire in Augustan England*, Berkeley, Los
Angeles, London, 1977.
[2] Charles Patin, *Histoire des médailles ou introduction à la connoissance de cette science*, Paris, 1695, p. 8.
Italics added. The same sentence in the 1691 edition is quoted by Arnaldo Momigliano, 'Ancient
history and the antiquarian,' in Momigliano, *Contributo alla storia degli studi classici*, Rome, 1955, p. 86, n.
31. Momigliano discusses at length the confidence of antiquarians in monuments, as well as the
distinctions one may draw between 'antiquarians' and 'historians.'
[3] L. Lévy-Bruhl, 'The Cartesian spirit and history,' in R. Klibansky and H. J. Paton, eds., *Philosophy
and history: essays presented to Ernst Cassirer*, Oxford, 1936, pp. 191–6. L. Davillé, *Leibniz historien*, Paris,
1909, pp. 337–40.
[4] L. A. Muratori, *Delle riflessioni sopra il buon gusto*, 2 vols., Venice, 1766, i, 184–6, and ii, 56–61. Works
in these volumes were originally published in the first decade of the century. Cf. Jean Mabillon, *Traité
des études monastiques*, Paris, 1691, especially p. 236, but also pp. 233–42. Leibniz to Thomas Burnett,
1/11 February 1697, in *Die philosophischen schriften von Gottfried Wilhelm Leibniz*, ed. C. J. Gerhardt, 7
vols., Berlin, 1875–99, iii, 193–4. This Burnett (sic) was a distant relative of the author of *The sacred
theory of the earth*.

monuments. The term included not only architectural remains and their inscriptions, but also medals, coins, charters, and laws, all of which could be accurately dated and forgeries detected. All had the additional virtue of being 'public' remains or documents, and they therefore could perpetuate no private bias. Even where bias did exist, monuments always had value if one knew how to use them; as Patin put it, medals may exaggerate the supposed virtues of rulers, but they also inadvertently supply reliable information about religious rites and gods, temples, tombs, ports, bridges, markets, libraries, theaters, palaces, games, clothing, marriage customs, etc. In short, monuments often contain data not intended to deceive or mislead, and they thus can furnish the 'proofs' necessary for judging the worth of a Livy or a Polybius. To call something a monument or medal, therefore, was to express confidence in its reliability. When one theologian described the soul as 'a living medal of Divinity,' Leibniz could only applaud.[5]

If modern geologists often think of their discipline as a physical science, their predecessors considered themselves to be both scientists and historians, and they regularly appealed to the concepts of both disciplines in moments of need. When, for example, naturalists like Robert Plot and Edward Lhwyd suggested that at least some fossils were *lusus naturae* or were produced by the germination of seeds or by some 'plastick virtue' within the rocks, a chorus of critics replied in traditional fashion that nature does nothing in vain, that nature is not capricious or intentionally misleading—in Newton's phrase, nature is 'always consonant to itself.' Explicit appeals to history were added by Robert Hooke and John Woodward, among others, who suggested that naturalists imitate historians in their analogical reasoning: if those who dig up coins do not hesitate to call them man-made, then naturalists ought to assume that fish-like fossils were once real fish.[6]

Apart from affinities in reasoning, geologists shared with historians the problem of evaluating the credibility of at least one crucial ancient text, the Bible. To Woodward and J. J. Scheuchzer, the biblical account of the

---

[5] Leibniz to Burnett, 29 December 1707, in *Schriften* (4), iii, 315; also, iii, 182, and Davillé, op. cit. (3), pp. 131, 465–9. C. B. Stark, *Systematik und geschichte des archäologie der kunst*, Leipzig, 1880, pp. 155–7. There is no study of the antiquarians, but Momigliano, op. cit. (2), has a useful bibliography. For a more sophisticated view of degrees of certainty in history, see the discussion of Fréret by Roger Mercier, 'Une controverse sur la vérité historique,' in *La Régence*, proceedings of a colloquium held at Aix-en-Provence, Paris, 1970, pp. 294–306.
[6] The quotation is from Newton's *Principia*, rule 3 of the Rules of Reasoning. See Robert Hooke, *Micrographia*, 1665, repr. New York, 1961, p. 112; G. W. Leibniz, *Protogée ou de la formation et des révolutions du globe*, originally published in Latin in 1749, tr. B. de St.-Germain, Paris, 1859, pp. 47–9; Hooke, *Posthumous works*, 1705 repr. with intro. by T. M. Brown, London, 1971, p. 321; and J. Woodward, *The natural history of the earth*, ed. and tr. B. Holloway, 2 vols., London, 1726, ii, 155. Also: Jean Astruc, in *Mémoires de Trévoux*, March 1708, pp. 515–16; J. B. A. Beringer, *The lying stones of Dr. Johann Bartholomew Adam Beringer*, originally published in Latin in 1726, tr. and ed. M. É. Jahn and D. J. Woolf, Berkeley, Los Angeles, 1963, pp. 59–60, 72–3, 183–4, 186–7; Paolo Boccone, *Recherches et observations naturelles*, Paris, 1671, p. 18; and B. de Fontenelle, in *Histoire de l'Académie royale des sciences*, 1706 (1731), pp. 10–11. For the problems posed by fossils, see M. J. S. Rudwick, *The meaning of fossils*, London, New York, 1972, chapters 1–2.

30

Flood was in danger of being considered by unbelievers to be the testimony of a mere writer of memoirs, and both naturalists set themselves to show that fossils provided those authentic monuments capable of supplying proof of the veracity of Moses. Neither author personally needed such assurance—and one contributor to the *Mémoires de Trévoux* considered their procedure to be both gratuitous and potentially dangerous—but they were speaking the language of some of their contemporaries when they recognized that a monument was more convincing than a text.[7]

Confidence in the reliability of natural monuments can be detected in contexts other than that of the Flood, as when early writers like Steno and Leibniz rejoiced to think that nature not only supports Scripture but also provides hard evidence to fill in the gaps in written records. More explicitly confident was Fontenelle, whose assertion that geologists deal with 'histories written by the hand of nature itself' faithfully echoes the often-heard distrust of human testimony. Decades later, when debates about historical certainty and the value of monuments as distinct from mere memoirs had long since abated, Torbern Bergman could still say that fossils 'resemble a series of ancient coins in the testimony they bear to the convulsions and revolutions of our globe, on which historical monuments [in contrast to natural ones] are wholly silent.' In 1777 Peter Simon Pallas remarked that the succession of geological formations 'provides the oldest chronicle of our globe, the least subject to falsifications.' Some thirty years later, Johann Friedrich Blumenbach still referred to fossils as 'the surest documents in the archives of nature,' while Georges Cuvier, in a famous passage, referred to his own reconstructions of extinct quadrupeds as requiring that he 'learn to decipher and to restore these monuments.'[8]

For many of these geological texts, and especially for the earlier ones, one may say that the vocabulary of monuments and medals was being used to suggest reliability, freedom from bias, and a resultant confidence in the conclusions drawn from such evidence. For some of the later texts, it is likely that conventional language dictated the use of terms like monuments and documents, since it was not then necessary to defend the study of history or geology, or to explain the affinities between the two. Readers of Cuvier's writings, for example, can only doubt that he seriously intended to defend his own work on the basis of its similarities to that of the antiquarians; still, even in Cuvier's day, his audience was being reminded

---

[7] Review of a French translation of Woodward, in *Mémoires de Trévoux*, February 1736, especially pp. 245, 247, 252. M. E. Jahn, 'Some notes on Dr. Scheuchzer and on *Homo diluvii testis*,' in C. J. Schneer, ed., *Toward a history of geology*, Cambridge, Mass., 1969, pp. 192–213. Also, R. Rappaport, 'Geology and orthodoxy: the case of Noah's flood in eighteenth-century thought,' *The British journal for the history of science*, 1978, *11*, 1–18.

[8] For Steno and Leibniz, see Leibniz, *Protogée* (6), p. 18. Fontenelle, in *Hist. Acad.*, 1722 (1724), p. 4. Bergman, *Essays physical and chemical*, Edinburgh, 1791, pp. 236–7. Pallas, *Observations sur la formation des montagnes et les changemens arrivés au globe*, St. Petersburg, 1777, p. 29. Blumenbach, *Beyträge zur naturgeschichte*, 2 vols., Göttingen, 1806–11, i, 113; translated in *The anthropological treatises of Blumenbach*, tr. T. Bendyshe, London, 1865, pp. 317–18. Cuvier, 'Discours préliminaire,' *Recherches sur les ossemens fossiles de quadrupèdes*, 4 vols., Paris, 1812, i, 1.

that new techniques in paleontology were as exact and reliable as the most secure branch of history, namely, the study of monuments.

Revolutions: the legacy of the abbé Vertot

Unlike 'monuments,' the word 'revolution' figures frequently in studies of the history of geology, where it is traditionally equated with catastrophic, widespread (often universal) change. Some years ago, however, Albert V. Carozzi undertook a close examination of that classic 'catastrophist' text, Cuvier's *Discours sur les révolutions du globe* (1812), and he discovered that Cuvier's use of 'revolution' was by no means consistent; in fact, revolutions might be large or small, sudden or gradual, depending in every instance upon the context. One may add to Carozzi's analysis the observation that Cuvier's usage was not idiosyncratic, and that 'revolution' had had behind it more than a century of ambiguity and versatility.[9] Because this word is so firmly embedded in modern minds as a synonym for large and sudden changes—or more localized but quite drastic ones—it will be helpful here to look first at the way 'revolution' was used in the decades before Cuvier by a variety of non-geological writers, so that geological usage may be put into an appropriate context.

From some of the dictionaries and better-known historical and philosophical works of the period, and from some modern treatments of the subject, one can extract three principal groups of meanings of 'revolution.'[10] The first, associated with astronomy, was movement in a circle; by extension, those who believed that history moves in repetitive or cyclical patterns used the word to mean 'return' or 'restoration.' A second group is represented by phrases like 'the revolutions of time' and 'the revolutions of empires.' Although a cyclical view of history was sometimes implied, these phrases often meant simply the passage of time and the fact that empires succeed each other historically. The third and most complex group has to do with unidirectional changes of various types. One kind of change might be 'revolutions of empires,' signifying either slow decline or sudden disaster or merely changes of dynasty in one state. Another species of the same genus was the concept of accumulated change, ultimately bringing about a new state of affairs; some writers used this idea in political history, while others described Isaac Newton's work as the completion of a 'scientific revolution' begun more than a century earlier.

---

[9] Carozzi, 'Une nouvelle interprétation du soi-disant catastrophisme de Cuvier,' *Archives des sciences*, Geneva, 1971, *24*, 367–77. Carozzi concludes, p. 374, that 'l'usage du mot 'révolution' par Cuvier ne peut pas être pris dans un sens général et à la lettre.'

[10] Especially valuable is Felix Gilbert, 'Revolution,' in *Dictionary of the history of ideas*, ed. Philip Wiener, 5 vols., New York, 1973, iv, 152–67. Pertinent quotations and extensive bibliography are in I. Bernard Cohen, 'The eighteenth-century origins of the concept of scientific revolution,' *Journal of the history of ideas*, 1976, *37*, 257–88. In addition to these and to works cited below, I have consulted the dictionaries of the Académie française (1694 and later editions), Nathan Bailey (1736, 1764), and Samuel Johnson (1755), as well as other pertinent passages in Dr Johnson's works, using as my guide W. K. Wimsatt, jr, *Philosophic words*, New Haven, 1948.

32

Given so rich an array of possibilities, it would be rash to state that one meaning was more common than another in the eighteenth century. David Hume, for example, uses 'revolution' in three different ways within the space of three consecutive paragraphs in his *Dialogues concerning natural religion*.[11] It is nevertheless my impression that the connotation of change-in-the-direction-of-novelty was gaining ground in this period, as was certainly true in the realm of political history. To cite but two political examples—examples of utterly different political complexions, but agreed in matters of definition—both the Diderot *Encyclopédie* and Edmund Burke considered a revolution to be 'an important change in the government of a State.'[12] Burke's meaning is especially clear when he denies that the English (1688) and American Revolutions were revolutions at all; they were restorations, with some necessary adjustments, of the way things ought to have been all along. The Encyclopedist, writing long before the turbulence of 1776 and 1789, cites as his authority on the meaning of political revolution a now obscure writer named the abbé Vertot.

Vertot, some of his contemporaries, and a host of successors all wrote about revolutions essentially political, and they popularized the notion that 'revolution' meant novelty of one sort or another. To judge from the numbers of editions and translations, Vertot's own works were by far the most popular in an otherwise unscholarly and unexciting array of narratives.[13] In an age when history is not easily divorced from scholarship, the modern reader can still understand why Vertot's works were so successful: they are unfailingly lively, informative, and intelligent, and Vertot's gift for narrative is not marred in any important way by demands upon the reader's mind.

What Vertot actually says about 'revolution' is meagre and hardly as clear as the Encyclopedist's interpretation. In his treatment of Swedish history, he does explicitly deny that 'revolution' means a cyclical return, since one of the Swedish revolutions he chronicles is the establishment of Lutheranism as the state religion; this, says Vertot, was viewed in the sixteenth century as a return to the Gospels, but he, Vertot, considers this change to be a gradual spoliation of the Catholic Church in Sweden. Vertot's political revolution in Sweden was the substitution of a hereditary monarchy for an elective one, and this, like the religious change, was a matter of piecemeal changes over a period of some fifty years. If one may sum up Vertot's unphilosophical and unreflective texts, it seems that

---

[11] David Hume, *Dialogues concerning natural religion*, ed. Henry D. Aiken, New York, 1957, part VI, p. 45. The meanings here are great alterations, cyclical changes, and the passage of time.

[12] Denis Diderot and Jean d'Alembert, *Encylopédie*, 28 vols., Paris, 1751–72, xiv (1765), 237: article 'Révolution,' unsigned.

[13] Among the many works with 'revolutions' in their titles, I have looked at those by Carlo Denina, P.-F. Guyot Desfontaines, Laurence Echard (or Eachard), and Benedetto Varchi. Standard catalogues (Bibliothèque nationale, British library, Heinsius) list not only many French editions of Vertot's works, but also translations ranging from the German to the Catalan; after 1800, certain of his works also appeared in Spanish and Russian.

revolutions are cumulative changes resulting eventually in the introduction of a new era in Swedish history.[14]

Neither Vertot nor his successors emphasized violence or suddenness as characteristics essential to revolutions. Nor, in fact, did they stress magnitude. A revolution may indeed possess all these features, but it may also consist of a series of minor changes of an additive sort. That this lack of precision was typical of minds better than Vertot's is evident when Gibbon could describe the decline and fall of Rome as both a 'series of revolutions' spread over thirteen centuries and a single revolution in the sense that these centuries ended with the disappearance of the Empire.[15]

With such verbal inconsistency prevailing among non-scientists, it is not surprising that geologists in search of a technical vocabulary found several uses for 'revolution.' One broad definition does, in fact, imply most of the variants detectable in geological texts, and this definition is succinctly stated in the Diderot *Encyclopédie*, article 'Révolutions de la terre' (1765): 'This is what naturalists call the natural events by which the surface of our earth has been and still is being changed in its sundry localities by fire, air, and water.'[16] Revolutions are merely events that succeed each other in time and that are local rather than universal in extent. Although rates of change are not made explicit, fire was commonly associated with the rapid and violent (as in volcanoes), water with both rapidity and slowness (floods and gradual encroachments of the sea), and air with 'insensible' change (subaerial erosion). While not every writer would have restricted 'revolution' only to local events, this definition summarizes accurately the two most common features of geological revolutions: successive events, varying in rate and extent.

How 'revolution' acquired so broad a significance in geology cannot be charted with precision, but one may point to some milestones antedating the Encyclopedist's definition. As early as 1684, Thomas Burnet was using the term to mean changes other than universal catastrophes. There had been, in his view, only one catastrophe since the Creation—the Flood—but the earth would undergo 'as many general Changes and Revolutions of Nature in the remaining part [of its history] as have already happen'd.' A second example comes from Fontenelle, who was doubtless familiar with the work of the abbé Vertot, his colleague in the Royal Academy of Inscriptions. In a series of texts dating from 1718–22, Fontenelle employed 'revolutions' to sum up what two naturalists, Réaumur and Antoine de Jussieu, described as local invasions of the

---

[14] René-Aubert Vertot, *Histoire des révolutions de Suède*, 2 vols., Paris, 1695, i, Avertissement. Henry Guerlac has called to my attention the fact that Vertot's work on Portugal ends with the *restoration* of the Braganzas. When first published in 1689, however, this work was entitled, *Histoire de la conjuration du Portugal*; revised and reissued in 1711, it was given a title of proven success, with 'révolutions' replacing 'conjuration.'

[15] Edward Gibbon, *The history of the decline and fall of the Roman empire*, 3rd ed., 6 vols., London, 1777–88, vol. i, pp. iii, 1–2.

[16] *Encyclopédie* (12), xiv, 237–8, article unsigned.

34

sea. Neither scientist wanted to use that universal catastrophe, the Flood, and they differed on whether the submersion of parts of France should be called a local flood or a slow encroachment. Fontenelle, however, blurred such distinctions, applying 'revolutions' to a variety of past changes. Not content to summarize the work of his colleagues, he went on to speculate whether revolutions in the past might have been greater in magnitude than at present, and he described geologists as historians of those revolutions the earth has undergone. Jussieu found this vocabulary sufficiently attractive to employ it himself shortly thereafter.[17]

Buffon, too, adopted 'revolution,' although his precise meaning is not always apparent. In one passage in his theory of the earth (1749), for example, he declared:

> There is no doubt . . . that there has taken place an infinite number of revolutions, upheavals (*bouleversements*), specific changes and alterations on the surface of the earth, due as much to the natural movement of the seas as to the action of rains, frosts, running water, winds, subterranean fires, earthquakes, floods, etc.[18]

If revolutions differ from upheavals and specific changes, then it is hard to see what role remains for the word. If these terms are intended to be roughly synonymous, then a revolution can be any kind of change whatever. Despite some ambiguity, Buffon nonetheless associates revolutions with local, successive changes, past and present.

In addition to such texts, one of Diderot's most important collaborators, the Baron d'Holbach, had before 1765 consistently applied *révolution* to a wide variety of geological events. The German text being translated by d'Holbach contained several vague and several specific references to geological *Veränderungen*, and d'Holbach meticulously followed J. G. Lehmann in his use of the noun and of qualifying adjectives; *Veränderungen* becomes *révolutions, kleine Veränderungen* are *révolutions particulières*, and the biblical Flood, 'eine gewisse allgemeine Veränderung,' is 'une révolution générale.' In the *Encyclopédie* itself, d'Holbach retained this broad usage when he said that although 'some scientists' consider mountains to be the product of post-diluvian, violent 'local revolutions,' it is truer to say that mountains and other 'inequalities' in the earth's crust were the result of 'revolutions which the earth has undergone and still undergoes today' and which take place 'either suddenly or little by little.'[19]

The Encyclopedist's broad definition of geological 'revolutions' was thus implicit in pre-1765 texts, and it can also be found in later writings.

[17] Thomas Burnet, *The theory of the earth*, London, 1684, p. 326. Fontenelle, in *Hist. Acad.*, 1718 (1719), p. 5; 1720 (1722), p. 8; 1721 (1723), pp. 1–4; 1722 (1724), p. 4. Jussieu, 'De l'origine des pierres appellées yeux de serpents et crapaudines,' *Mém. Acad.*, 1723 (1725), p. 205.
[18] *Oeuvres philosophiques de Buffon*, ed. Jean Piveteau, Paris, 1954, p. 104B. This passage comes from the 'Conclusion' following the 'Preuves de la théorie de la terre.'
[19] J. G. Lehmann, *Versuch einer geschichte von flötz-gebürgen*, Berlin, 1756, pp. 18, 20, 81, compared with Lehmann, *Traités de physique, d'histoire naturelle, de minéralogie et de métallurgie*, tr. d'Holbach, 3 vols., Paris, 1759, iii, 102, 104–5, 192. D'Holbach, 'Montagnes,' in *Encyclopédie* (12), x (1765), 672.

Those passages to be treated below have in common the idea that revolutions are local and successive, although rates of change vary, depending on the type of geological event under discussion. An admirably general use of 'revolutions' occurs in a classic memoir by Giovanni Arduino (1774): 'the visible parts of this Earth are not primordial in form and structure, but are the result of sundry revolutions and transformations (*varie rivoluzioni e metamorfosi*) which the Earth has undergone.'[20] Arduino then proceeds to rehearse the types of changes successively wrought by water and by fire, acting singly and in combination. Equally broad was the definition produced by Arduino's friend, the Swedish geologist J. J. Ferber:

> Every event (*événement*) gives rise to a new stratum, hill, or mountain, by the destruction of other strata and rocks. . . . Every revolution has produced and still produces one or several new strata on the earth's surface by the surbedding of materials brought from elsewhere. . . . These strata . . . are more or less thick, broad, and extended, depending on the type, energy, and duration of the revolution which produced them.[21]

Large, grand, and vague as are these statements, John Playfair's is still more insubstantial. Recalling when he and Sir James Hall had accompanied James Hutton on a field trip, Playfair described how Hutton moved his audience to consider the long history of changes suffered by the earth. This experience stirred Playfair to reflect that 'Revolutions still more remote appeared in the distance of this extraordinary perspective.'[22]

These post-1765 statements differ little from the writings of Fontenelle in their generality and ambiguity. At the same time, with so vague a term at their disposal, geologists could logically be expected to adapt 'revolution' to suit their individual specialities. The extent to which they did so becomes evident when one compares geologists whose interests lay in mountain-building on the one hand or sedimentation on the other.

Among those especially interested in mountains, the examples of Peter Simon Pallas and H.-B. de Saussure are instructive and dissimilar. Pallas seems to have thought of mountain-building as necessarily violent, and geological dynamics as a whole became a matter of explosions and floods, probably more violent in the past than at present. Saussure is more complex in his thinking, mainly because he was even more hesitant than

---

[20] Giovanni Arduino, 'Saggio fisico-mineralogico di lythogonia, e orognosia,' *Atti dell' Accademia delle scienze di Siena*, 1774, *5*, 230. An earlier statement of Arduino's ideas, but without the vocabulary of 'revolutions,' is in 'Due lettere del Sig. Giovanni Arduino sopra varie sue osservazioni naturali,' *Nuova raccolta d'opuscoli scientifici e filologici*, Venice, 1760, *6*, xcvii–clxxx.

[21] J. J. Ferber, 'Réflexions sur l'ancienneté relative des roches et des couches terreuses qui composent la croute du globe terrestre,' *Acta academiae scientiarum imperialis petropolitanae*, 1782 (1786), *6*, ii, 193–4, 198.

[22] Quoted from Playfair's life of Hutton (1805), in Gordon Y. Craig, ed., *James Hutton's theory of the earth: the lost drawings*, Edinburgh, 1978, p. 23.

36

Pallas to commit himself to any theoretical views; indeed, his geological testament, the 'agenda' he wrote late in life to direct the future research of naturalists, displays the uncertainties that can readily explain why he never wrote the 'theory of the earth' he had in mind. Saussure begins one section of his 'agenda' with the statement that *grandes révolutions* antedated the keeping of written records. Whether he associated these revolutions with violence remains unclear, since he goes on to speculate about a variety of mechanisms possibly responsible for uplift. As noted earlier in connection with d'Holbach, 'some scientists' did indeed associate mountain-building with violence, but Saussure refused to choose between violence and what he called the deceptive appearance of violence.[23]

Decidedly slow and tranquil were the revolutions invoked by Lamarck and perhaps by Ambrogio Soldani as well. Both were concerned mainly with sedimentary formations, and Soldani, when he raised the traditional question of how to explain the presence of marine fossils high in the Appenines and deep below the earth's surface, concluded that the earth has undergone one or more revolutions during which the sea changed its basin. In similar fashion, Lamarck—whose aversion to violent mechanisms is well known—asserted that fossils in general and marine forms in particular provide 'one of the principal means of determining the revolutions which took place over the earth's surface.'[24]

Some writers were careful to dissociate revolutions from disorder, insisting instead upon uniform causes and laws. Von Buch, for example, rejected the Plutonist interpretation of the formation of granite in part because fire was an agent of disorder, and 'nature, in its seemingly most terrible revolutions, always follows the same fixed, immutable, and beneficent laws.' Jean Darcet, a partisan of violent mountain-building, saw uniform laws behind those *bouleversements* which had produced the Alps and the Pyrenees, 'because why would not the same causes, given the same circumstances, produce the same effects?' To Desmarest, however, the term *bouleversement* was to be rejected as synonymous with disorder, and 'revolution' was the preferred alternative, signifying change produced by known and regular causes. Desmarest by no means denied that violence had played a role in the geological past, but he would have agreed with the abbé Fortis that the earth's history was not one of successive cataclysms; despite the violence of earthquakes and volcanic eruptions, the earth had

---

[23] Pallas, op cit. (8), pp. 35–6. Elsewhere, he refers to the greater intensity of revolutions during the early history of the earth: *Voyages de M. P. S. Pallas, en différentes provinces de l'empire de Russie, et dans l'Asie septentrionale*, tr. Gauthier de la Peyronie, 6 vols., Paris, 1788–93; i, 697. Saussure, *Voyages dans les Alpes*, 8 vols., Geneva, Neuchâtel, 1787–96, viii, 253, 271, 274–5. The 'agenda,' reprinted more than once, can also be found in the *Journal des mines*, floréal an IV [1796], *4*, no. 20, 1–70.

[24] Ambrogio Soldani, *Saggio orittografico*, Siena, 1780, p. 89. J. B. de Lamarck, *Hydrogéologie*, Paris, 1802, pp. 64–5; translation from Lamarck, *Hydrogeology*, tr. Albert V. Carozzi, Urbana, Ill., 1964, p. 58. Comparable statements are in Lamarck's *Système des animaux sans vertèbres*, Paris, 1801, p. 406, and in his 'Mémoires sur les fossiles des environs de Paris,' *Annales du Muséum national d'historie naturelle*, 1802, *1*, 299.

fundamentally had a history of successive events requiring long periods of time.[25] The foregoing examples do not exhaust the possibilities in the word 'revolution,' and some eighteenth-century geologists did use the term to mean universal catastrophes or cataclysms. Not surprisingly, some examples are furnished by those who sought to integrate a literal or figurative interpretation of the biblical Flood into the geological history of the earth. Others, like J. F. Blumenbach, might be uncertain about the universality of the Flood, but they did not object to the concept of universal catastrophes of different sorts. Noteworthy, however, is the language employed for world-wide events: adjectives like 'general' or 'universal' are usually provided, or 'catastrophe' or 'cataclysm' takes pride of place over 'revolution.'[26] One may, then, with some degree of confidence, distinguish the universal from the local revolution, and conclude that the latter meaning was far more common than has generally been acknowledged by historians of geology.

Why geologists found so useful a term so versatile as to defy exact definition remains a question with no definitive answer—but a tentative one may be suggested. Nicolaus Steno's *Prodromus* (1669) clearly offered an analysis of successive historical changes, although Steno had no word or phrase to summarize such a concept. Recognizing this lack—and adding dynamic elements of this own—Robert Hooke proposed a theory of 'earthquakes,' using that word to include not only seismic shocks, but also such phenomena as folding, faulting, and other displacements of the earth's crust. Hooke's theory was rejected by contemporaries who took 'earthquakes' to mean seismic tremors and who thus complained that history did not record enough events of this kind to account for the present state of the earth's crust.[27] Had Hooke developed this theory twenty years later, he might well have said 'revolutions' instead of 'earthquakes,' and with different results. In fact, since 'revolution' described events large and small, sudden and gradual, its chief function seems not to have been to characterize any event, but rather to indicate changes occurring in series. What geologists apparently wanted was a word signifying the successive

---

[25] *Leopold von Buch's gesammelte schriften*, ed. J. Ewald, J. Roth, H. Eck, and W. Dames, 4 vols. in 5, Berlin, 1867–85, i, 129. The quoted text, first published in 1800, occurs in a letter to Pictet on the controversy between Kirwan and Hall. Darcet, *Discours en forme de dissertation sur l'état actuel des montagnes des Pyrénées, et sur les causes de leur dégradation*, Paris, 1776, p. 21. Desmarest, *Encyclopédie méthodique: géographie physique*, 5 vols., Paris, 1794–1828, especially i (1794), 341; iii (1809), 197–8; iv (1811), 39–40. A. Fortis, *Mémoires pour servir à l'histoire naturelle et principalement à l'oryctographie de l'Italie, et des pays adjacens*, 2 vols., Paris, 1802, i, 194–5.

[26] Rappaport, op. cit. (7), especially p. 8, and above (19). Similarly qualified use of 'revolution' can be found in a historian aware of the content of geological literature: A. L. von Schlözer, *Weltgeschichte nach ihren haupttheilen im auszug und zusammenhange*, Göttingen, 1785, pp. 18–19, 36, 40.

[27] Hooke had developed his theory shortly before the publication of Steno's *Prodromus*, and he continued to lecture on 'earthquakes' for some years thereafter. For a discussion of this theory, see Carozzi in R. E. Raspe, *An introduction to the natural history of the terrestrial sphere*, tr. and ed. by A. N. Iversen and A. V. Carozzi, New York, 1970, pp. xxiv, 155–7. For objections by Hooke's contemporaries, see his own summary in *Posthumous works* (6), p. 404.

38

historical changes that have modified the earth's crust, and the term adopted was suitably versatile and increasingly familiar in the non-scientific literature of their day. Like 'monument,' 'revolution' possessed some novelty in the early decades of the century, and it became a conventional, normal, unexamined part of the vocabulary of Georges Cuvier and his older contemporaries.[28]

Accidents: the legacy of rationalism

My third geological term has long been familiar to students of eighteenth-century philosophy and historiography, although the concept of 'accident' has not always been associated with the particular word. To rationalist historians like Voltaire or Hume, the customs or beliefs prevailing at any given moment in history were 'accidents' attributable to time, place, and circumstance; such matters as the nature of man and, in Voltaire's phrase, 'the spirit of nations' were non-accidental and quite fundamental. The historian William Robertson could thus refer to several centuries of the Middle Ages as consisting of 'a succession of uninteresting events; a series of wars, the motives as well as the consequences of which were equally unimportant, fill and deform the annals of all the nations of Europe.'[29] That there is real history on the one hand and unfortunate aberrations on the other was also made plain by Dugald Stewart:

> it is of more importance to ascertain the progress that is most simple, than the progress that is most agreeable to fact; for . . . it is certainly true, that the real progress is not always the most natural. It may have been determined by particular accidents, which are not likely again to occur, and which cannot be considered as forming any part of that general provision which nature has made for the improvement of the [human] race.[30]

As a modern writer puts it, only in the eighteenth century could anyone have said that 'the matter of history could "deform" history itself.'[31]

The relevance of rationalist philosophy to the history of science has not gone unnoticed, although it has not received much emphasis. Well

---

[28] I do not know precisely when such words became old-fashioned or obsolete in geology, and, in any event, to seek precision in dating changes of usage is hopeless. One topic which could and should be clarified, however, is the transformation of 'revolution' from an ambiguous word into a 'catastrophic' one. Comparison of Cuvier's *Discours* (1812) with Kerr's translation (1813) would be an obvious starting point, but this should be followed by comparisons of British and Continental usage in the twenty years between Cuvier and Charles Lyell. Rudwick, op. cit. (6), chapter 3, is helpful in this respect.

[29] William Robertson, *The history of the reign of the Emperor Charles V*, 3 vols., London, 1769, i, 18.

[30] Dugald Stewart, 'Account of the life and writings of Adam Smith, LL.D.,' in Smith, *Essays on philosophical subjects*, ed. Joseph Black and James Hutton, London, 1795, p. xlvi. The same passage is cited by Burkhardt, below (32). Stewart also remarked: 'when we cannot trace the process by which an event *has been* produced, it is often of importance to be able to show how it *may have been* produced by natural causes,' and he praised the 'theoretical history' of Montesquieu who avoided 'bewildering himself among the erudition of scholiasts and of antiquaries.' Stewart, pp. xlii–xliii. Italics are in the original.

[31] Douglas M. Barnes, 'Edmund Burke and the history of the middle ages in the eighteenth century,' unpublished senior thesis in history, Vassar College, 1980, p. 56. I am indebted to Mr. Barnes for calling to my attention the passage in Robertson, quoted above (29).

known, however, is the idea common in the Enlightenment that the existing races of man are all accidental, in the sense that variations of skin color and physical conformation are temporal deviations from an original type. A recent and admirable study of Lamarck makes it plain, too, that Lamarck's treatment of the main lines of evolutionary development, with fruitless deviations identified, bears a striking resemblance to the philosophy of Dugald Stewart and other rationalist proponents of 'accidents.'[32]

In early geology, as in rationalist history, so-called accidents include virtually everything imaginable, depending upon what each author conceived to be the normal and the natural. But how did one know what was normal? Opinions might change from one day to the next, and the perils of such a position are implicit in a statement reportedly made by Rembrandt Peale: 'double-headed snakes were so frequently met with in America, that [Americans] considered them as species, and not as monsters.' A similar problem appears in Lavoisier's geological writings where it is plain that he thought sedimentary strata in any one locality ought to be either littoral or pelagic; alternations of the two types he considered 'accidental,' until he decided that such alternations were more common than he had earlier supposed.[33] Like Peale, Lavoisier does not tell us when a phenomenon is sufficiently common to be removed from the category of 'accidents' and transferred to the realm of normalcy.

The early use of 'accidental' to describe certain fossils—i.e., to distinguish organic remains ('accidents') from other curious stones ('natural' fossils) in the strata—indicates that the concept was hardly new in the eighteenth century.[34] But usage during the Enlightenment became considerably broader, in that geological 'accidents' signified such diverse matters as changes in the original position of strata, chemical changes in rocks and minerals, and almost any local (as distinct from universal) phenomenon. Daubenton, for example, provides one element in this broad picture when he says that, since minerals are constantly undergoing chemical alterations, there is a difference between 'the essential characteristics of plant and animal species and the accidental characteristics of minerals.' Arduino offers an analogous statement: 'bituminous substances are not originally part of the mineral kingdom, but are accidental and

---

[32] Richard W. Burkhardt, jr, *The spirit of system: Lamarck and evolutionary biology*, Cambridge, Mass., 1977, pp. 87, 142, 145–7. In traditional literature, the races of man all derived from the sons of Noah, but eighteenth-century writers emphasized not origins but rather the effects of climate, food, and way of life in producing racial variations in one or more original types.

[33] Peale is quoted in John Hunter, *Essays and observations on natural history, anatomy, physiology, psychology, and geology*, ed. Richard Owen, 2 vols., London, 1861, i, 251 n. 5; also, Hunter's own views, i, 239, 240, 251, in his essay entitled, 'On monsters.' R. Rappaport, 'Lavoisier's theory of the earth,' *The British journal for the history of science*, 1973, *6*, 252–3, 254–5.

[34] Early examples are legion, but Louis de Launay's terminology, in his *Mémoire sur l'origine des fossiles accidentels des Provinces Belgiques*, Brussels, 1779, seems a bit unusual by that late date. By the time Desmarest objected to such usage, it was no longer common, but he pointed out that fossil shells were so common in the strata that they could hardly be called 'accidents.' Desmarest, op. cit. (25), i, 415. Also, Rudwick, op. cit. (6), chapters 1–2.

40

derived on the whole from the plant and animal kingdoms.'[35] Local disturbances of the original positions of sedimentary strata were also accidental, as were local exceptions to the general order of primary, secondary, and tertiary formations. As Ferber put it:

> Nature remains faithful to its principles when it acts on a large scale; it is for us to grasp these principles and not to think nature has departed from them, at the first little observation that seems to us unusual if we do not examine it properly.[36]

In short, accidents must be reinterpreted to fit into the known pattern of earth history. That this should be the aim of geologists was later stated in broad terms by Alexander von Humboldt; since, he argued, the sciences rest mainly on inductions, 'the more complete these inductions, then the more local circumstances accompanying each phenomenon will be excluded from the formulation of general laws.'[37]

In retrospect, it is apparent that two kinds of philosophical confusion were inherent in the concept of 'accidents.' For one thing, the traditional search for natural laws was apparently being conducted without adequate attention to the problem of how one distinguishes normality from deviation. Avowed empiricists, geologists had not got the admittedly dense message of Francis Bacon that proper inductions ought to be based on careful enumerations, distinctions, and exclusions, so that one could reach conclusions in the affirmative.[38] Proper inductions are, alas, difficult. In the context of this article, a second area of confusion is more important: accidents and revolutions are in danger of becoming indistinguishable. Accidents are always local, revolutions usually so. But an accident is *merely* local and thus not very important, while revolutions are part of a recurrent, common pattern. This rather slippery distinction can be illustrated by a discussion of eighteenth-century opinion on lacustrine, as opposed to marine, deposits, and by some examination of the reception accorded to the study of the Paris Basin published by Cuvier and Brongniart in 1808. As we shall see, this monograph undermined the concept of 'accident,' while leaving 'revolution' untouched.

For most eighteenth-century geologists, there was one normal and natural way of thinking about the history of the more recent parts of the earth's crust, namely, that sedimentary formations had been laid down at the bottom of the sea. Decades of struggle against diluvialist explanations of sedimentary formations seem to have planted firmly in geologists' minds

---

[35] Daubenton, *Encyclopédie méthodique: histoire naturelle des animaux*, 10 vols., Paris, 1782–1825, I, iii–iv. Arduino, in *Atti* (20), p. 286; also, pp. 229, 293. A common word then for 'accidental' was 'adventitious,' which is also used by Arduino.
[36] Ferber, 'Réflexions' (continuation of article cited above, n. 21), *Nova acta academiae scientiarum imperialis petropolitanae*, 1784 (1788), 2, 178.
[37] Humboldt, *Essai géognostique sur le gisement des roches dans les deux hémisphères*, Paris, 1823, p. 63.
[38] Based mainly on Francis Bacon, *Novum organum*, Book I, aphorism 105, with apologies for the element of parody in my summary. Also, F. H. Anderson, *The philosophy of Francis Bacon*, Chicago, 1948, chapter 18.

the central principle of gradual, successive, marine deposition, and the classic memoir on this subject, Réaumur's early study of the faluns of Touraine, became a standard and unimpeachable part of geological literature.[39] With this principle firmly established, it is not surprising that such matters as volcanoes and earthquakes should have been treated as deviations from the normal; they were studied, but they were not to be given undue importance in the earth's history. The occasional specialist in things volcanic might hope that the traditional balance of water as more important than fire could be upset,[40] but the evidence, as in the case of Hooke's 'earthquakes,' seemed insufficient, even if a good many hitherto unrecognized volcanic sites were being identified in the latter half of the century.

If marine sedimentation supplied the normal pattern, then the intrusive, disturbing elements were not so much volcanoes as apparently *non-marine sediments*, i.e., sediments containing freshwater or terrestrial fossils. That specimens of the latter kinds were known throughout the century is clear enough, but how they were interpreted remains a tale as yet only partly told. The terrestrial forms, few in number, had been poorly identified—the classic case being Scheuchzer's *Homo diluvii testis*—but, even with more specimens available from about the 1760s, and even with Cuvier's reconstructions and identifications dating from the 1790s, it was clear that these would pose no serious problem in the general picture of the earth's relatively recent history. Decisive was the fact that many such specimens were found in recent alluvial terrains, clearly formed after the retreat of a universal ocean or after the last of several invasions of the sea. For some terrestrial specimens found in conjunction with marine shells, it was abundantly clear that the terrestrials had been washed into the sea. The great quadrupeds that established Cuvier's reputation were no threat to the accepted story of marine sedimentation.[41]

Freshwater molluscs posed a more delicate problem. Too little was known about these creatures, in part because collectors of shells had concentrated upon the more dramatically colored marine forms. Not only were freshwater shells less well-known than the marine, but paleontologists faced the additional problem that they possessed only the petrified shells and not the living animals. Was knowledge of the shell sufficient for knowledge of the environment of the animal? Was it possible that living freshwater molluscs might once have been adapted to live in the sea? Even

[39] References to the Touraine, often without mention of Réaumur's name, are too numerous to list. Some discussion is in Rappaport (7), p. 9.
[40] For example, William Hamilton, *Campi Phlegraei*, Naples, 1776, p. 6. Apparently more common was the attitude of two vulcanologists, Desmarest and Dolomieu, who continued to think of the history of the earth in essentially neptunist terms; valuable articles on these men, both by Kenneth L. Taylor, are in the *Dictionary of scientific biography*, ed. C. C. Gillispie, 16 vols., New York, 1970–80, iv, 70–3, 149–53.
[41] For the great quadrupeds in particular, see John C. Greene, *The Death of Adam*, Ames, Iowa, 1959, chapter 4, and Rudwick (6), chapter 3.

42

if one admitted that certain molluscs were indubitably freshwater, then and now, was it not possible that such forms had been carried into the sea, deposited there, and eventually petrified? This assemblage of questions occurs in no single text before 1808, because almost no one saw freshwater forms as a fundamental problem. Arduino, for example, refers in passing to deposits seemingly lacustrine, but he goes quickly on to say that the sea is the great source of sedimentary formations. More important is the testimony of Lamarck who, unlike Arduino, paid more attention to fossils than to rocks; but Lamarck, like Arduino, dismisses freshwater species as few in number and presumably carried into the sea by the rivers.[42] One of the few geologists to object to this common pattern of interpretation was the Chevalier de Lamanon, who insisted that he had discovered deposits left by freshwater lakes; more such examples would certainly be found, he argued, if naturalists would only look for them.[43]

Cuvier and Brongniart later mentioned Lamanon's name with respect when their own study of the Paris Basin revealed two formations containing freshwater forms and separated by others clearly marine in origin. Unlike Lamanon's work, this monograph did indeed provoke other geologists to make similar finds, but it also aroused opposition from what might be called rationalist critics. These not only assembled the questions outlined above, but they also accused Cuvier and Brongniart of having elevated a mere 'accident' into a position of undue importance in the history of the earth's crust.

One of the first to respond to the Cuvier–Brongniart study was their older colleague at the Muséum, Barthélemy Faujas de St.-Fond. As Faujas put it rather grandly, his confrères had studied 'geology in miniature and à l'eau douce; this is to confound very small effects with great causes.' In a similar vein, one of Faujas' disciples stressed the unlikelihood of such curious behavior by 'two different fluids coming and going several times in order to form some little hills whose maximum height is one hundred toises.'[44] Coupled with the philosophical criticism was examination of certain fossil and living species, so that the critics might raise doubts about the evidence on which Cuvier and Brongniart had based their interpretation. Brongniart replied in detail in 1810, indicating that he and Cuvier had not relied upon doubtful species, but on large numbers of well-known

[42] Arduino, in *Atti* (20), p. 287. Lamarck, 'Mémoires sur les fossiles des environs de Paris,' *Annales du Muséum national d'histoire naturelle*, 1804, *4*, 113, 298.
[43] Robert de Paul de Lamanon, 'Description de divers fossiles trouvés dans les carrières de Montmartre près Paris,' *Journal de physique*, March 1782, *19*, 193. Also, Lamanon, 'Mémoire sur un os d'une grosseur énorme qu'on a trouvé dans une couche de glaise au milieu de Paris,' *Journal de physique*, May 1781, *17*, 393–405. Lamanon died a few years later, having gone as naturalist with the famous Lapérouse expedition to circumnavigate the globe.
[44] Articles by Faujas and P. Brard, in *Annales du Muséum national d'histoire naturelle*, 1809, *14*, 324, and 1810, *15*, 421. For an excellent discussion of the Cuvier-Brongniart monograph, see Rudwick, 'Brongniart,' in *Dictionary* (40), ii, 493–7.

forms found in strata devoid of marine shells. Furthermore, his critics had committed that most serious of sins, he said: they had interpreted new and anomalous observations to fit into preconceived 'systems.' And he went on to offer what can only have been a tongue-in-cheek alternative to his and Cuvier's interpretation: if large assemblages of freshwater shells had in fact been deposited by the sea, then naturalists should find it of great interest to learn that the sea had at various times been inhabited solely by freshwater forms.[45]

This debate was not confined to the halls of the Muséum or the pages of the *Annales du Muséum*. In 1810 John Farey entered the lists, in 1811 Scipione Breislak, and in subsequent years Constant Prévost, Alexander von Humboldt, and others. On the whole, the problem of 'accidents' did not dominate such texts—although the issue was often mentioned or alluded to—and the authors instead debated such questions as the precise identification of species, the possibility that species now living in fresh water had been adapted to different environments in the past (or that the sea itself had been less salty in the past), and the question of whether rivers emptying into the sea could account for the disturbing observations of Cuvier and Brongniart and their successors. That the issues should have been put in these ways does suggest that the concept of 'accidents' was still alive and well, in that critics were doing their utmost to preserve the 'system' of marine sedimentation. Their real doubts about the evidence should not be minimized, but it is refreshing, after reading their works, to turn to Brongniart's blunt statement: when one sees a formation containing only freshwater shells, one concludes that they were deposited by fresh water.[46]

As I have tried to show, the 'accident' remained a living concept in both history and geology for a good many decades. That this idea came under attack in both disciplines at about the same time may well be coincidental, as it assuredly is in the case of Lamanon and J. G. Herder being contemporaries. But when Cuvier and Brongniart undermined 'accidents'—despite the emphasis of this article, many did greet their findings with enthusiasm—they were living in a period when the pioneering efforts of Herder had at last begun to arouse among historians the desire to recreate the past fully and in all its diversity. The 'theoretical history' praised by Dugald Stewart has its counterpart, so to speak, in the system of marine sedimentation, while the historicism of Barthold Niebuhr flourished alongside the monographic destruction of the concept of

---

[45] Brongniart, 'Sur des terrains qui paroissent avoir été formés sous l'eau douce,' *Annales du Muséum*, 1810, *15*, 357–405 (especially 357–9, 403–5). He indicates (p. 361) that the much enlarged, more detailed version of his and Cuvier's monograph was then in press.

[46] John Farey, 'Geological remarks and queries on Messrs. Cuvier and Brogniart's [sic] memoir on the mineral geography of the environs of Paris,' *Philosophical magazine*, 1810, *35*, 113–39. Scipione Breislak, *Introduction à la géologie*, tr. J. J. B. Bernard, Paris, 1812, pp. 257–8. (The original Italian edition was published in 1811.) G. P. Deshayes, in *Encyclopédie méthodique: histoire naturelle des vers*, 3 vols., Paris, 1789–1832, ii (1830), 355–6. Humboldt, op. cit. (37), pp. 37–8, 48–9. Brongniart, op. cit. (45), p. 358.

44

'accidents' by Cuvier and Brongniart. Whether there are causal connections between these changes occurring in history and geology remains unknown, but the two disciplines began simultaneously to share the idea that there are no deviations from nature, and that whatever 'happens to happen' is an integral part of the evidence to be used in a full reconstruction of the past.[47]

[47] The quotation is based on J. H. Hexter, *Reappraisals in history*, 2nd edition, Chicago and London, 1979, p. 16. Especially interesting studies of relationships between historians and scientists in this period are: Owsei Temkin, 'German concepts of ontogeny and history around 1800,' *Bulletin of the history of medicine*, 1950, *24*, 227–46, and M. J. S. Rudwick, 'Historical analogies in the geological work of Charles Lyell,' *Janus*, 1977, *64*, 89–107. Also, D. R. Oldroyd, 'Historicism and the rise of historical geology,' *History of science*, 1979, *17*, 191–213, 227–57.

XII

DANGEROUS WORDS:
DILUVIALISM, NEPTUNISM, CATASTROPHISM *

In *The Great Chain of Being* Arthur O. Lovejoy remarked about -isms that they are «trouble-breeding and usually thought-obscuring terms, which one sometimes wishes to see expunged from the vocabulary of the philosopher and the historian altogether». Such wishes being impractical, Lovejoy himself had already shown how these terms could be analyzed, as in his essay *On the Discrimination of Romanticisms* (1924). Decades later, Alfred Cobban would encounter incomprehension and resistance when he argued that «omnibus terms» and «over-full» categories – in this case, «bourgeoisie» and «peasantry» – obscured rather than clarified social analysis. More recently, Paolo Rossi has reminded us that labels like «progressives» and «reactionaries» are categories we impose on the past; the words express our own preferences and may well misrepresent how historical figures were viewed by their own contemporaries.[1]

Historians have long been aware of these perils, thanks to Herbert Butterfield's classic *Whig Interpretation of History* (1931), and various writers, including Butterfield, have examined the pros and cons of Whiggism in the history of science in particular.[2] Knowledge in the sciences does 'progress' in a way that politics or religion does not, and historians of science can thus more readily distinguish between the winners and losers of the past. If it has taken

---

* The main arguments of this essay were outlined in a speech delivered in August, 2003, and published in «Northeastern geology and environmental sciences», XXVI (2004), pp. 107-109.

    [1] A.O. LOVEJOY, *The great chain of being* (1936), repr. NY, Harper, 1960, pp. 5-6; the 1924 article was reprinted in LOVEJOY, *Essays in the history of ideas*, NY, Putnam, 1960. A. COBBAN, *Aspects of the French revolution*, NY, Braziller, 1968, chaps. 14-15. P. ROSSI, *I segni del tempo*, Milano, Feltrinelli, 1979, pp. 304-308.

    [2] Among relatively recent studies, D.L. HULL, *In defense of presentism*, «History and theory», XVIII (1979), pp. 1-15, is notable for the range of issues discussed.

time and disciplinary maturity for historians to devote attention to the critics
of obvious winners like Galileo and Newton, it may seem that historians of
geology have never needed such reminders, as, in a tradition dating back to
Charles Lyell, they have customarily paid attention to both sides of disputes.
How they have done so, however, has more than once called forth from François Ellenberger a warning against Manichaeanism in histories of geology: the
simplistic classification of early geologists into antagonistic «schools», and the
dismissal without adequate analysis of what is stigmatized as «non-modern».
It was premature in 1987 for Rachel Laudan to say that «we no longer need to
argue» that -isms «confuse as much as they clarify in the history of geology».[3]

In the next pages, the three -isms of this essay will be handled separately,
although historical treatments sometimes describe neptunism as a form of diluvialism and diluvialism as prototypically catastrophic. That such combinations occur is one reason for the selection here of these terms; in addition,
they are remarkably «over-full» categories, too readily used to characterize
old, wrong, bad science. Illustrative examples, drawn from recent works
(since about 1970), have been chosen to show the range of meanings attached
to each -ism. Consideration of 18th- and early 19th-century texts will be jux-
taposed to analysis of modern usage.

1. DILUVIALISM

Modern dictionaries indicate that «deluge» and «diluvial» refer to a flood
or floods and especially Noah's. Among so many options, historians of geology have long agreed that Thomas Burnet, John Woodward, and their disciples were diluvialists who, despite the differences among them, gave to the
Biblical Flood a major role in producing the earth's landforms; «major» in
these cases usually carried the explicit corollary that only negligible changes
had occurred since Noah's day. Problems then multiply in historical studies:
do other naturalists who mention the Flood assign to it a major role, and what
constitutes «major»? Should one call «diluvialists» those who considered the
Flood to be a recent event in a lengthened history of the earth? Does the use
of episodic floods or flooding warrant the label «diluvialism»?

---

[3] F. ELLENBERGER, *Le dilemme des montagnes au XVIIIe siècle: vers une réhabilitation des dilu-
vianistes*, «Revue d'histoire des sciences», XXXI (1978), pp. 43-52, remarks on p. 52; the same
author's *Histoire de la géologie*, II, Paris, Technique et documentation/Lavoisier, 1994, p. 265.
R. LAUDAN, *From mineralogy to geology: the foundations of a science, 1650-1830*, Chicago, Univ. of
Chicago Press, 1987, p. 223.

In an effort to summarize how this protean -ism has been used, Roy Porter in 1981 produced the following definition:

In geology, the doctrine that the Earth's surface has been overwhelmed by one or more great deluges, which have transformed the landscape, or even the structure of the Earth's crust. Until the 18th century, these deluges were usually identified with the Flood of Noah [...]. Such great controversy raged as to the physical causes and plausibility of this universal Flood that the belief became discredited. It was replaced in the [early 19th century] [...] by a theory of successive natural deluges, supposed to have caused the extinction of animal populations, and to have transported masses of superficial debris, such as pebbles [...] and erratic boulders [...].
Diluvialism is also used to describe the theory that the origin of all strata lay in precipitation out of a universal ocean. This view [is] sometimes called Neptunism.

Not entirely in accord with this text, another of Porter's articles in the same dictionary informs us that «most geologists» before about 1800 thought basic changes in the earth's crust «were the work of water, often attributed to the miraculous agency of the Biblical flood». This explanation would then be replaced by a series of deluges, «not necessarily miraculous».[4]

A single -ism thus signifies one universal flood or several non-universal floods, a miracle or a natural event or events, a theory of successive deposition in water (neptunism), and characteristic violence or tranquillity. Perhaps missing here, although it might be subsumed under «natural deluges», is the tendency of historians to call «diluvial» the powerful currents that some early geologists thought responsible for the excavation of certain valleys.

One feature of Porter's definition, the miraculous, had indeed aroused controversy by about 1700, when Burnet's critics accused him of reducing a miracle to a natural process and when Woodward's use of miracle laid him open to the charge of reducing nature to mystery. By 1750, Vallisneri, Celsius, Buffon, and others were setting the Flood aside, declaring it beyond their competence to analyze a miracle. Far from vanishing, however, the issue was still alive in 1813 when John Playfair criticized Georges Cuvier for combining the Flood with several other ancient flood traditions. Playfair argued that one could lengthen the «days» of Creation, but one did not have such liberties where the Flood was concerned. That is to say, Moses was an «inspired writer» recounting a miracle, and readers could not add geological details not in conformity with the brief, tranquil event described in Genesis.[5]

---

4 In W.F. BYNUM et al. (eds.), *Dictionary of the history of science*, Princeton, Princeton U. Press, 1981, s.v. diluvialism, catastrophism.
5 In Playfair's review of the 1813 English translation of Cuvier's *Discours*, «Edinburgh Review»,

Since some naturalists did continue to weave the Flood into their writings, just how they did so merits more attention than the topic has received. How many adopted Woodward's strategy of saying that the *cause* of the Flood had been miraculous but its *effects* detectable in nature? How many espoused (or debated) pre-1700 arguments that the Flood had not been universal?[6] More generally, although much has been written about the various attacks on miracles by David Hume, Denis Diderot, and others, we have no study of what might be called «the status of miracles» in this period. For example, N.-S. Bergier, often said to be the best of French religious apologists, perceived this new and (to him) deplorable philosophical trend by the 1760s, and he could only insist that miracles were of a different order from natural events, their truth established by reliable human testimony. Other theologians, however, sometimes tried to meet philosophers on their own ground, arguing that miracles were believable because they could be explained naturalistically. One such divine, George Campbell, a critic of Hume, even tried to do this for resurrections, while another, Henri Griffet, S.J., tackled the easier job of doing so for the Flood.[7]

If these examples show some diversity among theologians, we have too little information to conclude for any consensus even within a single Christian denomination. Nor have historians paid enough attention to the Flood to detect any consensus among naturalists. Unusual in this respect, geologist Davis

---

XXII, 1814, pp. 467-469 discuss the days of Creation and the Flood. For the Flood set aside as a miracle, see ELLENBERGER, *Histoire de la géologie*, cit., pp. 43-44, and the remarks on Celsius in T. FRÄNGSMYR (ed.), *Linnaeus: the man and his work*, Berkeley, U. California Press, 1983, p. 131. For debate about the Flood, see R. RAPPAPORT, *When geologists were historians, 1665-1750*, Ithaca, NY, Cornell U. Press, 1997, chap. 5, and L.E. PAGE, *Diluvialism and its critics in Great Britain in the early nineteenth century*, in C.J. SCHNEER (ed.), *Toward a history of geology*, Cambridge, MA, MIT Press, 1969, pp. 257-271.

[6] For the decades around 1700, see ROSSI, *I segni*, cit., and DON CAMERON ALLEN, *The legend of Noah* (1949), repr. Urbana, Univ. of Illinois Press, 1963. M.S. SEGUIN, *Science et religion dans la pensée française au XVIIIe siècle: le mythe du déluge universel*, Paris, Honoré Champion, 2001: an uneven volume which at times presents pre-1700 arguments in such detail that both France and the 18th century get minimal attention. R. PORTER, *The making of geology: earth science in Britain, 1660-1815*, Cambridge, Cambridge Univ. Press, 1977, offers conflicting assessments for the mid-18th century: a non-universal Flood was «increasingly popular», pp. 107-108, and the universality of the Flood was an issue that «still largely dominated theoretical inquiry», p. 116.

[7] N.-S. BERGIER, *Le déisme réfuté par lui-même*, 3rd ed., Paris, Humblot, 1768, esp. part I, pp. 74-121. In the *Encyclopédie méthodique: théologie*, 3 vols., Paris, Panckoucke, 1788-90, Bergier did try to reply to rational and natural questions, but sometimes lost patience, exclaiming that the Flood was not natural and the questions therefore irrelevant (I, 506a, art. «Déluge universel»). G. CAMPBELL, *Dissertation sur les miracles*, trans. Jean de Castillon, Utrecht, H. Spruyt, 1768, pp. 42-44; the original English edition appeared in 1762. H. GRIFFET, *L'insuffisance de la religion naturelle*, 2 vols., Liège, Bassompierre, 1770, II, pp. 44-76.

A. Young has devoted a recent book to the subject, showing how scientific evidence gradually diminished the role of the Flood until it became, first, a recent event, then non-universal, and ultimately irrelevant to geology. In an analysis devoted chiefly to British and American scientists, Young comments on associating the Flood with Cuvier's catastrophism or with George Greenough's transport of erratics. Unlike interpreters who are prone to see Flood-geologists whenever the Flood is mentioned, Young argues that Cuvier and Greenough «stopped short» of clearly singling out the Flood or giving it an explicit geological role.[8]

Other historians have ranged less widely than Young, tending to focus on specific writers who, in the years around 1800, did use the Flood as a geological event. The favorites here seem to be William Buckland, J.-A. Deluc, Richard Kirwan, and (more rarely) John Whitehurst and John Williams. Apart from the fact that four of the five were British (the fifth, Deluc, spent much of his career in England), the group is remarkably disparate. Unique among them, Whitehurst suggested that the Flood resulted from a fiery explosion of some kind. Williams, by contrast, has been called a latter-day Woodwardian, as his starting point was the complete dissolution of the antediluvian crust. Buckland, Deluc, and Kirwan all concentrated on what they deemed the recent, still-identifiable effects of the Flood: Kirwan on the resultant regional instabilities of the earth's crust, Deluc on the newness of today's continents, and Buckland on the superficial deposits (dubbed Diluvium) and the cave faunas which included species rendered extinct by the Flood. Deluc and Kirwan shared an antagonism to James Hutton's seeming eternalism and his emphasis on the earth's internal heat (which they criticized on the basis of chemical evidence); but they differed about time, Kirwan espousing a scale of the order of magnitude of the Biblical and Deluc a long or indefinite antediluvian time span.[9]

Does Buckland's claim that he had found evidence for the Flood make him a diluvialist? (He understood the Flood as a recent and transient event

---

[8] D.A. YOUNG, *The Biblical Flood: a case study of the church's response to extrabiblical evidence*, Grand Rapids, MI, William Eerdmans, 1995, p. 101. Contrast the article on Greenough by V.A. Eyles in the *Dictionary of scientific biography*, V 519a. Although Young is selective, he is in another sense too inclusive, ranging from antiquity to the present.

[9] For a detailed treatment of all five writers, see G.L. DAVIES, *The earth in decay: a history of British geomorphology, 1578-1878*, NY, American Elsevier, 1969. Also, M. RUDWICK, *Jean-André de Luc and nature's chronology*, in C.L.E. LEWIS and S.J. KNELL (eds.), *The age of the earth: from 4004 BC to AD 2002*, London, Geological Society, 2001, pp. 51-60, and PORTER, *Making of geology*, cit., pp. 196-202. For chemistry, LAUDAN, *From mineralogy*, cit., esp. pp. 122-123, and M.T. GREENE, *Geology in the nineteenth century*, Ithaca, NY, Cornell U. Press, 1982, pp. 48-53.

in a long history of the earth.) The need for definition becomes urgent if we notice Anthony Hallam's characterization of Buckland as «the leading diluvialist (or student of the Deluge)». Many «students» of the Flood can be found among Buckland's contemporaries, but they reached no consensus about whether the event had been natural or miraculous, tranquil or turbulent, or, indeed, whether it had had identifiable geological effects. Furthermore, Buckland's attribution of his cave faunas to a single flood had collapsed by the 1830s; are we to conclude that his diluvialism also collapsed, or can we now put him into another category, namely, «neodiluvial catastrophist»? [10]

The anglophone emphasis of the preceding discussion was to some extent inevitable, there being relatively large and productive programs in the history of science in Britain. More significant here is the emphasis at times on a «social history of science», resulting in studies of «British» geology (and other sciences). Too little effort has been made to compare Britain with the European continent, although a comparison is essential for the identification of anything distinctively British. Existing comparisons often take the form of unexamined simplifications: Continentals were «less constrained by [...] natural theology», or the French were polarized into Catholics and unbelievers. [11] Even within the English context, comparisons may be neglected. Buckland's biographer Nicolaas Rupke, for example, emphasizes the theological atmosphere Buckland encountered at Oxford, remarking on the same as true of Cambridge. And yet there is evidence that controversial German Biblical scholarship had made some inroads at Cambridge. More generally, those who focus on Britain find themselves at a loss for much of the 18th century, as remarkably little that can be called geological was produced there from about 1710 until 1775 or later. In somewhat different ways, Roy Porter and Davis Young both handle this gap by pointing to some influx of Continental publications that would eventually be absorbed into British writings. Neither, however, has confronted such issues as whether these importations included discussion of the Flood. [12]

---

[10] A. HALLAM, *Great geological controversies*, 2nd ed., Oxford, Oxford Univ. Press, 1989, p. 41.

[11] YOUNG, *Biblical Flood*, cit., pp. 85, 88. J. HEDLEY BROOKE, *Science and religion: some historical perspectives*, Cambridge, Cambridge Univ. Press, 1991, esp. pp. 180, 200. Aware of divergences among French philosophers, Brooke seems to assume that Catholicism was monolithic, although he hesitates on p. 238. Historians have long been aware of disagreements among (for example) the French Jesuits, the classic study being V. PINOT, *La Chine et la formation de l'esprit philosophique en France (1640-1740)*, 2 vols., Paris, Geuthner, 1932.

[12] Porter mentions some post-1750 British naturalists (e.g., Borlase, Brydone, Walker) but says little about their views; nor is there much substance in his remarks on British knowledge of Buffon, in *Making of geology*, cit., p. 159. Young (*Biblical Flood*, cit., pp. 86-87) discusses Johann Gottlob

Among Continental historians, a broader approach has remained customary. In his classic study of the life sciences (1963), Jacques Roger's title does announce a focus on France, but he observes that one must compare and analyze relevant persons and contexts in order to understand why different men could «look at the same object without seeing the same thing».[13] More recently, monographs by Ezio Vaccari and Luca Ciancio, dealing respectively with Giovanni Arduino and Alberto Fortis, follow these men wherever their careers lead: to local contexts (economy, technology, patronage) and to international ones (communication and collaboration with foreigners). Among specialists in Wernerian studies, Martin Guntau has long emphasized international social and intellectual contexts; in recent years, these contexts still remain, stripped of the never very functional Marxist terminology («the European bourgeoisie») of his earlier writings. As Ellenberger once asked, is it possible to write a «national» history of ideas?[14]

Continental studies, especially on the period after about 1750, make little reference to the Flood, and, like their British parallels, do little to explain why the decline of the topic took place more quickly in some areas than in others. Davis Young, though most concerned with varieties of Protestantism, especially in the Reformed tradition, asks no questions about scientists in the Reformed churches in Switzerland or The Netherlands.[15] As early as 1729, in fact, Louis Bourguet in Neuchâtel detected what he called a new «school» in geology: the different diluvialisms of Burnet and Woodward now had a rival in the idea that marine fossils signalled a «long presence» of the sea. Bour-

---

Lehmann without mentioning his giving the Flood a considerable role in geology. N. Rupke, *The great chain of history: William Buckland and the English school of geology (1814-1849)*, Oxford, Clarendon Press, 1983, pp. 21-26; contrast E.S. Shaffer, *Coleridge and natural philosophy: a review of recent literary and historical research*, «History of science», XII (1974), pp. 284-298, and Id., «*Kubla Khan» and the fall of Jerusalem: the mythological school in biblical criticism and secular literature, 1770-1880*, Cambridge, Cambridge Univ. Press, 1975, chaps. 1-3, for both Cambridge men and Dissenters excluded from Oxbridge.

[13] J. Roger, *Réflexions sur l'histoire de la biologie (XVIIe-XVIIIe siècle): problèmes de méthodes*, «Revue d'histoire des sciences», XVII (1964), pp. 25-40, quotation from p. 38. Also, Roger's analysis of various approaches to the history of science, including the national and social, in *Per una storia storica delle scienze*, «Giornale critico della filosofia italiana», ser. 6, IV (1984), pp. 285-314.

[14] Ellenberger's review of Porter, in «Revue d'histoire des sciences», XXXII (1979), pp. 177-182. E. Vaccari, *Giovanni Arduino (1714-1795)*, Firenze, Olschki, 1993. L. Ciancio, *Autopsie della terra: illuminismo e geologia in Alberto Fortis (1741-1803)*, Firenze, Olschki, 1995. Contrast M. Guntau, *Die Genesis der Geologie als Wissenschaft*, Berlin, Akademie Verlag, 1984, with his *Natural history of the earth*, in N. Jardine et al. (eds.), *Cultures of natural history*, Cambridge, Cambridge Univ. Press, 1996, chap. 13.

[15] Young, *Biblical Flood*, cit., pp. 99-100, does discuss questions raised by H.-B. de Saussure about the geological roles of tides, currents, and floods, saying that in these matters he «spoke for the geological community».

guet's short list of examples would be expanded in later decades by writers such as Arduino (1774) and J.-C. Valmont de Bomare (1767), Valmont proclaiming that this non-diluvial theory was «held by all philosophers, ancient and modern». By that time, reference to the Flood had not vanished from French texts, but the subject elicited both satirical remarks from philosophers and, on the part of some naturalists, efforts to put flood stories to use in geological and anthropological ways. Best studied in the latter genre are the writings of Buffon (1778), N.-A. Boulanger, and the Baron d'Holbach. Intimately familiar with interpretations of the Flood, these men developed alternative views of local (sometimes extensive) floods which both altered landforms and induced long-lived fear and myth-making among primitive survivors.[16] Notable, too, among these and other writers is the expanded time assumed for the earth's pre-human history; if ancient texts supplied clues to ancient mentalities and to some few events, Nicolas Desmarest would eventually declare himself «astonished» that some men still consulted «ancient authors» as relevant to studying the earth's crust.[17]

The apparent decline of references to the Flood in Continental geology requires more examination. We do possess pertinent studies of some individuals and groups – Werner and Wernerians, for example – and scholars have taken up relevant themes like secularism in German universities and scientific specialization in the Paris Academy of Sciences and the Museum of Natural History. In the Catholic tradition, theologians had a monopoly on the discussion of Biblical interpretation; J.-F. d'Aubuisson de Voisins would celebrate, in a conversation with Charles Lyell, the Catholic separation of secular from religious subjects.[18] Nonetheless, one wonders whether religious constraints affected Catholic naturalists. Until recently, the only relevant detailed studies

---

[16] For Bourguet and others (before 1750), see RAPPAPORT, *When geologists*, cit., esp. p. 180. Valmont de Bomare and others in ID., *Geology and orthodoxy: the case of Noah's flood in eighteenth-century thought*, «British journal for the history of science», XI (1978), pp. 1-18. VACCARI, *Giovanni Arduino*, cit., pp. 256-257. For Buffon, Boulanger, and d'Holbach, see G. CRISTANI, *D'Holbach e le rivoluzioni del globo*, Firenze, Olschki, 2003. Also, SEGUIN, *Science et religion*, cit., pp. 431-463, for anthropological and other uses of a flood or floods.

[17] Quoted by Cristani (*D'Holbach*, cit., pp. 166-167) from N. DESMAREST, *Encyclopédie méthodique: géographie physique*, 5 vols., Paris, H. Agasse, 1794/5-1828, I, pp. 477, 480 (article on Seneca). See below, n. 55.

[18] Institutional structures and affiliations are discussed by RUPKE, *Geology and paleontology from 1700 to 1900*, in G.B. FERNGREN (ed.), *The history of science and religion in the western tradition: an encyclopedia*, NY, Garland, 2000, pp. 405, 407. Institutions and other topics (including the Catholic tradition of reserving for theologians discussions of theology) are in T.A. APPEL, *The Cuvier-Geoffroy debate: French biology in the decades before Darwin*, NY & Oxford, Oxford Univ. Press, 1987, esp. pp. 56-58. D'Aubuisson's remarks are quoted by Lyell in a letter to his sister, Toulouse, 9 July 1830, in C. LYELL, *Life, letters and journals*, ed. K.M. Lyell, 2 vols., London, J. Murray, 1881, I, p. 276.

have been devoted to Buffon, but we now also have a notable analysis of Alberto Fortis and the religious and political factions of the Venetian Republic.[19] If, to theologians, the Flood had profound significance as God's punishment for sin, it is possible – and there is some evidence for this – that geologists (including clerical geologists) saw here something else: an historical event that could not be ignored.[20]

On a broader level, to notice that there were many clerical geologists should not mislead us: to be a cleric was often a matter of economic security – one was provided with a benefice, a 'living', and the opportunity then to pursue a career in literature or science. Without more study of such men, one cannot even assume that they had much education in theology, let alone a religious vocation or a commitment to defend particular beliefs against the possible inroads of science. One might indeed notice, and examine, W.H. Fitton's reassertion of the Augustinian view when he remarked, in 1823, that the *consensus in his own day* seemed to be that the Bible teaches religion and morals, not truths about nature.[21]

Noting the decline in references to the Flood among Continental geologists, Ellenberger concluded that diluvialism underwent a «revival» in the days of Buckland and others, chiefly for specific phenomena that would later be explained by glacial theory. Other historians have also detected here a «new diluvialism», sometimes using that phrase or alternatives like «diluvial catastrophism» or «neodiluvial catastrophism». On occasion, particular geologists are said to have used currents of «diluvial intensity» to explain certain features of the landscape.[22]

Historians who use such vocabulary do not always agree on whether the Flood itself was one of the cataclysms employed by particular writers – as in the case of Greenough, mentioned above. Peter Simon Pallas tried to explain the

---

[19] CIANCIO, *Geologia e ortodossia. L'eredità galileiana nella geologia veneta del secondo settecento,* in R. PASTA et al. (eds.), *La politica della scienza: Toscana e stati italiani nel tardo settecento,* Firenze, Olschki, 1996, pp. 491-507. For Buffon and the Sorbonne after 1749, see RAPPAPORT, *When geologists,* cit., pp. 252-253; for 1778 and the next years, see Roger, introduction to his critical ed. of BUFFON, *Les époques de la nature,* Mémoires du Muséum national d'histoire naturelle, n.s., sér. C, t. X, Paris, 1962, pp. CXXXII-CXXXVI. The case of Buffon is seriously and summarily misrepresented in D.R. DEAN, *James Hutton and the history of geology,* Ithaca, NY, Cornell Univ. Press, 1992, p. 5.

[20] For the Flood as history, see below at n. 55 and the discussion of Cuvier later in this essay.

[21] Fitton, review of Buckland's *Reliquiae diluvianae,* «Edinburgh review», XXXIX (1823-24), pp. 196-234: 196. Fitton went on to express admiration of Thomas Burnet on this point. Among clerics who virtually abandoned clerical functions in favor of a career in science, one thinks of Fortis, Soulavie, and Buffon's collaborators, Bexon and Needham. See also C. SPEAKMAN, *Adam Sedgwick, geologist and dalesman, 1785-1873,* Heathfield, Eng., Broad Oak Press, 1982, esp. p. 52.

[22] ELLENBERGER, *Histoire de la géologie,* cit., pp. 41-49. Also, DAVIES, *Earth in decay,* cit., pp. 191-192, 252-253; YOUNG, *Biblical Flood,* cit., chap. 7; and DEAN, *James Hutton,* cit., p. 184.

transport and extinction of Siberian mammoths as the result of a flood sweeping over the Continent after the sudden elevation of volcanic islands in the East Indies and elsewhere in the Pacific. Davis Young and Claudine Cohen assume that this flood was the Biblical one; this is possible, but Pallas also liked to refer to marine incursions (*débordements*), and he seems to have been relieved now and then to find ancient texts recording one or another of his floods, not just Noah's but also Deucalion's.[23] Sir James Hall's case is clearer by far, as he now and then invoked tsunamis; less clear is why Gordon Davies classes Hall among the «new» diluvialists. Nor is clarity promoted by Rupke's conflation of diluvialism with occasional «cataclysmal» floods, and, a few pages later, by his assertion that Buckland, by 1836, had «disavowed diluvialism» (meaning use of Noah's Flood).[24]

Texts cited in the last two paragraphs – and others that use the language of «neodiluvialism» – tend to couple the Flood, floods, and other surges of water with either «cataclysm» or «catastrophe». One message seems implied by this usage: the invocation of floods, even when not Biblical, is bad science. Another point, more important in the present context, has been noted by Martin Rudwick in his analysis of Charles Daubeny's writings about Auvergne: in the 1820s Daubeny expressed his awareness of using conventional language, without necessary reference to the Flood, when he classified lavas as «ante-diluvial» and «post-diluvial», or produced by flooding. As Rudwick indicates, Dolomieu had already expressed comparable sentiments. In the same period, B. Faujas de St. Fond dismissed those who used the Flood although he retained «diluvial» for floods and flooding.[25]

These last examples do provide some warrant for the diverse meanings historians attach to diluvialism. But to retain such usage for phenomena like tidal waves and episodic flooding does often hamper analysis and understanding. Furthermore, why characterize an early geologist's views by a focus on only one of the geological agents he employed?

---

[23] «Assume» is used here because neither Young nor Cohen cites or quotes Pallas; see YOUNG, *Biblical Flood*, cit., pp. 101-102, 107, and C. COHEN, *Le destin du mammouth*, Paris, Seuil, 1994, pp. 97, 125, 244. Pertinent passages by Pallas are in his *Observations sur la formation des montagnes et les changemens arrivés au globe*, St. Petersburg, Académie imperiale des sciences, 1777, pp. 35-36, and *Voyages de M. P. S. Pallas*, trans. M. Gauthier de la Peyronie, 6 vols., Paris, Lagrange, 1788-93, I, pp. xx, 215, 697; IV, pp. 133-134; V, pp. 188-190.

[24] For Hall, DAVIES, *Earth in decay*, cit., pp. 249-252. RUPKE, *Great chain of history*, cit., pp. 82, 95. Also, Young on Saussure, in *Biblical Flood*, cit., p. 100 and n. 3.

[25] RUDWICK, *Poulett Scrope on the volcanoes of Auvergne: Lyellian time and political economy*, «British journal for the history of science», VII (1974), pp. 205-242, at p. 210 and n. 24. B. FAUJAS DE ST. FOND, *Essai de géologie*, I, Paris, Levrault, Schoell et Cie, 1805, pp. 19-20; this seems to be a reprint of the original volume published in 1803.

2. NEPTUNISM

The word «neptunism» seems to have originated in the years around 1790, when the basalt controversy led some commentators to classify conflicting views as neptunist and vulcanist. Nowadays, although «vulcanist» is still used in its original sense, neptunism has been much expanded: not only is it commonly a synonym for the general and particular ideas of Abraham Gottlob Werner, but it has also been applied to any thinker who emphasized or analyzed the geological role of water. This curious expansion prompted Ellenberger to remark: «no one has ever claimed that igneous causes produced in their totality [the substances of the earth's crust]: thus, everyone was inevitably more or less a 'neptunist'».[26]

To illustrate briefly the different theories sometimes labeled neptunian, one need only mention that Thomas Burnet, John Woodward, *Telliamed* (1748), and Buffon (1749) have all been characterized in this way – as has Lucretius and even Thales of Miletus.[27] Burnet and Woodward have been reclassified from Biblical diluvialism to neptunism, while the other moderns offered a history of sedimentation without reference to the Flood – and one of the two put his theory in an anti-Biblical framework of eternal cosmological cycles. In short, neptunism is at times an epithet with little specific content and no claim to be called a «school» of thought.

Thanks to Alexander Ospovat, Rachel Laudan, and a Wernerian «industry» in Germany, an exception can be made for the kind of neptunism associated with Werner and his followers. This has by now achieved the respectable status of a scientific school meriting serious study, and myths about Werner himself have been examined and debunked.[28] Recent scholarship has not, of course, exhausted the subject, but has to some extent revealed topics still needing attention. Nor has it laid to rest certain suspicions still to be found in historical literature. For example, did not Wernerians use geological catastrophes? and was not Werner's theory welcomed by Biblical diluvialists? A recent writer has even summarized Werner's theory as the belief that «the

---

[26] ELLENBERGER, *Histoire de la géologie*, cit., p. 265.

[27] BYNUM, *Dictionary*, cit., s.v. neptunism. SEGUIN, *Science et religion*, cit., p. 151, n. 40. G. GOHAU, *Histoire de la géologie*, Paris, Editions La Découverte, 1987, pp. 29-30 (with some cautionary words). A.V. CAROZZI, *De Maillet's Telliamed (1748): an ultra-neptunian theory of the earth*, in SCHNEER, *Toward a history of geology*, cit., pp. 80-99.

[28] Perhaps the best study of Werner himself is the editorial apparatus in A.G. WERNER, *Short classification and description of the various rocks*, trans. with introduction and notes by Ospovat, NY, Hafner, 1971. In addition to Laudan (*From mineralogy*, cit.) Greene's, *Geology in the nineteenth century*, cit., chaps. 1-2, includes an excellent analysis of Werner and his followers.

surface features of the Earth were formed primarily by water erosion during a single, catastrophic flood».[29] And one may still find it said that Wernerian refusal to recognize the volcanic origin of basalt marks the theory not only as wrong, but also establishes the modernity of those students who, thanks to their fieldwork, broke away from their teacher's errors.

The association of Werner's universal ocean with Biblical diluvialism is part of an old historiographical tradition, but historians have at last become more cautious in the way they express this connection. In 1959, for example, Francis Haber could say that neptunism «probably derived from» Moses, but a dozen years later Colin Russell would betray a long-common assumption and attempt to evaluate it: «During the early years of the nineteenth century there *must have been many* who identified Werner's universal ocean with the Flood of Noah. Clear cases are hard to find». Subsequent writers have found ways to soften this connection, saying that neptunism offered «comforting parallels» to Genesis, or that Christians (unlike deists) were «receptive» to neptunism.[30] These remarks have been made almost exclusively about British geologists, and one understands the underlying impatience in Rachel Laudan's insistence that Wernerianism was important whether or not individuals associated it with the Flood.[31]

As Anthony Hallam once noted, neptunism is also «often [...] associated» with catastrophism, and both have had «a bad press».[32] Common elements here seem to be not only the use of violent change, but also a supposed tendency to invoke or imply supernatural causes and the adoption of a short geological timescale. With respect to supernaturalism, Werner himself and one of his Scottish disciples, Robert Jameson, both acknowledged that they could not explain why the universal ocean had apparently been shrinking – nor, indeed, account for periods of resurgence and alternations of calm and turbulence. But the supernatural and unknowable did not form part of their causal repertory or serve to fill gaps in their own knowledge.[33] The short timescale

---

[29] K. OSLUND, *Imagining Iceland: narratives of nature and history in the north Atlantic*, «British journal for the history of science», XXXV (2002), pp. 313-334: 321.

[30] F.C. HABER, *The age of the world* (1959), repr. Westport, CT, Greenwood Press, 1978, p. 161. C. RUSSELL, *Noah's Flood: Noah and the neptunists*, «Faith and thought», C (1972-73), pp. 143-158, at p. 147 (italics added). Editor's introduction in C.C. ALBRITTON jr. (ed.), *Philosophy of geohistory: 1785-1970*, Benchmark papers in geology no. 13, Stroudsburg, PA, Dowden, Hutchison, Ross, 1975, p. 2. DEAN, *James Hutton*, cit., pp. 93-94. LAUDAN, *From mineralogy*, cit., p. 115.

[31] *Ibid.*, p. 110.

[32] HALLAM, *Great geological controversies*, cit., p. 30.

[33] LAUDAN, *From mineralogy*, cit., p. 93.

does occur if neptunism is used for the 17th or the earlier 18th century. In Werner's day, however, «long» time (duration usually unspecified) had become commonplace, often because thick piles of sedimentary strata evoked from observers reflections on the «incalculable» time essential for their successive deposition.[34]

Simplifications of other kinds plague the literature on Wernerian neptunism, perhaps because some topics and contexts have not received enough attention. Three such areas will be selected here, all of them relevant to neptunism as well as to other debates of the period.

First, there is the so-called lowering of the Baltic Sea, as this was measured and then debated in the Academy of Stockholm in the 1740s. As Alexander Ospovat liked to point out, neptunism did not originate with Werner; the key notion of a universal ocean, diminishing in time, might owe its origin to Anders Cels (Celsius) and some of his colleagues. At present, we have excellent studies of the Swedish debate, but only fragmentary information on the diffusion of this research outside of Sweden.[35] We have even less information about the diffusion of observations known to the Swedish debaters, namely, the evidence that the Adriatic Sea was not only fluctuating in level but apparently *rising*. Although some historians have noticed Wernerian use of the Baltic, they seem to be unaware of the conflicting data coming from Italy. This conflict did not escape the attention of Nicolas Desmarest, Paolo Frisi, and John Playfair.[36]

A second topic has been noticed especially by Ellenberger, who indicates that, from about 1750, geologists had almost nothing to say about tectonics.[37] Robert Hooke's theory of earthquakes having aroused little enthusiasm in his

---

[34] Nicoletta Morello (*La macchina della terra*, Torino, Loescher, 1979) considers Wernerian time to be short and traditional, p. 25, but her own translation from Werner's *Kurze klassifikation* shows otherwise, p. 181. See ELLENBERGER, *Histoire de la géologie*, cit., pp. 35-39, for an inventory of those (including Werner) who adopted long timescales.

[35] For the debate in Sweden, see Frängsmyr's chapter in FRÄNGSMYR (ed.), *Linnaeus*, cit.; also, E. WEGMANN, *Changing ideas about moving shorelines*, in SCHNEER, *Toward a history of geology*, cit., pp. 386-414. Some information about diffusion outside Sweden is in RAPPAPORT, *When geologists*, cit., pp. 226-228, 232-233.

[36] *Ibid.*, p. 233, n. 85 for Frisi and Desmarest. J. PLAYFAIR, *Illustrations of the Huttonian theory of the earth* (1802), facsimile reprint with introduction by G.W. White, NY, Dover, 1964, sections 387-401. See also ELLENBERGER, *Histoire de la géologie*, cit., pp. 28-31, and GOHAU, *Les sciences de la terre aux XVIIe et XVIIIe siècles*, Paris, Albin Michel, 1990, pp. 170-173. Those who focus on the Baltic and not the conflicting evidence include Wegmann (above, n. 35) and LAUDAN, *From mineralogy*, cit., pp. 59, 188. Ciancio (*Autopsie della terra*, cit., pp. 104, 186), mentions the debate, but without explicit attention to the Adriatic.

[37] ELLENBERGER, *Histoire de la géologie*, cit., pp. 48, 135, 165, and other entries in the index, s.v. tectonique; also, pp. 290-294 on Saussure.

own day, and the Biblical diluvialism of Woodward et al. having been largely abandoned, how could one account for the oblique position of formerly horizontal strata and, more generally, for the origin of mountains? In 1749 Buffon had insisted that all landforms originated on the sea floor, but he had to declare his inability to explain the emergence of land from under the sea. Those who posited subterranean forces, as Buffon did in 1778, were at a loss to explain their causes or relied on what was regarded as unproven and speculative: a «central» heat in the earth.

In this context, Wernerian neptunism possessed a decided advantage: the unevenness of «primitive», largely granitic, terrains resulted from a process of crystallization in a fluid, and these terrains formed the scaffolding upon which subsequent sediments were deposited. In other words, mountains and continents were not the product of unknown forces, but had merely been *exposed* as the seas retreated. As Eugène Wegmann observed in connection with the Baltic evidence, water is more *mobile* than land. From Saussure to Lyell, those who suggested or argued for tectonic movements either had no causal explanations, or hesitated about possible explanations, or aroused doubt among contemporaries by invoking «heat» within the earth.[38]

What can be called «the status of heat» merits more analysis than it has received in studies of critics of James Hutton. Whether the earth possessed internal heat in non-volcanic regions became a subject of investigation in the days of Robert Boyle; subsequent natural philosophers concluded that such heat did seem to exist, but could not specify its nature, depth, or extent.[39] Nor did volcanoes offer important clues to the earth's history, in part because, as Sir William Hamilton complained, they were commonly thought to play a «destructive» rather than a «constructive» role. More fundamentally, the sulfurous odor accompanying eruptions, and the sulfur compounds found in ejecta, led most naturalists to conclude that volcanoes were relatively recent structures, the product of the burning of bitumens. In other words, as Ken-

---

[38] WEGMANN, *Changing Ideas*, cit., p. 290; also, GREENE, *Geology in the nineteenth century*, cit., p. 50, as well as pp. 75-76, 94-97. For Lyell's struggle with the problem of the elevation of land in the Scandinavian case, see his *Principles of geology* (1830-33), facsimile of first ed., with introduction by Rudwick, 3 vols., Chicago, U. Chicago Press, 1990-91, I, esp. 231-232. In her section on «Wernerian theories of elevation», Laudan (*From mineralogy*, cit., pp. 181-197), essentially shows that they had none, until Leopold von Buch began to expand the role of volcanoes.

[39] For the period to about 1750, RAPPAPORT, *When geologists*, cit., pp. 180-189. In *Observations sur la physique*, VIII, août 1776, pp. 81-82, it was said that continuing disagreements suggested the need for an academic prize question on central heat; in the same journal, XVII, avril 1781, p. 325, J.-B. Romé de l'Isle was said to have effectively demolished the notion of central heat in his critique of Buffon's *Epoques de la nature*.

neth Taylor has shown, volcanoes were often regarded as «accidents» rather than widespread, law-like features of the earth's history. Under these circumstances, to say that Hutton «recognized» the importance of heat is to do less than justice to the context in which he worked.[40]

The basalt controversy has been given undue prominence and has to some extent been misinterpreted. For vulcanologists like Desmarest and Dolomieu, basalts in volcanic areas were volcanic products, but volcanoes were recent. Insofar as either man espoused a theory of the earth, it was neptunian. The same may be said of supposedly break-away Wernerians who visited Auvergne, their neptunism being flexible enough to accommodate their changed interpretation of some basalts.[41] Missing from such discussions, however, is the important subject of Italian terrains as studied both by Italians and by foreigners. As Luca Ciancio shows in his study centering on Fortis, «neptunism» does not adequately apply to the baffling conclusion reached by those familiar with Italian volcanoes, sometimes described as «half cooked, half raw», i.e., partly volcanic, partly calcareous. These and other observations would lead at least one prominent Italian, Arduino, very tentatively in the direction of a «central» heat, but more generally to the conclusion that volcanoes were older than commonly thought. Although there is evidence that northerners had knowledge of studies by Arduino and others, little has thus far been done to examine more than selected cases even of those who travelled to Italy.[42]

This brief – and far from exhaustive – survey illustrates several points. First, «neptunism» when used for philosophers from antiquity through Wer-

---

[40] W. HAMILTON, *Campi phlegraei*, Naples, Paolo De Simone, 1776, pp. 3, 13. K. TAYLOR, *Volcanoes as accidents: how 'natural' were volcanoes to 18th-century naturalists?* in MORELLO (ed.), *Volcanoes and history*, Genova, Brigati, 1998, pp. 595-618. Omitted from this discussion is the complex subject of chemistry: could heat produce the regular crystals of some basalts and granites? The essential starting-point probably remains C. STANLEY SMITH, *Porcelain and plutonism*, in SCHNEER, *Toward a history of geology*, cit., pp. 317-338, and good discussions can also be found in the volumes by G. Davies, M. Greene, and R. Laudan, as cited above. Except for the chemical experiments of Richard Kirwan in opposition to Hutton, the problem of heat enters Dean's *James Hutton*, cit., pp. 108-109, 190-194, only in about the 1820s and then almost wholly in Britain.

[41] The problem was less basalts «in general» or even prismatic columns in particular, but basalts found in areas seemingly devoid of known volcanic traces. See GREENE, *Geology in the nineteenth century*, cit., pp. 62-63, and RUDWICK, *Minerals, strata and fossils*, in JARDINE (ed.), *Cultures of natural history*, cit., chap. 16, esp. p. 270 and notes.

[42] CIANCIO, *Autopsie della terra*, cit., esp. chap. 2; here and elsewhere in this volume, Ciancio has much to say about Sir John Strange (1732-99), traveller in Italy who would become diplomatic «resident» in Venice. Relevant works by Arduino, and his hesitations, are indicated in RAPPAPORT, *The earth sciences*, in PORTER (ed.), *The Cambridge history of science*, vol. 4: *Eighteenth-century science*, Cambridge, Cambridge Univ. Press, 2003, p. 428 and n. 27. See also TAYLOR, *Nicolas Desmarest and Italian geology*, in G. GIGLIA et al. (eds.), *Rocks, fossils and history*, Firenze, Festina Lente, 1995, pp. 95-109.

nerians qualifies as one of Alfred Cobban's «over-full» categories. Even when used more restrictively, the term, as Ernst Hamm remarks, «is not very helpful».[43] It has acquired a load of dubious baggage (association with diluvialism, catastrophism, short timescale), and its adherents struggled with a range of topics current in their day but not given adequate attention by historians.

3. CATASTROPHISM

In general catastrophism has a significance not possessed by diluvialism and neptunism: it has long remained a part of geological vocabulary, usually as a synonym for bad science. When, in the 1950s, Reijer Hooykaas began his pioneering efforts to re-examine catastrophism (and uniformitarianism), he had difficulty finding a publisher for so bizarre a subject. Some years later, Martin Rudwick would propose that «catastrophism» be abandoned by historians of science, unless they returned to its original meaning as defined by William Whewell in the 1830s. More recently, thanks to such literature and to what is sometimes called neo-catastrophism, some efforts have been made by geologists to re-examine more analytically than before the thinking of their predecessors.[44]

In reviewing successive volumes of Lyell's *Principles of Geology*, Whewell at one point coined the words «uniformitarianism» and «catastrophism», asking:

Have the changes which lead us from one geological state to another been, on a long average, uniform in their intensity, or have they consisted of epochs of paroxysmal and catastrophic action, interposed between periods of comparative tranquillity?

And, he added, the earth's great antiquity and our limited knowledge of its past (geology being a new science) lead us to suppose, not unreasonably,

---

[43] E.P. HAMM, *Unpacking Goethe's collections: the public and the private in natural-historical collecting*, «British journal for the history of science», XXXIV (2001), pp. 275-300, at pp. 285-286.

[44] For the resistance of modern geologists to one explicit, limited use of catastrophe, see V.R. BAKER, *The Spokane flood controversy and the Martian outflow channels*, «Science», CCII (1978), pp. 1249-1256. Recent reconsiderations, including attention to the history of geology, may be found in HALLAM, *Great geological controversies*, cit.; R. HUGGETT, *Catastrophism: systems of earth history*, London, Edward Arnold, 1990; and V.R. BAKER, *Catastrophism and uniformitarianism: logical roots and current relevance in geology*, in D.J. BLUNDELL and A.C. SCOTT (eds.), *Lyell: the past is the key to the present*, Geological Society special publication 143, London, Geological Society, 1998, pp. 171-182. See Rudwick's review of Hooykaas's first book, in «History of science», I (1962), pp. 82-86, and ID., *Uniformity and progression: reflections on the structure of geological theory in the age of Lyell*, in D.H.D. ROLLER (ed.), *Perspectives in the history of science and technology*, Norman, OK, Univ. of Oklahoma Press, 1971, pp. 209-227. Also, RUDWICK, *The glacial theory*, «History of science», VIII (1969), pp. 136-157, esp. p. 137.

that we may be ignorant of some of the «agents» in action in remote times.[45] For a glimpse of what this thoughtful definition had become by 1990, we have Richard Huggett's description of what he thought his students had been encountering in geological textbooks:

> Catastrophism is no longer the dirty word that it has been for the last century and a half [...]. Gone [now] is the invocation of supernatural events; gone is the dissolution of the entire globe in Flood waters; gone is a series of new creations of life; gone is the insistance on catastrophes as the only potent force of terrestrial change. But despite these differences, the catastrophisms old and new have one cardinal point in common – they both claim that biological and geological rate and state are non-uniform.

This sort of thing can indeed be found in textbooks, but not in Whewell.[46]

Nor have historians and philosophers of science – with some notable exceptions – been much better than the textbooks. When Roy Porter in 1981 summed up what scholars had been saying, he began with an inaccurate version of Whewell: there had been «one or more convulsive, sudden, and fundamental revolutions [...] in conditions on the Earth's crust». The meaning of «fundamental» is here not specified, and Porter goes on to say:

> Most geologists before the 19th century were Catastrophists. They believed that only a devastating force, or forces, quite *outside the ordinary operations* of Nature, could account for such features as the irregular shapes of mountains and coastlines [...]. This seemed cogent because the Earth did not appear to have existed for *long enough* for more gradual causes to have produced such effects. Most believed these changes were the work of water, often attributed to *the miraculous agency* of the Biblical flood.

The italicized phrases are not only dubious, misleading, or plainly wrong if applied to the decades around 1800, but they also have little relationship to Whewell's emphasis on causes acting with greater «intensity» at times in the past. Instead, Porter refers to «a series of (not necessarily miraculous) revolu-

---

[45] In «Quarterly review», XLVII (1832), pp. 103-132. Discussion and quotations are in S. JAY GOULD, *Dinosaur in a haystack*, NY, Crown Trade Paperbacks, 1997, pp. 165-168. Whewell would later indicate that he did not reject the possibility that some past causes might no longer be in operation; W. WHEWELL, *History of the inductive sciences*, 3 vols., London, J.N. Parker, 1837, III, pp. 606-609. Worth consulting are studies in M. FISCH and S. SCHAFFER (eds.), *William Whewell: a composite portrait*, Oxford, Clarendon Press, 1991.
[46] HUGGETT, *Catastrophism*, cit., preface, p. VII. S.J. Gould offers examples of what he calls «textbook cardboard» in his *Time's arrow, time's cycle*, Cambridge, MA, Harvard U. Press, 1987, pp. 112-114.

tions», in contrast to the Huttonian and Lyellian insistence on «normal, natural laws [operating] over a long enough period of time». Does this mean that catastrophists did not employ «normal, natural laws»? and what is «long enough»?[47] Buried in Porter's prose are three themes typical of the historical literature. First, a «short» geological timescale required the use of catastrophes. Second, there being no observable, present-day processes analogous to catastrophes, the latter were produced by causes no longer in action. Third, as a variant of the latter, at least some catastrophes were the work of supernatural intervention. Thanks chiefly to Hooykaas and Rudwick, an alternative to these common interpretations has resulted in the (originally French) term «actualism» making its way into anglophone analyses of catastrophism. An «actualist» makes no assumption about uniformity of «rate and state», but uses present-day («actual») processes to infer that the same processes could in the past have acted at times with greater intensity. (This, indeed, was Whewell's point.) As Anthony Hallam once put it, «it is perfectly possible to use actualistic methods and come to 'catastrophist' conclusions». In addition, Rudwick and others have argued that, during the early 19th century, increasing evidence for a «central heat» within the earth gave to catastrophism another key characteristic: if the earth had a history of progressive cooling, then it had within itself a geological agent that could have acted with greater energy in the past.[48]

The rest of this essay will be devoted to examining selected «catastrophist» texts of the late 18th century and a more detailed discussion of Georges Cuvier. Treating some of Cuvier's predecessors can illuminate such basic topics as the various uses of the words «revolution» and «catastrophe», the geological timescale, and the inference from «actual» causes to operations in the past.

Missing from most historical studies is any awareness that the traditional meaning of «catastrophe» – in Greek drama, an end or dénouement – might be relevant to early geological texts, their authors educated in the ancient classics. Whewell probably coined his -ism in full awareness of its classical overtones, and John Playfair would use the term in just this way when he wrote of the end of the world, a catastrophe not within the realm of natural causes. In

---

[47] In BYNUM, *Dictionary*, cit., s.v. catastrophism. Italics added.

[48] HALLAM, *Great geological controversies*, cit., p. 30. RUDWICK, *Uniformity and progression*, cit. Also, GREENE, *Geology in the nineteenth century*, cit., esp. pp. 73-76, 90-91. Alberto Elena (*The imaginary Lyellian revolution*, «Earth sciences history», VII (1988), pp. 126-133), offers some evidence that the anglophone world opted for «uniformitarianism» and Continentals for the somewhat different «actualism». That the two terms are at times used interchangeably may be seen in the French and English abstracts preceding Ellenberger's article, *Le dilemme des montagnes*, cit., p. 43.

an opening address to *lycée* students, Alexandre Brongniart would tackle the balance (*économie*) of nature that had apparently been maintained for «millions of years», but in his peroration announced that nature had always struggled to delay the triumph of imbalance that would one day result in a «terrible catastrophe», namely, an earth becoming «different» from the one we know.[49] These two examples cannot lead to any general conclusion, except that historians should pay more attention to a vocabulary with meanings less familiar now than in the past.

Semantic dangers are more evident for «revolution», generally used by historians as a synonym for «catastrophe». But in 1830 and later, Charles Lyell and some of his contemporaries were still using «revolution» in the older sense of successive changes, not necessarily violent in nature.[50] Early language dictionaries usually do not apply «revolution» to events in nature – except, of course, for the revolutions of the planets – but rather to human affairs; here, significantly, a favorite example was the non-violent accession of William and Mary to the English throne (1688-89).[51]

Despite the dictionaries, Fontenelle would, as early as 1716, find «revolution» useful to describe geological changes. Summarizing and commenting upon the writings of colleagues and correspondents, Fontenelle apparently began with the Cartesian model of the earth's history to suggest that the initially smooth crust had collapsed here and there to form mountains. Such collapses, «revolutions», had become rarer as the earth itself became less malleable in the course of time. (The Newtonian version was readily reconcilable with the initial premise: to writers as different as Buffon, Whitehurst, and Deluc, the earth's early malleability was shown by its spheroidal shape.) Other members of the Academy of Sciences, and non-members as well, gradually adopted Fontenel-

---

[49] For Whewell, see RUDWICK, *Uniformity and progression*, cit., p. 218, and LAUDAN in J.L. HEILBRON (ed.), *The Oxford Companion to the history of modern science*, Oxford, Oxford Univ. Press, 2003, pp. 810-811. PLAYFAIR, *Illustrations*, cit., pp. 119-120. Bibliothèque centrale du Muséum national d'histoire naturelle, Paris, MS 2323: A. BRONGNIART, *Discours d'ouverture du cours d'histoire naturelle faite au Lycée republicain en L'an 6 [1798-99]*, where examples of struggle include denudation vs. the soil-holding strength of shrubs and trees. Why this balance would end he does not say, but there is no allusion to the supernatural. Contrast the Rev. William Hamilton, *Letters concerning the northern coast of the county of Antrim*, Dublin, L. White, 1786, who argued that destructive processes like denudation were slowed by God's «saving hand», p. 191; this last Letter was omitted from the French translation of 1790, presumably because much of it is an attack on the French and especially their rejection of final causes.

[50] LYELL, *Principles of geology*, cit., I, 124, and other examples in W.F. CANNON, *The uniformitarian-catastrophist debate*, «Isis», LI (1960), pp. 38-55, at pp. 44 (Murchison), 53 (Lyell).

[51] These remarks are based on the dictionaries of the Académie française (editions of 1694, 1695, 1740, 1762), Antoine Furetière (1690, 1727), the Jesuit *Dictionnaire de Trévoux* (1721), Nathan Bailey (1730), and Samuel Johnson (1755).

le's term as an inclusive way of referring to a range of changes – by 1759, for example, the Baron d'Holbach was using it to translate the several meanings of the German *Veränderungen*. Eventually, Nicolas Desmarest would contrast regular and orderly revolutions with disorder and upheaval. And in 1802, Alberto Fortis would refer to revolutions as transformations «by slow and regular causes [...] even if this requires a thousand million years».[52]

The examples of Desmarest, Fortis, and Lyell suggest that, contrary to a common assumption, «revolution» did not necessarily become synonymous with violence as a result of the several French revolutions (1789 through 1830). In fact, Lamarck was still using the term in 1801 and 1802 to mean non-violent changes. As an interesting variant, Lavoisier began to envisage what he called a «chemical revolution» as early as 1773, and in 1790 he would declare that, although not all chemists had yet been converted, the revolution was «complete». That year, 1790, he would compare what had been happening in chemistry with «our political revolution»: this, too, was a *process* nearing «completion».[53] By way of contrast, naturalist J.-G. Bruguière, writing in 1792, objected to one of Fontenelle's texts because he interpreted «revolution» to mean «some violent perturbation of our globe». In that early text, Fontenelle had referred to «great floods» transporting plants from the Indies to Europe, but Bruguière somehow transformed this into an event seemingly global and (by implication) inexplicable.[54]

If «revolution» was an ambiguous and versatile term, «catastrophe» was not: here we do have a synonym for drastic and violent change, although on a scale from anything between a global flood and the excavation of a single valley. As commonly understood, catastrophes – unlike ordinary upheavals

---

[52] FONTENELLE, *Histoire de l'académie royale des sciences*, 1716 (1718), pp. 14-15; for the authorship of this text, see RAPPAPORT, *When geologists*, cit., p. 214, n. 35, and below, n. 54. ID., *Borrowed words: problems of vocabulary in eighteenth-century geology*, «British journal for the history of science», XV (1982), pp. 27-44, p. 34 for d'Holbach. Cristani (*D'Holbach*, cit.) shows that the baron himself employed two kinds of revolutions, one of them long and slow (shifts in the earth's axis, producing climatic changes), the other fast and violent (the burning of bitumens). Desmarest (*Encyclopédie méthodique: géographie physique*, cit., III, 197-198, art. «Bouleversemens») contrasting such disorders with revolutions. Fortis in CIANCIO, *Autopsie della terra*, cit., p. 275, n. 110.

[53] J.-B. LAMARCK, *Système des animaux sans vertèbres*, Paris, Chez Deterville, 1801, p. 406, and his *Hydrogeology* (1802), trans. A.V. Carozzi, Urbana, IL, U. of Illinois Press, 1964, p. 58. In both texts, the study of fossils allows us to determine the revolutions the earth has undergone. For Lavoisier, see H. GUERLAC, *The chemical revolution: a word from Monsieur Fourcroy*, «Ambix», XXIII (1976), pp. 1-4.

[54] Bruguière (or Bruguières) is quoted in R.W. BURKHARDT jr., *The spirit of system: Lamarck and evolutionary biology*, Cambridge, MA, Harvard Univ. Press, 1977, p. 109. Although Bruguière here referred to a 1718 paper by Antoine de Jussieu, he had probably read Fontenelle's summary in which there was more violent language than Jussieu used; see RAPPAPORT, *When geologists*, cit., pp. 215, 216.

such as earthquakes – have never been recorded by human observers. Why, then, did early geologists conclude that such events had occurred? One explanation, but applicable to only one catastrophe, was summarized by Valmont de Bomare (and later echoed by Desmarest): «At first glance, nothing seems more natural than to have recourse [...] to the greatest, oldest, and most general catastrophe *mentioned in History*». But Valmont immediately expressed his preference for the long, slow operations of the sea in producing sedimentary strata and their enclosed fossils. Others would at times argue explicitly that catastrophes of various sorts were analogous to, albeit greater than, ordinary and observable events. As Buckland put it, we can see «ravages» produced by «the bursting of a dyke in Holland», and we thus get clues to what larger masses of water can do.[55]

John Whitehurst, called by Gordon Davies «a thorough-going catastrophist», was unusual in confronting problems of tectonics. At first glance, as he put it, the existence of mountains, broken strata, and erratics (and other «debris») all seemed unlike «the effects of a regular uniform law». But, he protested, history does testify to violent events – earthquakes, volcanic eruptions, the appearance of new islands – that have produced uneven and debris-laden terrains. (Whitehurst quotes at length accounts of such events, modern and ancient, starting with the Lisbon earthquake of 1755). If such causes are not now sufficient to elevate the Alps, we do have some reason to think that the same causes operated more violently in the past; here Whitehurst uses new volcanic islands to argue that «subterraneous fires actually exist under the bottom of the ocean» in at least some parts of the world, and that other such fires «may have been extinguished time immemorial». Thus, wherever we find debris resembling what modern volcanoes produce, we should infer volcanic causes even in areas now having no other «vestiges of ancient volcanos». Finally, Whitehurst remarks that the earth's volcanoes, fissures, and broken strata have been serving as vents, so that subterranean pressures can no longer build up and produce such huge effects.[56]

---

[55] J.-C. VALMONT DE BOMARE, *Dictionnaire raisonné universel d'histoire naturelle*, 3rd ed., 9 vols., Lyon, Jean-Marie Bruyset, 1776, VIII, p. 502 (art. «Terre, *terra*»), italics added. DESMAREST, *Encyclopédie méthodique: géographie physique*, cit., III, p. 606 (art. «Déluge»). BUCKLAND, *Reliquiae diluvianae*, 2nd ed., London, J. Murray, 1824, Appendix, pp. 235-237.

[56] J. WHITEHURST, *An inquiry into the original state and formation of the earth*, 2nd ed., London, W. Bent, 1786, pp. 61, 67-86, 129. Neither Davies (*Earth in decay*, cit., pp. 132-133), nor Porter (*Making of geology*, cit., pp. 124-127), indicates the chain of reasoning used by Whitehurst. The catalogues of the British Library and the Bibliothèque nationale list no translations of Whitehurst's work, but he was known to Swedish naturalist J.J. Ferber who visited him in Derbyshire in 1769 and to Faujas de St. Fond who visited him on Benjamin Franklin's recommendation in 1784.

Whitehurst based his catastrophism on thoughtful actualism. Much the same is true of one of his better-known contemporaries, Déodat Dolomieu, who was said by one of his early readers to have invoked causes «utterly foreign to the ordinary course of nature». These the same reader identified as extra-large tidal waves.[57] Like Whitehurst, Dolomieu had found it necessary to magnify the ordinary because, as he put it in a letter to Saussure, «the sea today does not form strata like ours, nor carve valleys». What Dolomieu needed was not more time, but more *force*. Little evidence seems to survive about Dolomieu's views on geological time, except that the scale was longer than the Biblical and shorter than that of many of his contemporaries. In the same letter to Saussure, he referred to those who unduly lengthened the earth's history and the history of mankind as if (certainly in the latter case) such ideas were merely part of the arsenal of «our modern philosophers», seeking to substitute this «prejudice» for others.[58]

Whitehurst and Dolomieu shared with others – including Saussure and Sir James Hall – a desire to inject more force into geological processes. Each man singled out particular features of the earth's crust that seemed to require such explanation. Summarizing his own uncertainties in the famous «Agenda» published at the end of his *Voyages dans les Alpes*, Saussure wondered if strata «in general» show signs of having their original orientations changed by violence or by «collapses» simply due to the weight of the superincumbent rocks.[59] For all that these men shared an interest in force, Dolomieu himself seems to have been unusual with respect to the timescale. In fact, the common wisdom of his day was expressed in phrases like «thousands of centuries» (and «centuries» could mean «ages» of unspecified length), the chief evidence for which was the sheer thickness of sedimentary formations; but slow accumulation of horizontal and parallel strata could not account for the subsequent deformations of such strata.[60]

---

[57] N. ANDRÉ, *Théorie de la surface actuelle de la terre*, Paris, Société typographique, 1806, pp. 286-289. André approved of this line of reasoning, saying he had himself already published something similar.

[58] Dolomieu to Saussure, Paris, 26 April 1792, in A. LACROIX, *Déodat Dolomieu*, 2 vols., Paris, Libr. Acad. Perrin, 1921, II, pp. 40-43. Dolomieu's statement on time and force dates from a 1791 article quoted and discussed by TAYLOR, *The geology of Déodat de Dolomieu*, in *Actes du XIIe congrès international d'histoire des sciences* (Paris, 1968), Paris, A. Blanchard, 1971, VII, pp. 49-53. See also, GOHAU, *Dolomieu et les idées de son temps sur la formation des montagnes*, «Histoire et nature», XIX-XX (1981-82), pp. 83-97.

[59] H.B. DE SAUSSURE, *Voyages dans les Alpes*, 8 vols., Genève, Barde, Manget et Cie, 1787-96, VIII, pp. 274-276 (chapters 10 and 11 of the «Agenda»). The «Agenda» was also published as a separate brochure, 1796, and in the «Journal des mines», IV, floréal an IV, pp. 1-70.

[60] The common phrases for «long time» are given by ELLENBERGER, *Histoire de la géologie*, cit.,

There is some consensus that Cuvier was, as Goulven Laurent has put it, the founder, father, and leader of the «school» of catastrophists in the early 19th century.[61] Analysis of Cuvier might be relatively simple using Whewell's definition as the test, but the conventional view of catastrophism has come to include additional elements: a short timescale, global revolutions, and the use of supernatural intervention (or, at least, some «religious» motivation or agenda). At one interpretive extreme, Franck Bourdier focused his discussion of Cuvier's science on a list of his errors: adoption of Biblical chronology, advocacy of global cataclysms followed by «creations» of new species, and defense of the fixity of species. Parts of this condemnation have been repeated, modified, or denied by everyone who writes on Cuvier.[62] As Dorinda Outram has pointed out, those who see a religious motivation or agenda behind Cuvier's most famous text, the *Discours* of 1812, have «rarely if ever» tried «to discover just what his religious ideas really were».[63] In fact, rarely do historians look any further than that text – not even at its subsequent editions – to seek either clarifications of Cuvier's ideas or evidence that they changed during his career. Recently, however, Martin Rudwick has produced an invaluable collection of Cuvier's writings, through 1812, accompanied by noteworthy com-

---

pp. 35-40; examples included here are Cuvier's «milliers de siècles» (1812) and its English translation in 1813 as «thousands of ages». One of the then-unusual attempts to devise chronometers for the recent past is analyzed by RUDWICK, *Jean-André de Luc and nature's chronology*, cit. Contrast the emphasis on Biblical time (the examples are British) by L.G. WILSON, *Uniformitarianism and actualism*, in FERNGREN (ed.), *The history of science*, cit., p. 409.

  [61] In G. CUVIER, *Discours sur les révolutions de la surface du globe*, préface de Hubert Thomas, postface de Goulven Laurent, Paris, Christian Bourgois, 1985, pp. 318, 322-323. Laurent's reasons are in part the fame (many editions and translations) of the *Discours* and in part the fact that when others (he names two) called Cuvier «le maître le plus eminent» of catastrophism, Cuvier, a man able and willing to argue, did not protest. Thomas Burnet, Biblical diluvialist, is at times listed among neptunists and catastrophists – surely an odd bedfellow for a Werner or a Cuvier!

  [62] BOURDIER, «Cuvier», in *Dictionary of scientific biography*, III, pp. 521-528. Bourdier and others have been put into particular anti-Cuvier contexts by D. OUTRAM, *Scientific biography and the case of Georges Cuvier: with a critical bibliography*, «History of science», XIV (1976), pp. 101-137. The usual clichés about Cuvier and catastrophism may be found in the Thomas preface and the Laurent postscript in the edition cited above, n. 61. LAUDAN, *From mineralogy*, cit., p. 140, and COHEN, *Le destin du mammouth*, cit., p. 20, both associate Cuvier with global cataclysms, while for Huggett (*Catastrophism*, cit.) Cuvier's catastrophes are global (p. 42), not global (pp. 93-94), and «near-global» (p. 173). The best studies of Cuvier as scientist are W. COLEMAN, *Georges Cuvier zoologist*, Cambridge, MA, Harvard U. Press, 1964, and RUDWICK, *The meaning of fossils*, London, Macdonald, and NY, American Elsevier, 1972, chap. 3.

  [63] OUTRAM, *Georges Cuvier: vocation, science and authority in post-revolutionary France*, Manchester, Manchester Univ. Press, 1984, pp. 142-143. Outram's analysis of this issue relies chiefly on the testimony of people who knew Cuvier, but also on a few explicit remarks in his writings. As a striking example, p. 144, when Cuvier's daughter prayed for her father's «conversion», we do not know to what. In general, Cuvier was as reticent about his beliefs (or lack thereof?) as about aspects of his science.

mentaries. These include remarks on Cuvier's vocabulary and style, his knowledge of German Biblical criticism, and his habitual caution in specifying causal agents.[64]

Using Rudwick's collection and some supplementary material, one may begin by asking whether Cuvier equated the Biblical Flood with the most recent geological catastrophe. In 1806, he had dismissed the error of a century earlier: authors who used the Flood to account for fossils «forgot that the Deluge is presented in Genesis as a miracle» and is thus not to be woven into naturalistic studies.[65] The careful phrasing («is presented in») and Cuvier's use of the Flood in 1812 suggest that he thought the event no miracle. In one of his earlier texts, he was already saying (in 1800) that legends of floods and human giants were virtually universal, and that «peoples», like the sciences themselves, «pass from poetry to history».[66] The latter phrase, following so soon after the coupling of tales of giants and floods, may be much more than a bit of eloquence on Cuvier's part, as late 18th-century scholarly and philosophical works had shown, on the one hand, that parts of the Pentateuch (and other books of the Old Testament) were poetry, and had suggested, on the other, that poetry was the earliest form of human expression.

That such arguments could be unsettling to traditional apologists is manifest in the case of the abbé Bergier, author of the theology volumes of the *Encyclopédie méthodique*. Bergier agreed with those who said that the poetry of Moses, being inspired, was unrivalled in its «perfection». What Bergier did not ask was whether Hebrew poetry could be read as fairly straightforward narrative, and he preferred to call Moses historian, «not poet or orator». How much Cuvier knew about such debates is not clear, although he did in 1812 refer briefly to one of the German Biblical scholars who considered the first chapters of Genesis to be «authentic records of the experiences of the earliest human beings, once allowance was made for the primitive or oriental» mentality entering the text.[67]

---

[64] RUDWICK, *Georges Cuvier, fossil bones, and geological catastrophes*, Chicago, Univ. of Chicago Press, 1997. Cuvier's *Discours préliminaire* first appeared as an introduction to his *Recherches sur les ossemens fossiles de quadrupèdes*, 4 vols., Paris, Déterville, 1812. It was given its «definitive» title in the 3rd ed., printed as a separate volume in 1825 and reprinted in 1985 (above, n. 61). In the notes that follow, *Discours*, 1812/1997 will be used to designate the text in Rudwick, and *Discours*, 1825/1985 for the 3rd ed.; also indicated will be the subtitles of the sections being cited, as these did not change after 1812.

[65] Report on Noël André's theory of the earth, 1806, in RUDWICK, *Georges Cuvier*, cit., pp. 103-104; for authorship of the report, *ibid.*, p. 99. Rudwick uses one of the 1807 printings of the report; as was traditional in the Academy, the report was originally published at the end of ANDRÉ, *Théorie* (1806), cit., pp. 315-336.

[66] In RUDWICK, *Georges Cuvier*, cit., pp. 46, 47.

[67] Quotation from J.W. ROGERSON, *Old Testament criticism in the nineteenth century: England*

Whatever his knowledge of German scholarship, Cuvier was familiar with British writings on the history of ancient India and referred to Champollion in a post-1812 edition of his *Discours*. In 1818, when elected to the Académie française, he began his inaugural speech with ideas reminiscent of the brief phrases he had used in 1800, now suggesting that the first of the «ages of man» was an «age of inspiration», characterized in part by the use of poetry; without the subsequent development of prose, he declared, there would have been no history and no science, and men would have remained in «the circle of an eternal infancy». In short, Cuvier kept expanding the final section of his *Discours* not to refine or clarify the relevance of the Bible or of human history to geology, but because he found fascinating the latest research on remote history.[68]

Cuvier did not take ancient accounts of floods as sources of «facts», but as reinforcing indications of a flood-event in the relatively recent past. Since his own geological revolutions were caused by the coming and going of bodies of water, salt and fresh, ancient writings confirmed what he had already inferred; furthermore, they gave an approximate date for this most recent revolution. As he also noted in that final section of his *Discours*, the study of «events closest to us» in time offered the best hope for further understanding «of more ancient events and of their causes».[69]

---

and Germany, Philadelphia, Fortress Press, 1985, pp. 17-18, summarizing the views of several German scholars, including Johann Gottfried Eichhorn who is mentioned in CUVIER, *Discours*, 1812/1997, p. 240, n. 126 (subtitle: «All known traditions...», or «L'histoire des peuples»). This discussion would be lengthened and Eichhorn's name omitted, *Discours*, 1825/1985, pp. 146-147. Cuvier's library contained some works by Eichhorn, but not the one he mentioned in 1812. Although he owned a remarkable array of poetic works and studies of comparative linguistics, Cuvier's only copy of Bishop Lowth's seminal writings on Old Testament poetry was an 1813 French translation. (I am indebted to the librarian at the Ecole normale supérieure, Paris, for allowing me to see, on microfilm, the manuscript catalogue of Cuvier's non-scientific library). Cf. BERGIER, *Encyclopédie méthodique: théologie*, cit., III, 223b-224b (art. «Poésie des Hébreux», with reference to Lowth's arguments, concluding for Moses as inspired poet), and I, 499b-508b (art. «Déluge universel», where Moses is historian, «not poet or orator», 500b).

[68] Cuvier's 1818 speech, printed more than once, can be found in the appendix to CUVIER, *Recueil des éloges historiques*, 2 vols., Paris, F.G. Levrault, 1819, II, pp. 443-467, followed, as was traditional, by the reply of the Academy's *directeur*. For his interest in ancient India, see RUDWICK, *Georges Cuvier*, cit., pp. 80, 181, and esp. 243; also, OUTRAM, *Georges Cuvier*, cit., p. 151. Among others aware of this British research, J.-A. Deluc used the occasion to reassert the traditional view: such ancient tales of floods were myths, the true, factual source of all of them being Genesis. DELUC, *Lettres sur l'histoire physique de la terre, adressées à M. le Professeur Blumenbach*, Paris, Nyon aîné an VI [1798/99], letter VI, dated September, 1794. Playfair's papers on ancient Indian astronomy were reprinted in his *Works*, ed. James G. Playfair, 4 vols., Edinburgh, A. Constable, 1822, vol. III; he became aware of and began to use the new research after his earliest papers had been published.

[69] *Discours*, 1812/1997, p. 249 («Ideas for research still to be carried out»); also in *Discours*, 1825/1985, p. 227.

For Cuvier the order of magnitude of the Biblical scale applied only to the most recent geological revolution. In his rare references to time in general, he used the customary locution of his day: «thousands of centuries». In a text of 1804, the phrase refers to the age of (fairly recent) fossils in the gypsum beds near Paris; in 1812, the same phrase is a rhetorical flourish at the very end of the *Discours*, and it would vanish from the 1825 edition when Cuvier instead added pages treating paleontological work of the intervening years.[70] Elsewhere, in critical remarks on system-builders, Cuvier dismissed extreme views: «some want billions of years for the formation of the Secondary formations, while others claim they were formed in one year», or «some naturalists rely [...] on the thousands of centuries that they pile up with a stroke of the pen; but in such matters we can hardly judge what a long time would produce, except by multiplying in thought what a lesser time produces». To illustrate the latter maxim, Cuvier offered one example: if over «a lesser time» living animals had not come to differ significantly from the mummified animals of ancient Egypt, then species had not changed in a much longer time.[71]

As Rudwick noticed, «catastrophe» was not Cuvier's preferred locution – perhaps the word had too explicitly disastrous connotations for a man educated in the classics? Indeed, he now and then changed passages in the *Discours* to eliminate the word. Thus, where in 1812 he has «catastrophes» tilting sedimentary strata, the same passage in 1825 refers to «some cause or other». And the remark in 1812 that «catastrophes have produced revolutions» subsequently disappeared.[72]

Apart from showing that Cuvier's texts require systematic comparison, the last example indicates that in Cuvier's usage revolutions were not being equated with catastrophes. Like his predecessors, Cuvier found «revolution» to be conveniently flexible and ambiguous. As it happens, Cuvier owned a volume on «the revolutions of the globe» by F.-X. Burtin, an antiquarian and fossil collector, who offered no causal explanations of revolutions but charted

---

[70] Memoir on a fossil marsupial, 1804, and *Discours* («Ideas for research [...]»), in RUDWICK, *Georges Cuvier*, cit., pp. 70, 252, and comments by Rudwick, pp. 68, 174. Gohau (*Les sciences de la terre*, cit., p. 323) suggests that Cuvier suppressed his 1812 reference to long time because «ces durées faisaient encore peur». Possibly so, but the new last pages of 1825 had no place for rhetorical flourishes.

[71] The first quotation is from Cuvier's report on André, 1806, p. 105 (in Rudwick's volume), and the rest from *Discours*, 1812/1997, pp. 228-229 («Lost species are not varieties»). The latter was unchanged in 1825.

[72] *Discours*, 1812/1997, p. 188 («Initial evidence of revolutions»), pp. 188-189 («Proofs that these revolutions have been numerous»). *Discours*, 1825/1985, pp. 38, 39. Also, RUDWICK, *Meaning of fossils*, cit., pp. 131-132.

their frequency in the earth's history. For Burtin, successive changes in lithology and in fossil populations signified intervening revolutions, and he enumerated many such local alterations. In addition, he argued repeatedly that more fossil species were extinct than his contemporaries realized. The resemblances to Cuvier's line of argument are evident, as is the choice of language. This language naturally misleads the modern reader, but it allowed Cuvier to say in 1812 that revolutions were «not [...] slow at all», and then, in 1825, that «not all» revolutions were slow.[73]

Although he changed his terms (and his ideas) about catastrophes, Cuvier did not eliminate all references to them. On one occasion, he referred to disturbances «which initially perhaps shook the entire crust of the earth to a great depth, but which have since become steadily less deep and less general».[74] But this remark would later vanish, and he would continue to use a range of adjectives for both revolutions and catastrophes: sudden, violent, gradual, local. («Local», to be sure, could be very extensive, affecting a whole continent).

Cuvier had no evidence for any ancient source of energy within the earth; if, as he allowed, volcanoes had formerly been more widespread and more powerful, they nonetheless did not «disrupt the other beds», and «they have not contributed at all to the elevation of high nonvolcanic mountains». And he remarked that the much-admired Saussure had «destroyed» the idea of a central heat in the earth by showing that the water of seas and lakes gets colder with depth.[75] If Cuvier did favor the Wernerian version of a formerly universal ocean, his own interest lay not in the unfossiliferous «Primary» rocks; rather, he noticed conflicting observations on modern changes in sea levels – the supposed diminution of the Baltic, reports of a rising sea in parts of the Mediterranean and elsewhere, and evidence for local fluctuations of sea levels. Such mixed signals from observers, plus Cuvier's own knowledge of

---

[73] *Discours*, 1812/1997, p. 190 («Proofs that these revolutions have been sudden»); *Discours*, 1825/1985, p. 41. BURTIN, *Réponse a la question physique, proposée par la Société de Teyler, sur les revolutions generales, qu'a subies la surface de la terre, et sur l'ancienneté de notre globe*, «Verhandelingen, uitgegeeven door Teyler's tweede genootschap», VIII (1790). Text in both French and Dutch. Cuvier's copy, now in the Bibliothèque centrale du Muséum national d'histoire naturelle, contains no marginalia, but nor do other books known to have been owned by him. Burtin did allow one «general» revolution, but it was not to be confused with the Flood (esp. his chap. 7); he disliked non-literal readings of Genesis in efforts to have nature give unneeded support to Scripture.

[74] *Discours*, 1812/1997, p. 190 («Proofs that these revolutions have been sudden»).

[75] Report on the progress of the sciences since 1789 (published 1810), in RUDWICK, *Georges Cuvier*, cit., p. 122. *Discours*, 1812/1997, p. 197 («Volcanoes»). In *Discours*, 1825/1985, Cuvier modified his text and added a note, p. 45, on the origin of granite («Proofs that there were revolutions before organisms existed»). Eulogy of Saussure, 1810, in CUVIER, *Recueil des éloges*, cit., I, p. 427.

local fossil populations, may well have led him to doubt greater and more widespread changes in the earth's more remote past. In any event, Cuvier's catastrophes served mainly to cause extinctions of local populations of organisms, and he knew it would be rash to pronounce on the extent of any catastrophe when evidence was still being assembled about fossils in various parts of the globe.[76]

On the whole, Cuvier could find no causal mechanisms for the particular changes of most interest to him, such as the alternations of marine and freshwater deposits that he and Brongniart had investigated. Nor could he explain the sudden appearances of new species. For the latter, Whewell would invoke God's hand, and some historians have assumed that Cuvier's silence on this and other topics mean that he, too, resorted to supernatural intervention.[77] A famous passage by Cuvier seems clear and simple until one notices the range of interpretations it has provoked: «The thread of operations is broken; nature has changed course, and none of the agents she employs today would have been sufficient to produce her former works».[78]

For Goulven Laurent, convinced that Cuvier advocated global catastrophes, extreme breaks made «everything return to the hand of God, who alone could remake» a harmonious natural world. At the opposite pole, A.V. Carozzi considers Cuvier's passage to be one of those «vexatious exaggerations» that mar the text, as Cuvier could not possibly have meant that the

---

[76] Doubts about the Baltic, *Discours*, 1812/1997, pp. 196-197 («Incrustations»), and more information added to *Discours*, 1825/1985, p. 57 and notes. On the ocean having formed the Primary rocks, see RUDWICK, *Georges Cuvier*, cit., p. 76, and Report on André, p. 105. Cuvier's early concern with the distribution of the world's fossils, *ibid.*, chap. 5. In his eulogy of Pallas, 1813 (in *Recueil des éloges*, cit., II, 138), Cuvier's only comment on the idea that a great irruption of water had carried rhinoceros and elephants into Siberia from southeast Asia was: «Il est bien démontré aujourd'hui que les animaux fossiles [de Sibérie] sont très-différens de ceux de l'Inde».

[77] An extreme statement can be found in S. TOULMIN and J. GOODFIELD, *The discovery of time*, London, Hutchinson, 1965, p. 166: «Failing any natural analogy within our experience, the source of these convulsions must be supernatural». More recent writers often focus on new species and faunas; see, for example, GOHAU, *Histoire de la géologie*, cit., pp. 140-142; ID., *Les sciences de la terre*, cit., pp. 285-286, 309-310; THOMAS and LAURENT in CUVIER, *Discours*, cit., pp. 14-15, 323-329. Also, the thoughtful discussion by APPEL, *Cuvier-Geoffroy debate*, cit., pp. 44, 133. For the use of the word «creation», see RUDWICK, *Georges Cuvier*, cit., p. 77, n. 3, and ELLENBERGER, *Histoire de la géologie*, cit., pp. 301-303. Ellenberger remarks about Cuvier, p. 302, that he added a disclaimer to the 1830 edition of the *Discours*: «Je ne prétends pas qu'il ait fallu une création nouvelle pour produire les espèces aujourd'hui existantes». In fact, that sentence was already there in *Discours*, 1812/1997, p. 229 («Tabulation of the results of the present work»). Gohau in particular insists on the stratigraphic necessity of worldwide destructions of species and thus the need for new «creations»; this, in turn, led to a theological embarrassment, as God would have to engage once more in creative activity. Logical as this argument seems, Cuvier did not commit himself either to worldwide extinctions or to worldwide catastrophes.

[78] *Discours*, 1812/1997, p. 193 («Examination of the causes that today still operate»).

laws of physics and chemistry had changed after the most recent geological revolution. Carozzi's solution is essentially a return to Whewell: Cuvier must have meant that the same causes now at work had acted with greater intensity at times in the past. Rudwick, for his part, has noticed that Cuvier's statement is immediately preceded by a reference to continuity, «the causes that still operate» on earth (sedimentation, erosion, and so on); the troublesome passage thus presumably refers to unexplained but natural causes «of another kind».[79] Another possibility is that since no catastrophes have occurred in the last few thousand years, «nature has changed course» may mean only that we now observe the ordinary processes that are not «sufficient» to produce catastrophes. Are there «sufficient» causes «of another kind», or are they perhaps like the gigantic tidal waves used by a Dolomieu? We do not know.

Given all the uncertainties, hesitancies, silences, and changes in Cuvier's writings, how can we define his catastrophism? Stephen Jay Gould once offered a characterization meriting serious attention: «empirical literalism» was the method of catastrophists, who «tended to accept what they saw as reality: abrupt transitions of sediments and fossils indicated rapid change of climates and faunas». And, Gould added, anti-catastrophism requires that one «interpolate» into an imperfect geological record some explanation of what had happened during these gaps.[80]

Gould's view does much to make sense of Cuvier's reticence and hesitations. Unable to observe what had occurred during «abrupt transitions», he could only insist that something-or-other had «changed» (hence his recourse to the convenient term «revolutions»). Subterranean forces he sometimes regarded as merely hypothetical or, as in the case of volcanoes, irrelevant to the sedimentary terrains in which he specialized; on the whole, his only mechanism of change was the movement of bodies of water. On occasion, he insisted that a particular change had been sudden, as the condition of the fossils showed that species had died before they could escape whatever presumably threatened them. All we can be sure of about Cuvier's catastrophes is that they were sufficient to account for local disappearances of species.

Cuvier's catastrophism may have been a method more than a theory. It may fit Whewell's definition in that the intermittent events were larger than ordinary ones, even if Cuvier did not clearly accept greater «energy» or «in-

---

[79] LAURENT in *Discours*, cit., p. 329; also, pp. 321, 323. CAROZZI, *Une nouvelle interprétation du soi-disant catastrophisme de Cuvier*, «Archives des sciences», XXIV (1971), pp. 367-377. RUDWICK, *Georges Cuvier*, cit., pp. 175-176.

[80] GOULD, *Time's arrow*, cit., p. 133. Also, his essay on Cuvier, in ID., *Hen's teeth and horse's toes*, NY, Norton, 1983, pp. 94-106.

tensity» of action at times in the past. Under these circumstances, what kind of «school» could he have founded and led? Consider William Buckland, who began with only one catastrophe, and it was global: the Flood. After he abandoned the geological role of the Flood, he became more Cuvierian in his references to local «irruptions of of water» as the cause of at least some local extinctions.[81] Very different catastrophists were Leopold von Buch and Léonce Elie de Beaumont, both of them especially concerned with the tectonics that Cuvier had almost entirely ignored.

When Cuvier encountered Buch's argument that the Swedish landmass was rising (thus accounting for the supposed «falling» level of the Baltic), he objected that the supposition was premature, as the accumulated data on the Baltic had not been published in any detail. The man often identified as Cuvier's disciple, Elie de Beaumont, began to publish in 1829 a series of papers on the origin of mountain ranges. Using the latest findings on the existence of a central heat in the earth, he argued that progressive cooling had resulted in stresses within the crust, which were relieved by episodes of violent adjustment; thus (to simplify the argument here), the earth's history had been punctuated by paroxysms that had produced mountain ranges and consequent floods and tidal waves. Cuvier's only recorded reaction: the «ingenious» author provided a way to determine the relative ages of tilted strata and the undisturbed horizontal ones near them.[82]

By the time Cuvier produced these minimally informative comments on the work of younger contemporaries, he had ceased to do much original scientific work. We do not know if he regarded himself as founder or leader of a school, or if he thought such a school existed. As Mott Greene once pointed out, catastrophists differed among themselves on some key points, and they acquired a degree of cohesion when they found inadequate Lyell's insistence on uniformity of rate and state.[83] As the above examples suggest, this cohesion probably did not exist in Cuvier's day.

---

[81] BUCKLAND, *Geology and mineralogy considered with reference to natural theology*, 2nd ed., 2 vols., Philadelphia, Carey, Lea & Blanchard, 1837, I, 81. Also, GOULD, *The flamingo's smile*, NY, Norton, 1985, esp. p. 123, and CANNON, «Buckland», in *Dictionary of scientific biography*, II, 566-572.

[82] Cuvier referred to Buch in *Discours*, 1825/1985, p. 57, n. 1 («Incrustation»); one does not know if he was aware of Buch's volcanic studies. The reference to Elie de Beaumont occurs in the last edition published during Cuvier's lifetime, *Discours*, 6th ed., Paris, Edmond d'Ocagne, 1830, pp. 11-12 («Premières preuves de révolutions»). Playfair had in fact noticed in his review of the first English translation of the *Discours*, in «Edinburgh review», XXII (1814), at pp. 460, 463, that Cuvier used the motions of water, not land. An excellent analysis of Buch and Elie de Beaumont is in GREENE, *Geology in the nineteenth century*, cit., chaps. 3, 4.

[83] *Ibid.*, pp. 73-75. Rupke (*Great chain of history*, cit., pp. 193-200), argues that there was no

DANGEROUS WORDS: DILUVIALISM, NEPTUNISM, CATASTROPHISM

Although Greene and other historians of geology have analyzed components of catastrophism – notably actualism and the progressive or directional aspects of the earth's history – this -ism continues to be saddled with a load of dubious baggage. The clichés of traditional historiography have not yet died: did these men not emphasize or at least resort to mystery and even the supernatural? did they not employ global catastrophes? Phrased in this way, these categories provide a way of characterizing non-modern «schools» of thought, of distinguishing the goats from the admirable sheep, the reactionaries from the progressives. And one common trait among non-moderns is the introduction of religion into science.

When Cobban insisted on the need for analysis of an «omnibus» term like «bourgeoisie», one reviewer replied that its meaning was «well understood by all undergraduates».[84] But Cobban, Rossi, and Lovejoy argue for better analysis, awareness of historical contexts, and discriminating use of historical vocabulary in ways that «all undergraduates» would find unfamiliar. In 1989, Gordon Davies, arguing for a change of attitude among historians of geology, criticized his own book, *The Earth in Decay* (1969), for its judgments of earlier scientists.[85] Are we so sure, he asked, that modern science is a repository of unalterable truth? After all, the rocks themselves are mute, and it is men, now and in the past, who interpret their meaning. Our task as historians is to study when and why men thought as they did.

English «school» of catastrophism, the English having «no preoccupation» with devising causal explanations of catastrophes, p. 194. As a brief response, one may wonder if Continentals had such a preoccupation.

[84] Jacques Godechot, quoted in COBBAN, *Aspects of the French revolution*, cit., p. 266. Cobban pointed out that he and Godechot agreed in thinking the Revolution chiefly a *political* upheaval, but that this was why Godechot thought the old, «well understood» *social* categories would do well enough.

[85] G.L. HERRIES DAVIES, *On the nature of geo-history, with reflections on the historiography of geomorphology*, in K.J. TINKLER (ed.), *History of geomorphology: from Hutton to Hack*, Boston, Unwin Hyman, 1989, pp. 1-10. His earlier book was not devoid of historical contexts, but one reviewer had responded by doubting the importance of «the modern approach of interpreting scientists within the context of their time». After all, Hutton's deism was insignificant when compared with the fact that his ideas were «sound, modern, and right». Quotations from review of Davies by CAROZZI, in «Studies in the history and philosophy of science», IV (1973), pp. 289-299.

XIII

ITALY AND EUROPE: THE CASE OF ANTONIO VALLISNERI (1661–1730)

Deservedly famous during his lifetime, Antonio Vallisneri* has never been wholly forgotten by historians of the biological and geological sciences. Jacques Roger, for example, has examined his views on generation, Francesco Rodolico his geological writings, and Massimo Baldini and others his philosophical and methodological pronouncements. In 1961 the University of Padua celebrated the tercentenary of Vallisneri's birth, inviting distinguished lecturers to discuss *Il metodo sperimentale in biologia da Vallisneri ad oggi*. And Vallisneri's name has also entered into studies of early journalism, early museums (*cabinets*), Italian intellectual life, and even surveys of the history of biology, medicine, and geology.[1]

With some few exceptions, most such studies, valuable as they are, have shortcomings of fundamental kinds: they tend to neglect the various editions of Vallisneri's works, the articles he published in periodicals, and the correspondence extant in Italian and other archives. Such neglect will soon become impossible, thanks to recent developments in Italian scholarship. In the next pages, I shall survey some of these developments and how they may affect Vallisneri scholarship in particular; in addition, I shall indicate, for selected aspects of Vallisneri's career, what light may be shed on Italian and European intellectual life by further study of this distinguished Paduan professor.

In a recent article Maurizio Torrini analysed the philosophical and ideological debates which have characterized the history of science in Italy since early in this century. With evident relief, Torrini concluded: "we have started after almost a century, to publish sources, archives, [and] scientific correspondences, ... without which we may be able to discuss methods and perspectives, but we cannot write history."[2] Torrini by no means denied that such publication of archival material had occurred from time to time in the past, and historians of science will think immediately of Antonio Favaro's lifetime of scholarship on Galileo or Howard Adelmann's on Marcello Malpighi. In

---

* The spelling of Vallisneri's name varies (the alternative being Vallisnieri), he himself having used both versions. Although standard reference works, such as the catalogue of the British Library, usually opt for including the *i* (Vallisnieri), I have chosen to follow the spelling adopted by such recent Italian scholars as Dario Generali.

areas indirectly related to the history of science, there have also been valuable publications sponsored by such organizations as the Centro di Studi Vichiani and the Centro di Studi Muratoriani, as well as the important texts published in the series *Illuministi italiani* (begun in 1958), under the direction of Franco Venturi. What Torrini alludes to as having been "started after almost a century", however, is the systematically organized (and funded) inventorying and editing of manuscripts, and especially of the correspondence of scientists of the seventeenth and eighteenth centuries.

The "Archivio della corrispondenza degli scienziati italiani", founded in 1982 under the direction of Paolo Galluzzi of the Institute and Museum of the History of Science in Florence, was conceived chiefly as a centre of "documentation", that is to say, a centre for the collection of photocopies drawn from Italian and foreign archives, and organized and catalogued. The focus on correspondence seemed to the founders especially important for revealing the thought processes normally concealed in polished publications; and the focus on the seventeenth and eighteenth centuries was deemed particularly necessary because scientists of that period could not always express themselves fully and freely in print. Thanks to the interest expressed by the publishing house of Leo S. Olschki (Florence), this initial design quickly broadened to include the publication of editions of correspondence. The first volume of an "Archivio" series appeared in 1985: the correspondence of Jacopo Riccati with Antonio Vallisneri.[3]

Establishment of the "Archivio" — both the centre of documentation and the series of publications — was evidently symptomatic of an interest and outlook shared by a good many Italian scholars in the early 1980s. In 1984, for example, at La Sapienza in Rome, another such centre was established to deal with "Corrispondenze letterarie, scientifiche, erudite dal Rinascimento all'età moderna". Under the aegis of this group, there has begun to appear an edition of the letters of Jean LeClerc (1657–1736), edited by Mario Sina.[4] In 1982, at a conference in Bologna on "Scienza e letteratura nella cultura italiana del Settecento", approximately one-third of the papers presented treated such matters as unpublished correspondence and the need to edit both letters and other texts. One of the participants at Bologna, Bruno Basile, is in fact co-editor with M. L. Altieri Biagi of two anthologies of scientific texts: *Scienziati del Seicento* (1980) and *Scienziati del Settecento* (1983). Although these volumes offer mainly excerpts from published texts, the excerpts are not only substantial but are also accompanied by a superb editorial apparatus (introductions, bibliographies, footnotes) that can serve as a model of what critical editions ought to be.[5]

From these and other projects, Antonio Vallisneri has been one of the first beneficiaries. I have already mentioned the Riccati–Vallisneri letters, published recently, and other scholars have in the past edited other portions of the correspondence of Vallisneri with such contemporaries as Giovanni Bianchi, Giacinto (or Diacinto) Cestoni, Scipione Maffei, Ludovico Antonio Muratori, Johann Jakob Scheuchzer (excerpts), and Apostolo Zeno.[6] At the Bologna conference in 1982, however, Dario Generali of the University of Milan announced a projected edition of the complete correspondence of Vallisneri: more than 8000 letters received and about 1400 replies by Vallisneri. Since the largest published collection, the letters exchanged with Muratori, numbers only 234 items, Generali's edition will obviously contain material largely unknown to historians.[7] (Any historian, sufficiently interested or enterprising, could of course track the Vallisneri correspondence to various repositories. As far as I know, only one collection of unpublished letters, those from Vallisneri to the Swiss naturalist Louis Bourguet, has been from time to time examined by scholars. An unusually large — 180 items — collection, and an exceptionally valuable one, not a single letter has been published in full.[8])

Apparently not as high on most agendas is the editing of texts already available in their eighteenth century printings. Dario Generali, however, has made an admirable start in this direction by producing a scholarly edition of the first biography of Vallisneri, written by a friend on the basis of information supplied by the subject himself. First published in 1733, this biography by Giovanni Artico di Porcia has regularly been consulted by historians, for it offers details available in no other printed sources; Generali has corrected some details, identified allusions, filled in bibliographical information, and at times supplied complementary information drawn from letters and other manuscripts.[9] If Porcia's biography is no sophisticated piece of analysis, in Generali's critical edition it has become not only a guide but an invaluable research tool for study of Vallisneri's life and works.

As for works published by Vallisneri himself, any effort to produce a critical edition would instantly encounter severe difficulties. Generali's notes to the Porcia *Vita* suggest that Vallisneri's books were only way-stations, so to speak, in the development of his ideas, arguments, and evidence. The books are often collections of 'letters', they were preceded by articles in journals, and they were sometimes followed by further articles either in periodicals or as appendages to new editions of the books themselves. Whether Vallisneri significantly modified his ideas in the course of time has rarely been asked by historians, many of whom have ignored the articles, ignored differences in editions of his books, and relied upon that convenient assemblage of texts published in 1733: the *Opere fisico-mediche*, edited by Vallisneri's son, with the prefatory biography by Porcia. If Vallisneri knew well the critical editions produced by his erudite friends, his son, Antonio junior, was not a good editor — he was sloppy about indicating what editions he chose to reprint, he left unidentified the short pieces and undated the letters he selected for printing, and he even altered his father's prose to render it more Tuscan.[10] The *Opere*, to be sure, contains some texts not published by Vallisneri during his lifetime — such as his dictionary of Italian scientific terms and some medical *consulti* — and in these areas no historian can ignore the *Opere*. For books and articles published by Vallisneri himself, however, the *Opere*, for all its convenience, cannot be considered a reliable text.

Any critical edition of a Vallisneri text must include, as I have suggested, consideration of periodical articles. Here one encounters the problem of anonymity, so common in journals of the period. Dario Generali's research in the correspondence has revealed at least one pseudonym used by Vallisneri ("Ettore della Valle") and has also shown — this is not very surprising — Vallisneri to be the author of an anonymous article advocating the use by Italian scholars of the Italian language.[11] Perhaps the correspondence will also confirm my own impression that Vallisneri wrote a number of book reviews, unsigned, for *La Galleria di Minerva* — a subject to which I shall return later. More generally, to establish a Vallisneri bibliography must also involve analysis of his editorial role in the *Giornale de' letterati d'Italia* (Venice, 1710–40), of which he was one of the founders. Periodicals of this type usually made professions of impartiality: they would provide "extracts" of the books reviewed, without "judging" their contents. Few if any journals lived up to such standards. As a founder and editor of the *Giornale*, Vallisneri assumed responsibility for the choice of books to be reviewed in such large areas as medicine, botany, anatomy, natural history, and "philosophy". Whether the "extracts", if compared with the books reviewed, contain any critical comments remains unknown.[12] Further study of this periodical should eventually reveal whether any reviews in the *Giornale* ought to be included in a bibliography of Vallisneri's works. In short, even to establish a Vallisneri bibliography entails formidable difficulties, and yet such a bibliography is exactly what historians need if they are to shed their dependence on the *Opere* and to study Vallisneri and his era in a scholarly fashion.

Even in this preliminary stage in Vallisneri scholarship, historians of science should be alerted to this scholarly area where new and exceptionally rich sources will soon be available. Furthermore, what is already known about Vallisneri's career guarantees that these new materials will be of value to those, like myself, who specialize in regions of Europe north of the Alps. In the next pages, therefore, I shall use some of the older and newer publications, as well as a few of the unpublished letters, to discuss three elements in Vallisneri's career: his journalistic activities, his cultural nationalism, and his geological ideas. In addition to suggesting a larger European context for each of these topics, I shall offer necessarily tentative comments about areas where we may expect further illumination from sources not yet published.

Some of the salient features of Vallisneri's life may first be indicated briefly.

Having studied with Malpighi in Bologna in the 1680s, Vallisneri went on to establish himself in medical practice in Reggio. His first known publication, on "la curiosa origine di molti insetti" (1696), framed in the form of a dialogue between Malpighi and Pliny, is generally described as an extension of Francesco Redi's experiments which undermined traditional belief in spontaneous generation.[13] By 1700 Vallisneri's reputation was such that he was appointed professor of practical medicine at the University of Padua, and he was later, in 1710, named professor of theoretical medicine. From a very early date, perhaps during the years in Bologna, he established a habit of regional exploration; begun as a matter of botanical 'simpling', his travels came to include visits to medicinal springs and more general study of natural curiosities of all kinds. Study with Malpighi apparently did not immediately label Vallisneri a 'modern', and the audience for his inaugural lecture at Padua may well have believed him sincere in his argument that medical modernity did not conflict with, but actually "confirmed", traditional doctrines. As a later eulogist noted, with marked reservations, these conciliatory words proved false, it soon becoming evident that Vallisneri was indeed pro-modern.[14] Even before that lecture of 1700, however, Vallisneri had embarked on a typically modern activity: he joined an "informal academy" founded by the Venetian printer-bookseller Girolamo Albrizzi and meant in part to provide consultants for Albrizzi's new journal, the *Galleria di Minerva* (Venice, 1696–1717). It was in the first volume of this journal that Vallisneri began to publish his dialogues on the origin of insects.[15]

Elements in this biographical sketch provide themes which may be traced through the rest of Vallisneri's life. The Malpighian modern who in his youth had translated Descartes's *Meditationes* into Tuscan became a mechanist whose geological writings insist upon the Earth as a machine subject to "the ordinary laws of nature". A critical mechanist, however, Vallisneri found Cartesianism inadequate to account for certain biological phenomena; if the mechanical philosophy proved deficient in this realm, Vallisneri retained the hope that laws comparable to those of physics would be discoverable in biology.[16] The modern who had joined Albrizzi's "academy" in Venice would become a member of the Royal Society of London (1703), the German Academia Naturae Curiosorum (1707), and some half dozen or more Italian academies. Later, beginning in about 1720, Vallisneri would serve as a consultant to the Countess Clelia Grillo Borromeo who wanted to establish in her Milan home an academy of experimental philosophers.[17] Early experience as a contributor to the *Galleria di Minerva* would be followed in 1710 by Vallisneri's becoming one of the founders and editors of the *Giornale de' letterati d'Italia* (Venice, 1710–40) and by other journalistic ventures. And the collecting activities begun in his youth would eventually result in the formation of a great "cabinet" in which natural objects were joined by human artifacts.[18]

Throughout this distinguished career, presented here in outline, large and small puzzles abound. One would like to know, for example, how Vallisneri acquired so great a reputation, in Italy and abroad, so early in his life. His success as a physician — and about this we know virtually nothing — may well have given him local or regional fame. But we know nothing about the origins of his association with Albrizzi, Apostolo Zeno ("secretary" of Albrizzi's academy), and the *Galleria di Minerva*. Unlike his future collaborators, Zeno and Scipione Maffei, whose letters have been published at least in large part, Vallisneri's early friendships cannot yet be described or even readily identified. Even after the obscure period of the 1690s, we remain uninformed about the origins in 1701 of his intimate friendship with L. A. Muratori and, by about 1707, with Maffei.[19] Once having published the dialogues on insects, however, Vallisneri's name came to the attention of one of the *riformatori* (governing magistrates) of the University of Padua, and one may suppose he was now on his way to greater fame in Italy and abroad. We know nothing, however, about the circulation of those dialogues — either in the *Galleria* or in the separate reprint of 1700 — outside of Italy, and we can at present only speculate about how Vallisneri acquired fame north of the Alps. Is it possible that his position at Padua, rather than any impressive publication, accounts for his remarkably early (1701) friendship with J. J. Scheuchzer in Zurich and for his membership (1707) in the German academy which specialized in medicine? As for the Royal Society of London, Generali's edition of the correspondence will certainly clarify Vallisneri's own belief that he had proper credentials to be a Fellow of that institution. Despite these fragmentary indications of European fame, one still wonders why, in 1709, when news of the imminent publication of the Venetian *Giornale de' letterati* circulated in Paris, it was assumed (erroneously) that Vallisneri would be the chief editor.[20] Since most historians begin their accounts of Vallisneri's work with his books published in 1710 or 1715, such pre-1710 questions have rarely been asked.

Vallisneri's more or less active role in connection with a number of periodicals can readily be placed in both Italian and European contexts. Late in the seventeenth century, in much of Europe, the learned journal was a new phenomenon, the first models being the *Journal des savants* and the *Philosophical transactions* of the Royal Society. The former soon evolved into what was primarily a book-review journal, and the latter specialized in original articles, with only occasional notices of new books. Among the several imitations of the *Journal des savants*, the first in Italy was the Roman *Giornale de' letterati*, first published in 1668 and soon to be followed by other short-lived ventures, often with the same title, in Ferrara, Parma, Venice, and elsewhere. Like the

Paris prototype, and like the several journals begun in the Netherlands late in the century, the Italian journals generally did not serve as organs for any established academy; instead, they were private ventures undertaken by small groups of scholars, or commercial enterprises founded by printers, and they attempted, insofar as they could, to survey the whole international world of learning.[21]

In all these ways, the *Galleria di Minerva* was hardly a new experiment when its first number appeared in 1696: it was founded by a printer, Girolamo Albrizzi, with an advisory "academy" and a "secretary" to generate both original articles and notices of books.[22] And it proposed to deal with international literature, ancient and modern, in all subjects. Vallisneri began immediately to contribute to the *Galleria*, but we do not yet have a complete bibliography of his articles — including any anonymous or pseudonymous pieces. Nor do we know how this so-called academy, said to have contained six hundred members, functioned: was it essentially just a long list of people willing to send information now and then to its "secretary", Apostolo Zeno? Soon after 1700, Zeno began to express dissatisfaction with the *Galleria*, and in about 1707 Vallisneri was beginning to suggest the need for a new and better journal.

Reasons for these complaints will, one hopes, emerge more clearly from further study of the letters of both Zeno and Vallisneri. Meanwhile, fragmentary information suggests that Albrizzi himself had adopted policies offensive to at least these two collaborators. Regarding the *Galleria* with the eyes of an entrepreneur, Albrizzi saw in the periodical a useful way of announcing to the public works he would later print in book form. (In 1700, he would reprint Vallisneri's dialogues on insects as a small book, without offering the author a chance to revise or expand.) To consider the periodical an ephemeral piece of promotion for future books may well explain why, as Vallisneri later complained, the *Galleria* was also carelessly printed and full of errors. Furthermore, it eventually became clear that the *Galleria*, for whatever reason, could not live up to its early promise of regular publication (only five volumes appeared in the first eleven years, 1696–1707, and then only two more before the journal expired in 1717).[23]

When Vallisneri began in about 1707 to discuss with friends the need for a new journal, he may have envisaged something as substantial — for both length and originality of articles — as Angelo Calogerà's later *Raccolta di opuscoli scientifici e filologici* (1728).[24] Ultimately, however, Vallisneri combined with Apostolo Zeno and Scipione Maffei to create a new *Giornale de' letterati d'Italia*, with Zeno as chief editor. This new Venetian journal proved less ephemeral than its predecessors, although it began a long decline when Zeno departed for Vienna in 1718. Nonetheless, its solidly admirable qualities have rightly been celebrated by historians, and it served as an important vehicle for Vallisneri's publications and for articles in which Vallisneri's views were invoked to support or combat issues raised in works by his contemporaries. Once again, as in the case of the *Galleria*, we do not yet have a complete bibliography of Vallisneri's contributions to this journal.[25]

If the *Giornale* has received a certain amount of attention, we continue to lack a good deal of information about it. For example, there seems to be uncertainty about the roles of two of the three founder-editors, Vallisneri and Maffei: did they write notices of books, or did they more often select books and assign the task of reviewing to colleagues (some of whom have been identified)? Did Vallisneri continue his association with the *Giornale* after about 1720? It is known that at some stage in the 1720s, Vallisneri began to "encourage" the young Angelo Calogerà, contributor to one short-lived periodical, editor of two more ephemeral journals, and eventually editor of the successful *Raccolta di opuscoli scientifici e filologici* (1728). Can one say that post-1720 changes in the *Giornale* prompted Vallisneri to look elsewhere and even to take a hand in the planning of one or more new periodicals?[26]

An especially noteworthy aspect of the *Giornale* was its abandonment of any effort to be international in its coverage: from its inception an avowedly "national" periodical, the *Giornale* was meant to be a vehicle for communicating news of Italian learning to other Italians, and for indicating to foreign scholars that Italy had a respectable intellectual life. Vallisneri's friends and colleagues — including Maffei, Muratori, Antonio Conti, and Luigi Ferdinando Marsigli — both defended and called for the revitalization of Italian culture, Maffei complaining in his introduction to the first volume of the *Giornale* that foreign book-review journals had unfairly ignored valuable work being produced in Italy.[27] Vallisneri, too, had powerful feelings on this subject, but his interest in promoting Italian cultural pride would be coupled with attacks on French arrogance and French cultural hegemony.

Like so much about Vallisneri, the origins of his cultural nationalism remain unclear, even when one acknowledges that he shared these sentiments with many other Italian intellectuals. Recalling the period of the 1690s, he tells us that he then learned French "because of the beauty of the books constantly being published" in that language. And he adds that his study of insects had benefited from his knowledge of French views. Later, in 1710, he informed Muratori that he had decided to write even technical works in Italian, not Latin.[28] To be sure, Vallisneri's early dialogues on insects had been written in Italian, and he would continue to use Latin in communications to foreign scholars; but he had come to think of Latin as the language not only of scholastic philosophy but also as a cloak used by physicians to disguise their ignorance. If these anti-Latin sentiments are understandable, Vallisneri's vigorous condemnations of the French need both dating and explanation. As early as 1703, and more forcefully in 1708, Muratori himself had written eloquently on the need to cultivate good learning in Italy so that one might properly take pride in Italian culture; but Muratori, an admirer and disciple of the Maurists, could hardly go so far as to condemn French learning.[29] Perhaps Muratori's efforts to promote "buon gusto" in Italy inspired his Paduan friend — their earliest correspondence is no longer extant — and the added anti-French element in Vallisneri may date from about 1710, when there began the long dispute over generation with French physician Nicolas Andry. Certainly it was to Muratori, in the years 1710–12, that Vallisneri began to express his dislike of the French. Another of Vallisneri's friends, Antonio Conti, would later provide additional fuel when he reported from Paris that Fontenelle, in an *éloge* delivered in 1714, had declared that Italy was "a country where the old philosophy still predominates because it is old".[30]

Fontenelle's remark continued to rankle in Vallisneri's mind when he prepared *De' corpi marini* (1721), with curious results for some of the content of that book — a subject to which I shall return later. Here it seems more pertinent to raise the old question of the repute of Italian learning north of the Alps. Did part of the communication program of the *Giornale*, to inform Europeans about Italian learning, have any success? Vallisneri's correspondence may provide the beginnings of an answer to this question, and it will surely clarify his own relations with French savants.

Until that correspondence appears, it may be helpful here to indicate briefly the difficulties of assessing Italian relations with *oltramontani*. We now have remarkably little information about the circulation of Italian publications outside of Italy; nor do we know much about the international book trade in general. In 1728, the editors of the new *Bibliothèque italique*, published in Geneva, would assert that northerners had indeed been neglecting Italian learning, in part because of problems of the book trade: limited communication of Italian booksellers with foreign countries, and the high price of Italian books. These comments require evaluation, for they fail to explain why the Jesuits, with their excellent international network, allegedly neglected Italian works in their *Mémoires de Trévoux*, while the *Acta eruditorum* in Leipzig supposedly came close to doing justice to Italian publications.[31] The Genevan editors also cited as a problem the lack of knowledge of the Italian language, and the same complaint can occasionally be found elsewhere. Here, too, evaluation seems necessary when one recalls, for example, the familiarity of men of letters, Fontenelle and Voltaire among them, with Italian poetry and drama or the frequency with which French men of letters became members of Italian academies.[32] As a third element, those thoughtful Genevan editors suspected that there existed among Europeans "the prejudice that fewer good books are published there [in Italy] than in places where there is greater freedom to publish what one thinks". An interesting judgement, this remark is impressionistic and hard to evaluate. At the simplest level, for example, it must have been apparent to northerners that not every scholarly or scientific subject pursued by Italians lay under a cloud of theological suspicion.[33]

However perceptive the Genevans may have been, their analysis must be regarded as representing contemporary impressions, not to be accepted without investigation. (On the other side of the coin, the same can be said about Italian complaints in this period about the difficulty of obtaining foreign books.) Only in some few instances can historians now judge these matters with any degree of confidence — it is known, for example, that the works of Malpighi and Muratori were widely appreciated outside of Italy, and that Fontenelle, despite that remark so resented by Vallisneri, actually expressed a good deal of ambivalence in assessing the quality of Italian learning.[34] In fact, the case of Fontenelle may provide a clue to the attitude of other northerners as well: European intellectuals seem to have expected great things from the country of the Renaissance and of Galileo, and they rejoiced when expectations were fulfilled, but condemned the peninsula when they were disappointed. In the end, I think we must admit that we have little reliable or systematic information about relations between Italy and the rest of Europe, and the Vallisneri correspondence will doubtless add something of value in this much-debated but little-understood area of intellectual history.

In the privacy of his correspondence, Vallisneri did confess that Italian learning was not all it might be. To Bourguet and Scheuchzer in Switzerland, he complained that few Italians studied natural history and that scholasticism remained a burden hard to shake off.[35] To Muratori and others, he expressed his distrust of Jesuits ("cold friends and hot enemies"), his awareness that certain issues might provoke the hostility of the religious orders ("the blacks and the whites"), and his general antagonism to "priestly" interference with freedom of expression.[36] Whatever the handicaps of Italian learning, however, Vallisneri had no hesitation in acting as an intermediary for the dissemination of Italian publications to his foreign friends.

However limited the Italian book trade with Europe may have been, it is likely that such formal channels of exchange were less significant than the cultivation of foreign friends willing and able to supply books. In Vallisneri's case, he and Johann Jakob Scheuchzer served one another in this way for a number of years, until Vallisneri began to dislike his role as agent for the Zurich naturalist's own numerous publications. By the time a group of Zurich intellectuals, with whom Scheuchzer was associated, began to publish their short-lived *Neue Zeitungen aus der Gelehrten Welt* (1724–25), Vallisneri's long friendship with Scheuchzer had cooled sufficiently for the Zurich editors to be unable to rely on this source of information; for Italian book news, they instead leaned heavily on the *Giornale de' letterati.*[37] With Louis Bourguet in Neuchâtel, Vallisneri remained on excellent terms, so that when Bourguet and others began to plan for and then to publish their *Bibliothèque italique*, they had, for at least a year, a valuable correspondent in Padua.[38] As I have suggested, too little is known about Vallisneri's role in the several journals with which he had some association, but one historian familiar with the Scheuchzer, Bourguet, and Vallisneri letters located in Zurich has remarked: "among all the Italian scholars (*eruditi*), Vallisneri seems really to have been the one with most influence on Swiss culture."[39] A large claim, this comment nonetheless opens up possibilities that further study of archives may reveal surprising roles for the Paduan in other areas of Europe as well.[40]

If much has been written about Vallisneri as biologist, rather less attention has been given his admittedly slender production as a geologist. With rare exceptions, historians of geology have confined themselves to analyses of one or both of his major publications: the *Lezione accademica intorno all' origine delle fontane* (1715; expanded second edition, 1726), and *De' corpi marini che su' monti si trovano* (1721; second edition, 1728).[41] But Vallisneri's geological writing was not limited to two books, one of them a tract on the origin of springs, and the collecting activities, begun early in his career, suggest a lifelong interest in natural curiosities.

As early as 1703, Vallisneri's interest in petrified shells seems to have been taken for granted by one of his correspondents, the Livornese pharmacist Giacinto (or Diacinto) Cestoni. The next year, in fact, Vallisneri is said to have sent to the Royal Society of London a Latin account of some of his early travels and observations.[42] What may be his first geological publication was a review of Scheuchzer's Latin translation of John Woodward's theory of the Earth, published in the *Galleria di Minerva* in 1708. The review being unsigned, its author cannot be identified with certainty; but Vallisneri was a regular contributor to the *Galleria*, and the whole review explicitly relies on his observations and conclusions — he is cited always in the third person — to undermine every part of Woodward's book. Vallisneri would later refer to Woodward with a degree of respect, but his real friend was Scheuchzer, himself a convert to Woodwardian theory. The same volume of the *Galleria* in which Woodward was demolished contains a mildly critical review of one of Scheuchzer's Woodwardian books. One may plausibly conclude that Vallisneri wrote both reviews, handling his Swiss friend more gently than the Englishman.[43]

Even if Vallisneri did not write these reviews — a remote possibility — the reviewer somehow knew a good deal about Vallisneri's observations during his travels, his views on how fossils and strata had been deposited, and his aversion to the use of miracles (the Flood) in natural philosophy. More detail on all these matters, supposedly "found" in manuscripts Vallisneri had written "for his own use", appeared in 1712 in the *Giornale de' letterati*.[44] Meanwhile, for reasons now unknown, Louis Bourguet, as early as 1710, had demanded of Vallisneri his views on fossils and strata. And Vallisneri's replies of 1710 rehearse in detail the arguments he would expand later in *De' corpi marini*. Repeatedly Vallisneri explained to Bourguet why the Biblical Flood could not or should not be used in geology, sometimes indicating that fuller exposition of both sides of the argument would make the whole debate more intelligible. On one occasion, setting aside the specific difficulties of interpreting the Flood in a naturalistic way, Vallisneri burst into a wholly Galilean explanation of why the Bible could not teach science.

> ... Nature is infallible and is the voice of God, with this difference, that the language of Holy Scripture can and should be interpreted in many ways (otherwise it would say many things contrary to the evidence of the senses), but the language of Nature is always the same, without metaphor, without allegory, without hyperbole, without doubtful, obscure, mysterious meanings. Nature speaks clearly to him who knows how to understand her, and has no need of interpretation.[45]

The exegetical part of this remarkable passage might be called traditionally Augustinian or Thomistic; but the juxtaposition of such remarks with an insistence on the inexorable nature of law and the openness of the book of nature ("no need of interpretation") instantly suggests parallels with Galileo's "Letter to Christina of Lorraine" (1615). Vallisneri did know this text — it had been reprinted clandestinely in Naples in 1710 — when he wrote to Bourguet.[46] More significantly, Vallisneri did not indulge in such explicit statements about Biblical exegesis in *De' corpi marini*. Indeed, on the one occasion in that book when he suggested that the Biblical account of the Flood might be interpreted figuratively, he immediately refused to meddle with such subjects; instead, he insisted that the Flood could not account for his own geological observations, that the event should be considered a miracle, and that any other course of action would merely produce "an indigestible mixture of science and morality".[47]

In public and in private, then, Vallisneri had a reputation as a non-diluvial geologist for some years before the publication of *De' corpi marini*. This reputation may help to explain why, in 1720, a German visitor turned up in Padua, showing (delivering?) to Vallisneri yet another book which would prove that fossils were really relics of the Flood. In a mood of hilarity, Vallisneri wrote to Muratori about the whole affair. The German author thought that "anyone who contradicted [the diluvial interpretation of fossils] was the wickedest of men, an atheist, a deist, an impious person", deserving of punishment in this life and the next. "Oh, that cowardly German! And what will he say when he sees in my little book, soon to be published, that I have assembled the strongest arguments to prove that marine shells etc. were not deposited by the Flood?"[48] With admirable tact, Vallisneri omitted all such commentary in writing to Bourguet about the new book; to the like-minded Muratori he could express himself more frankly.[49]

This is not the place to examine the main arguments of *De' corpi marini*, but certain aspects of the book stand out more clearly when one has some familiarity with Vallisneri's earlier career and his correspondence. One such element is his attack on French intellectual pride. In the early volumes of the *Histoire* of the Paris Academy of Sciences, Vallisneri found the suggestion that fossil forms might have "grown" from "seeds" in the rocks where they are now found; and in a popular French travel account, he found the argument that fossils are mere "tricks of nature". (Such ideas, of course, were not peculiarly French, and Vallisneri must have known as much.) At a time when, according to Vallisneri, the French — and here he refers to Fontenelle's infamous remark — were mocking Italian learning as superstitious and barbarous, as comparable to the learning of Lapps and Iroquois, such French errors required exposure and refutation.[50] Neglecting to point to later, quite different passages in the *Histoire*, which he proceeded to use, and failing to identify as French another author cited more than once to provide evidence for his own views, Vallisneri blithely moved on to devote most of his book to an examination of the theories of two Englishmen, Thomas Burnet and John Woodward — only to conclude that he believed he had succeeded in refuting "French errors" and in tackling many "other" problems as well.[51]

A second curious feature of *De' corpi marini* becomes evident when one realizes that Vallisneri had been in correspondence with Scheuchzer for twenty years. Woodward's disciple and translator, Scheuchzer is nowhere mentioned in Vallisneri's book. Relations between the two men had already begun to deteriorate, as Vallisneri was showing some reluctance to continue acting as agent for Scheuchzer's publications. In 1719, Scheuchzer's unsuccessful bid to obtain a professorial chair in Padua may have further strained cordiality between them, but more divisive, without doubt, were their disagreements about the Flood. When *De' corpi marini* was in press, Vallisneri briefly described the book to Scheuchzer as a sceptical tract, raising but not answering questions about the Flood. A year later, his book now in print, he would repeat this description, adding that he had purposely omitted all reference to Scheuchzer, valuing his friendship and his many admirable qualities. Instead, he had chosen to focus on Woodward, "one of the main leaders of that sect [of diluvialists]".[52] Not surprisingly, correspondence between the two men almost ceased entirely after that date.[53]

One of the most interesting questions one may ask about the value of the Vallisneri correspondence is whether his letters clarify the true intentions of *De' corpi marini*. As far as we know, the manuscript submitted to the censors encountered no difficulties, and one wonders if Vallisneri's apparent defence of miracles — and his exclusion of them from the province of the natural philosopher — disarmed potential criticism. In a letter to Jacopo Riccati in which he explained the sceptical stance he would adopt in print, Vallisneri remarked that he would speak clearly and yet pacify the clergy by inserting "le necessarie proteste".[54] Historians have rightly noticed about the book, however, that it implies a rejection of the historical veracity of Moses by subjecting every detail of the Flood story to damaging scrutiny. In addition, the text insists on nature's obedience to God's uniform laws, separating the physical history of the Earth from questions of mankind's sinfulness.[55] Might one conclude, then, that the defence of miracles was a sop to the censors, or that Vallisneri really intended to damage irretrievably the historicity of the Bible?

In some respects, Vallisneri's correspondence does little to clarify *De' corpi marini* — that is to say, many issues are handled in identical fashion in the letters and the book. One finds, for example, the same insistence on the uniformity of law and the same, often ironic, critique of a naturalistic interpretation of the Flood. In the book he announced his sceptical stance by declaring it easy "to destroy an ill-made structure [or theory]", but hard to erect a better one.[56] Similarly, he informed Riccati that it would be so difficult "to save the universal Flood" in a naturalistic fashion that all he could do was to adopt a sceptical attitude towards all such theories. Indeed, Vallisneri repeatedly told his correspondents that his book was a sceptical tract, raising but not answering questions.[57]

On the other hand, the correspondence also contains a few striking and suggestive remarks, confirming in explicit fashion the fact that Vallisneri exercised considerable self-restraint in *De' corpi marini*. To Riccati, for example, he deplored the way Protestants ("heretics") insisted on using Biblical texts to explain nature; Moses, he added, was "a great philosopher in his era, a great political leader" who held the people in check by teaching them to fear God — and Moses was also "a great prophet when he did not try to explain natural phenomena". To a friendly "heretic", Bourguet, he declared that the only evidence for a universal Flood was the Bible, "confirmed" by faith but in no other way.[58] In other words, Vallisneri did indeed reject Genesis as a source of historical information about nature, as he did in his letter to Bourguet in which he summarized the Galilean position on the autonomy of science.[59]

If Moses was no historian of the Earth, did Vallisneri also intend to attack traditional views of the age of the Earth? In general, *De' corpi marini* might be interpreted in such a way, for Vallisneri there insisted upon a (presumably long) history of small, successive episodes of flooding or of changes in sea level; but he verged on explicit comment only once when he referred to geological changes occurring "in the most remote and obscure times (and God knows when)".[60] Two friends, Antonio Conti and Giovanni Artico di Porcia, apparently detected in the book — perhaps aided by letters not yet published — an implied lengthening of the time scale, for both responded with letters discussing the possibility that the Earth had had a far longer history than that recorded in ancient writings or, more specifically, in the Old Testament.[61] In letters to Bourguet, Vallisneri did become more explicit: not only was the world older than commonly supposed, but the evidence entailed extrapolation from the small geological changes emphasized in *De' corpi marini* — for example, it had taken as long as two centuries for relatively small changes to occur in the sea level at Venice. And to Muratori, Vallisneri admitted that there existed contradictions, "indissoluble knots", if one compared his own geological ideas with the Biblical time scale.[62]

Among Vallisneri's friends, Louis Bourguet deserves special attention, for it was to Bourguet, possibly more than to any other correspondent, that Vallisneri had for years written detailed explanations of his geological views. And yet in a letter to a Genevan friend, Firmin Abauzit, Bourguet wondered whether Vallisneri thought the Flood had been universal or local, miraculous or natural.[63] That Bourguet raised such questions suggests that Vallisneri could baffle his closest friends. Bourguet himself, however, wanted very much to discuss the Flood in naturalistic terms, and he told Abauzit that he had plausible replies to some of Vallisneri's published arguments. If Vallisneri had made obeisance in the direction of miracles in *De' corpi marini*, he had also remarked privately to Bourguet: "When we have recourse to miracles, everything is answered in natural history."[64] This Galilean remark probably seemed to Bourguet pious and philosophical — and puzzling. If we are not to use miracles, and if the Flood was a miracle, what, then, ought we to do with the Flood? For Bourguet, there could be no way merely to exclude the Flood from geology, the Flood being a well-attested historical event.[65]

North of the Alps, among those who had not corresponded with Vallisneri, some reviewers apparently found *De' corpi marini* merely puzzling and paradoxical. The *Acta eruditorum*, for example, indicated briefly that Vallisneri had disputed various received ideas, but had offered no solutions. In the equally brief, but somewhat hostile, notice published in the *Mémoires de Trévoux*, we are told that "he gives no new explanations of these problems, and he leaves the whole matter as perplexed as before".[66] These non-reviews would merit little attention were it not for the fact that the Trévoux journalists really began to understand the issues little more than a year later, when the periodical reviewed a French translation of Woodward's theory of the Earth. Ironically, the reviewer of Woodward unconsciously retreated to the position of Vallisneri: bad science was no firm support for the truths of Scripture, and it would be wiser to accept the Bible on faith and without dubious efforts to make Nature support Scripture.[67]

Before one can generalize about how readers interpreted Vallisneri's book, further research in the manuscripts and printed works of the 1720s and 1730s will be essential. Vallisneri himself remarked to Bourguet that the book had made him "many enemies", but he identified none of them; nor is it clear that "enemies" were aroused by the geological part of *De' corpi marini* rather than by the appended portions on the more vexing biological issue of generation.[68] At present, provisionally, one may say that readers like Conti and Porcia detected implications that they did not find alarming, whereas some readers evidently felt dissatisfaction with a book which professed to have no answers to the questions it raised.[69] Above all, perhaps, Bourguet represents an increasingly powerful disposition among Europeans to discover in nature proofs of religious truths, and a naturalistic interpretation of the Flood was essential to such a programme. In Germany, diluvialist Peter Wolfart — the "cowardly German" of Vallisneri's letter to Muratori — warned that the alternative to accepting the veracity of Moses would be atheism and deism. In England, Thomas Burnet simply could not comprehend John Keill's argument that the Flood should be considered a miracle and thus wholly outside the realm of natural philosophy. The mere existence of a John Keill suggests that Vallisneri was not alone in this thinking — comparable sentiments were also being expressed by some of Vallisneri's villainous Frenchmen.[70] On the whole, however, I suspect that most readers of *De' corpi marini* remained as optimistic as Bourguet that they could find ways "to save the universal Flood" as a valuable part of the geological history of the Earth.

REFERENCES

1. Jacques Roger, *Les sciences de la vie dans la pensée française du XVIIIe siècle* (Paris, 1963). Francesco Rodolico, *La Toscana descritta dai naturalisti del Settecento* (Florence, 1945), and *L'Esplorazione naturalistica dell'Appennino* (Florence, 1963). Marino Berengo (ed.), *Giornali veneziani del Settecento* (Milan, 1962), esp. the editor's introduction. F. B. Crucitti Ullrich, *La 'Bibliothèque italique': Cultura 'italianisante' e giornalismo letterario* (Milan and Naples, 1974). Walter Kurmann, *Presenze italiane nei giornali elvetici del primo Settecento* (Bern and Frankfurt a. M., 1976). Krzysztof Pomian, *Collectionneurs, amateurs et curieux. Paris, Venise: XVIe–XVIIIe siècle* (Paris, 1987), 81–142, 213–91. Other works devoting considerable space to Vallisneri include Massimo Baldini, *Teoria e storia della scienza* (Rome, 1975), and Vincenzo Ferrone, *Scienza, natura, religione: Mondo newtoniano e cultura italiana nel primo Settecento* (Naples, 1982). Brief discussions can be found in older works, such as the histories of geology by Archibald Geikie and Frank Dawson Adams and the history of medicine by Arturo Castiglioni.

2. Maurizio Torrini, "Observations on the history of science in Italy", *The British journal for the history of science*, xxi (1988), 427–46, p. 446.

3. Paolo Galluzzi, "Presentazione", in Jacopo Riccati, *Carteggio (1719–1729). Jacopo Riccati, Antonio Vallisneri*, ed. by M. L. Soppelsa (Olschki, Florence, 1985), 1–3.

4. The first of three volumes of the LeClerc *Epistolario*, published by Olschki in Florence, appeared in 1987. Neither the Rome nor the Florence centre has any planned list of publications, but will sponsor editions that individual scholars prepare. See also Paul

Dibon, "Communication épistolaire et mouvement des idées au XVIIème siècle", and Mario Sina, "L'epistolario di Jean LeClerc", in G. Canziani and G. Paganini (eds), *Le edizioni dei testi filosofici e scientifici dell '500 e del '600* (Milan, 1986), 73–88, 221–8.

5. The volumes edited by Altieri Biagi and Basile are in the series "La Letteratura italiana, storia e testi", published by Riccardo Ricciardi, Milan and Naples. Papers delivered at the Bologna conference were edited by Renzo Cremante and Walter Tega under the title, *Scienza e letteratura nella cultura italiana del Settecento* (Bologna, 1984).

6. G. Bianchi, *Carteggio inedito di Antonio Vallisneri con Giovanni Bianchi (Jano Planco)*, ed. by Alessandro Simili (Turin, 1965). Giacinto Cestoni, *Epistolario ad Antonio Vallisneri*, ed. by Silvestro Baglioni (2 vols, Rome, 1940–41). Scipione Maffei, *Epistolario (1700–1755)*, ed. by Celestino Garibotto (2 vols, Milan, 1955). L. A. Muratori, *Carteggi con Ubaldini ... Vannoni*, ed. by M. L. Nichetti Spanio (vol. xliv of the Edizione Nazionale del Carteggio di L. A. Muratori; Florence, 1978). Excerpts, often quite substantial, from the Vallisneri–Scheuchzer correspondence are in Kurmann (ref. 1). The only edition of the letters of Zeno is neither scholarly nor complete, the letters at times abridged: *Lettere di Apostolo Zeno cittadino veneziano* (6 vols, Venice, 1785).

7. See Dario Generali in Cremante and Tega (eds), *op. cit.* (ref. 5), 487–510. Also, the article by Generali in Canziani and Paganini (eds), *op. cit.* (ref. 4), 193–207.

8. The letters from Vallisneri to Bourguet are in the Bibliothèque Publique et Universitaire de Neuchâtel, Fonds Bourguet, MS. 1282. I would like here to express my gratitude to Madame Maryse Schmidt-Surdez, Conservateur des manuscrits, for the help and courtesy she and her staff extended during my visit to Neuchâtel. Although there is no published catalogue of the Fonds Bourguet, the inventory available at the library itself is detailed, careful, and excellent. Some indication of the riches of the Fonds Bourguet is provided by F. Ellenberger, "Bourguet", *Dictionary of scientific biography*, xv, 57–58. Portions of these MSS. have been used by Kennard Bork, "The geological insights of Louis Bourguet (1678–1742)", *Journal of the Scientific Laboratories, Denison University*, lv (1974), 49–77, and especially in the superb article by F. S. Tucci, "'Il parlare della S. Scrittura e l'operare della natura': Gli interrogativi della geologia storica nella riflessione di Antonio Vallisnieri", *Contributi*, vii (1983), 5–37. In addition to Tucci's excellent analysis, he used not only the letters to Bourguet (in Neuchâtel) but also found three — unfortunately, only three — responses by Bourguet in Italian archives. The only historian before Dario Generali to have a large acquaintance with Vallisneri's letters was Bruno Brunelli Bonetti who formed a vast private collection (now in public archives) of the letters, and published a charming, not very analytical, study based on his collection: *Figurine e costumi nella corrispondenza di un medico del Settecento, Antonio Vallisnieri* (Milan, 1938).

9. G. A. di Porcia, *Notizie della vita, e degli studi del kavalier Antonio Vallisneri*, ed. by Dario Generali (Bologna, 1986). This biography was first published in Vallisneri, *Opere fisico-mediche*, ed. by Antonio Vallisneri, Jr (3 vols, Venice, 1733), i, pp. xli–lxxx. All references in this article will be to the edition by Generali. Brief and excellent sketches of Vallisneri's life may also be found in Muratori, *Carteggi* (ref. 6), 102–24, and, with excellent bibliography, in Altieri Biagi and Basile, *Settecento* (ref. 5), 3–10.

10. Bruno Basile and Dario Generali, in Cremante and Tega (eds), *op. cit.* (ref. 5), 462, 505–10. My conclusions about the 'stages' in Vallisneri's writings are based on the footnotes by Generali in Porcia, *Notizie* (ref. 9), 90–149; unhappily, Generali's otherwise excellent edition does not offer a Vallisneri bibliography, and it is difficult to arrive at a coherent, chronological sequence of writings because of some of the conventions used in the footnotes. Two of Vallisneri's works have recently been reprinted, but without any critical apparatus: Silvia Scotti Morgana, *Esordi della lessicografia scientifica italiana: Il*

"*Saggio alfabetico d'Istoria medica e naturale*" *di Antonio Vallisnieri* (Florence, 1983), reprints the *Saggio* published posthumously in the *Opere* (ref. 9) of 1733, and Massimo Baldini, *Vallisneri e la scoperta dell'origine delle fontane perenni* (Brescia, 1981), reprints the *Lezione accademica intorno all'origine delle fontane*. Not having seen Baldini's volume, I do not know if he reprints the first (1715) edition or the much expanded second edition of 1726.

11. Porcia, *Notizie* (ref. 9), 103 and n. 203, referring to the text in *Supplementi al Giornale de' letterati d'Italia*, i (1722), 252–330.

12. For an especially valuable study of the *Giornale*, see Brendan Dooley, "The *Giornale de' letterati d'Italia* (1710–40): Journalism and 'modern' culture in the early eighteenth century Veneto", *Studi Veneziani*, n.s., vi (1982), 229–70. Also, Berengo (ed.), *op. cit.* (ref. 1), pp. XI–XII, and Dario Generali, "Il 'Giornale de' letterati d'Italia' e la cultura veneta del primo Settecento", *Rivista di storia della filosofia*, xxxix (1984), 243–81. For some effort to identify an editorial point of view in articles in the *Giornale*, see G. Ricuperati, "Giornali e società nell'Italia dell' 'ancien regime' (1668–1789)", in Carlo Capra *et al.*, *La Stampa italiana dal cinquecento all'ottocento* (Bari, 1986), esp. 133–48 for the first decade of the *Giornale*. In looking through the *Giornale* for reviews of books on natural history, I have found only the mildly critical statement about Giuseppe Monti's diluvial geology that his book describes fossils which are "volgarmente" called diluvial; see *Giornale*, xxxii (1719), 534. No critical comments are detectable in the discussion of the origin of fossils, treated in the review of Buonanni's *Musaeum Kircherianum* (1709), in *Giornale*, vii (1711), esp. 258–9. For Vallisneri's responsibilities in connection with the *Giornale*, see Porcia, *Notizie* (ref. 9), 78, and compare Ricuperati, *op. cit.* (ref. 12), as cited below, ref. 26.

13. For example, Giuseppe Montalenti, "Vallisnieri", in *Dictionary of scientific biography*, xiii, 563. Also, Porcia, *Notizie* (ref. 9), 56–57.

14. The text of Vallisneri's lecture is no longer extant, but its content is discussed in some detail in Porcia, *Notizie* (ref. 9), 65–74. The reserved judgement is in the "Éloge historique de M. Antoine Vallisneri", *Mémoires de Trévoux*, September 1734, 1595–1608, pp. 1600–2.

15. For Albrizzi's journal, see Berengo (ed.), *op. cit.* (ref. 1), pp. XI and ll, n. 6; Ricuperati, *op. cit.* (ref. 12), 108–11; and Luigi Piccioni, *Il giornalismo letterario in Italia* (Turin, 1894), 58–65. In the bibliography of early Italian periodicals compiled by Marco Cuaz, in Capra *et al.*, *op. cit.* (ref. 2), 371, 374, Vallisneri is listed as a "chief collaborator" on the *Galleria di Minerva*.

16. Vallisneri's translation of Descartes is no longer extant. Geology will be dٍ cussed below. For analyses of his biological ideas, in addition to Roger, *Les sciences de la vie* (ref. 1), see M. F. Spallanzani, "Esperienza e natura in Antonio Vallisneri", *Contributi*, i (1977), 5–36, and L. Geymonat, "Problemi metodologici e filosofici suggeriti dall'opera di Antonio Vallisneri", in *Il metodo sperimentale in biologia da Vallisneri ad oggi* (Padua, 1962), 157–74.

17. The Countess is discussed in Brunelli, *op. cit.* (ref. 8), 37–44, and her desire to establish an academy is mentioned repeatedly in letters of the early 1720s from Vallisneri to Bourguet, Fonds Bourguet (see ref. 8), MS. 1282. An incomplete bibliography of Vallisneri's articles in the *Ephemerides* of the Academia Naturae Curiosorum is in Porcia, *Notizie* (ref. 9), 144–8. Generali (*ibid.*, 84, n. 165) mentions a letter from Vallisneri to Hans Sloane, 16 February 1705, in which Vallisneri requests membership in the Royal Society; but the Journal Books of the Society record his election much earlier, 30 November 1703. Perhaps news of his election had been slow to reach Vallisneri in wartime?

18. Journalism will be discussed below. Vallisneri's *cabinet* is described by Porcia, *Notizie* (ref. 9), 86–101, and discussed by Pomian, *op. cit.* (ref. 1).

19. Some of Vallisneri's medical *consulti* were published in *Opere* (ref. 9), iii, 483–558, and

additional information is in Porcia, *Notizie* (ref. 9), 149 and nn. 420–3. Jacopo Riccati was one of his patients, as is evident in their correspondence (ref. 3). To contrast what is known about Vallisneri on the one hand and Zeno and Maffei on the other, see Ricuperati, *op. cit.* (ref. 12), 126–8, who sketches the careers of the three men before 1710.

20. The repute of the early dialogues is discussed in Porcia, *Notizie* (ref. 9), 59–64. For the Royal Society, see the references to letters from Vallisneri to Hans Sloane, *ibid.*, 84, n. 165. Vallisneri's relations with Scheuchzer are analysed by Kurmann, *op. cit.* (ref. 1). For the German academy, its founding and early history (to about 1701), see Rolf Winau, "Zur Frühgeschichte der Academia Naturae Curiosorum", in Fritz Hartmann and Rudolf Vierhaus (eds), *Der Akademiegedanke im 17. und 18. Jahrhundert* (Bremen and Wolfenbüttel, 1977), 117–37. The rumours in Paris are mentioned by Berengo, *op. cit.* (ref. 1), p. XII, n. 3, citing an unpublished letter from the abbé Bignon to Apostolo Zeno, 28 February 1709.

21. The best study of Italian periodicals is by Ricuperati, *op. cit.* (ref. 12), but Berengo (ed.), *op. cit.* (ref. 1) is very useful. In addition, Claude Bellanger *et al.*, *Histoire générale de la presse française* (5 vols, Paris, 1969–76), i, and the several studies of individual journals and journalists, such as: Annie Barnes, *Jean LeClerc (1657–1736) et la république des lettres* (Paris, 1938), Alfred Desautels, s.j., *Les Mémoires de Trévoux et le mouvement des idées au XVIIIᵉ siècle, 1701–1734* (Rome, 1956), Elisabeth Labrousse, *Pierre Bayle* (2 vols, The Hague, 1963–64), Betty T. Morgan, *Histoire du Journal des sçavants depuis 1665 jusqu'en 1701* (Paris, 1928), and Jean Ehrard and Jacques Roger, "Deux périodiques français du 18ᵉ siècle: 'Le Journal des savants' et 'les Mémoires de Trévoux'", in Geneviève Bollème *et al.* (eds), *Livre et société dans la France du XVIIIᵉ siècle* (Paris and The Hague, 1965), 33–59. For the publications emanating from learned societies in particular, see James E. McClellan, *Science reorganized: Scientific societies in the eighteenth century* (New York, 1985). There is no study of the availability of northern journals in Italy, but see Ugo Baldini and Luigi Besana, "Organizzazione e funzione delle accademie", in *Storia d'Italia, Annali 3* (Turin, 1980), esp. 1319–20.

22. See works cited above, ref. 15.

23. Scipione Maffei, "Introduzione" to the first volume (1710) of the *Giornale de' letterati d'Italia*, reprinted in Berengo (ed.), *op. cit.* (ref. 1), 11; also, p. XII, n. 1, for Maffei's authorship. Porcia, *Notizie* (ref. 9), 62–63. Letter from Vallisneri to Riccati, 22 August 1721, in Riccati, *Carteggio* (ref. 3), 114, where he refers to his early dialogues as having been "assassinati con errori intollerabili dall'Albrizzi".

24. According to Dooley, *op. cit.* (ref. 12), 235, unpublished letters indicate that in 1707 Vallisneri was thinking about a new journal devoted to what Dooley describes as "scientific *opuscoli*". Since he quotes none of the letters, it is not clear whether 'science' should be interpreted in the modern sense or in the common eighteenth century manner (i.e., as the equivalent of any 'discipline'). For Calogerà, see below, ref. 26.

25. Three studies of the *Giornale* are mentioned above, ref. 12. Ricuperati emphasizes the cultural orientation of the editors and hence of the periodical; Generali discusses the range of topics and issues to be found in the *Giornale*; and Dooley offers analysis of a variety of subjects, from culture to Venetian politics, from censorship to the problems which induced Zeno to leave Italy (and so affected the subsequent history of the *Giornale*). The *Giornale* is very well indexed (vols xxv, xxxviii), and I have checked all references to Vallisneri; it is on this basis that I have described articles in the *Giornale* as invoking "Vallisneri's views ... to support or combat issues ...". See also, the example cited below, ref. 44.

26. For the roles of the editors, compare Porcia, *Notizie* (ref. 9), 78, on Vallisneri's responsibilities, with Ricuperati's assertion, *op. cit.* (ref. 12), 129, that the different sciences were the

responsibility of several men, Vallisneri dealing only with medicine. The hiatus of 1718–20 and change of editors, discussed in Ricuperati, 149–50, leaves unclear Vallisneri's subsequent connection with the *Giornale*. Vallisneri's encouragement of Calogerà is discussed in Porcia, *Notizie* (ref. 9), 78–79. The Calogerà correspondence, including many Vallisneri letters for the period 1727–29, has been inventoried but not published; see Cesare De Michelis, "L'epistolario di Angelo Calogerà", *Studi veneziani*, x (1968), 621–704. For the several journals with which Calogerà was associated, see Ricuperati, *op. cit.* (ref. 12), 158–9.

27. For the expressed aims of the *Giornale*, see Maffei "Introduzione", *Giornale*, i (1710), 13–67, esp. pp. 47–67. Dooley, *op. cit.* (ref. 12), 233, 249, offers information on the circulation of the *Giornale* in Italy, but not in other countries. For the several defensive and reform-minded proposals of Vallisneri's contemporaries, see Antonio Rotondò, "La censura ecclesiastica e la cultura", and Antonio La Penna, "Università e istruzione pubblica", in *Storia d'Italia*, v (Turin, 1973), 1443–5, 1758–73. Muratori occupied a special place in this movement for cultural revival; see below, ref. 29.

28. Vallisneri, "Ricordi autobiografici (1679–1701)", transcribed by Dino Mariotti, in *Metodo sperimentale* (ref. 16), 310–11. Vallisneri to Muratori, 11 May 1710, in Muratori, *Carteggi* (ref. 6), 145–6. As he explained to Muratori, the handicaps were fundamental: he was so accustomed to lecturing in Latin at the University of Padua that he tended to forget rules governing the vernacular language.

29. Writing under the pseudonym "Lamindo Pritanio", Muratori had published *I Primi disegni della repubblica letteraria d'Italia* (1703) and *Riflessioni sopra il buon gusto* (1708); I have consulted both in the later edition, *Delle riflessioni sopra il buon gusto nelle scienze e nell'arti* (2 vols, Venice, 1766). An excellent analysis by Gaetano Righi, "L'idea enciclopedica del sapere in L. A. Muratori", is in *Miscellanea di Studi Muratoriani. Atti e Memorie della R. Deputazione di Storia Patria per le Provincie Modenesi*, Serie VII, viii (Modena, 1933), 61–102. Also, Bruno Neveu, "Muratori et l'historiographie gallicane", in *L. A. Muratori Storiografo* (Florence, 1975), 241–304.

30. The anti-French theme recurs often in Vallisneri's letters to Muratori, *Carteggi* (ref. 6), c. 1710–12. It is hard to date precisely the start of the dispute with Andry, which may have begun in periodicals before it entered into Vallisneri's *Considerazioni ed esperienze intorno alla generazione de' vermi ordinarj del corpo umano* (Padua, 1710). Fontenelle's remark would later be cited by Vallisneri in *De' corpi marini, che su' monti si trovano*, 2nd edn (Venice, 1728), 16. (All references to *De' corpi marini* will be to the second edition, the text of the geological part of the work being virtually identical to that of the first edition, 1721. Even pagination is the same, with small variations in first and last lines of each page, showing merely that the type was reset for the second edition.) The Fontenellian *gaffe* is quoted and his views analysed in Enea Balmas, "Fontenelle en Italie", in *Fontenelle. Actes du colloque tenu à Rouen du 6 au 10 octobre 1987*, ed. by Alain Niderst (Paris, 1989), 577–88, p. 584. The anti-French theme is evident especially in letters to and from Riccati, mainly in 1721, and Vallisneri on occasion equated Andry with "i francesi" and with the whole Establishment, the Académie Royale des Sciences. See Riccati, *Carteggio* (ref. 3), 101, 111, and 125, but especially the letter of 22 August 1721 (pp. 113–14) where Vallisneri declares that several of his works were "una continuata critica de' francesi e della loro Real Accademia". To Giovanni Bianchi, in a letter dated 17 June 1722, he would declare: "Non abbiamo una Nazione più nemica degli Italiani, della Francese", and he went on to single out those "tanto venerati Accademici" as people unjustly accepted as "Oracoli delle Scienze, e de' Costumi etc". Bianchi, *Carteggio* (ref. 6), 20–22. Vallisneri's prose, especially in his letters, is usually vigorous and colourful, his

views roundly delivered, his opponents often condemned as idiots. Even so, the anti-French remarks seem to me to be unusually impassioned and sometimes ferocious.

31. "Préface", in *Bibliothèque italique*, i (1728), esp. pp. xiv–xviii. The author of this preface has been identified as a *lausannois*, C. G. Loys de Bochat (1695–1754), but his writing was vetted by the other editors over so long a period of time that one may call the preface an editorial statement. See T. Cavadini-Canonica, *Le lettere di Scipione Maffei e la Bibliothèque italique* (Lugano-Friburgo, 1970), 16–18. The Genevan editors single out the *Acta eruditorum* as reliable but too scholarly for ordinary consumption, and Maffei, *op. cit.* (ref. 27), 25, indicates his view that the *Mémoires de Trévoux* neglected Italy. Basic differences in outlook, separating Italian ('baroque') from French ('classical') Jesuits, are suggested by Ricuperati, *op. cit.* (ref. 12), 120–1; the characterization of different kinds of Jesuits has great merit — there were other cleavages among the Jesuits during this period, such as the controversy over the antiquity of the Chinese Empire — but I am not really convinced that such differences materially affected the transmission of news about the publication of new books.

32. On alleged ignorance of the Italian language, see also the remarks in Cavadini-Canonica, *op. cit.* (ref. 31), 22–24. For Franco-Italian cultural relations, in addition to the study by Balmas on Fontenelle (ref. 30), see Gabriel Maugain, *Étude sur l'évolution intellectuelle de l'Italie de 1657 à 1750 environ* (Paris, 1909), and the same author's "Fontenelle et l'Italie", *Revue de littérature comparée*, iii (1923), 541–603. Maugain's book (1909) is still often acknowledged to be the best *synthesis* for a period for which a great deal of research, conducted since 1909, has made better syntheses less feasible or more dangerous — the more information we possess, the more hazardous is any attempt to generalize.

33. As a more-or-less random 'confirmation' of the suspicions of the Genevans, one may cite John Ray's surprise to find in Naples "such a knot of ingenious persons and of that latitude and freedom of judgment in so remote a part of *Europe*, and in the communion of such a Church". See John Ray, *Observations topographical, moral, & physiological* (London, 1673), 272. Recent studies indicate a regular and frequent kind of Anglo-Italian communication in matters of science. See, for example, Marie Boas Hall, "La scienza italiana vista dalla Royal Society", in Cremante and Tega (eds), *op. cit.* (ref. 5), 47–64; Marta Cavazza, "Bologna and the Royal Society in the seventeenth century", *Notes and records of the Royal Society*, xxxv (1980), 105–23; W. E. Knowles Middleton, *The Experimenters: A study of the Accademia del Cimento* (Baltimore, 1971), ch. 6; Howard B. Adelmann, *Marcello Malpighi and the evolution of embryology* (5 vols, Ithaca, N.Y., 1966), i, ch. 21.

34. In addition to works on Malpighi (Adelmann, *op. cit.* (ref. 33)) and Fontenelle (Balmas, *op. cit.* (ref. 30); Maugain, *op. cit.* (ref. 32)), see the comment by Ehrard and Roger, *op. cit.* (ref. 21), 38–39, that their effort to assess the Italian content of the *Journal des savants* was skewed by the predominance of a single Italian author: Muratori.

35. For example, Vallisneri to Bourguet, 25 April 1721, in Fonds Bourguet (ref. 8), MS. 1282, fols 245–6. and Vallisneri to Scheuchzer, 10 January 1719, in Kurmann, *op. cit.* (ref. 1), 208. In letters like these, Vallisneri at times bemoaned not the lack of will or talent among Italian *savants*, but rather the lack of patronage that would allow them to publish works as lavishly illustrated as that by Peter Wolfart (see below, ref. 48).

36. For example, Vallisneri to Muratori, 24 May 1714 (in Muratori, *Carteggi* (ref. 6), 182), Muratori to Vallisneri, 8 May 1721 (*ibid.*, 229), and Vallisneri to Muratori, 3 May 1723 (*ibid.*, 272). In the last letter, the prose is worth quoting: "tutti i frati e bianchi, e bigi, e neri, e tutta la buona turba de' preticelli." Also, letters from Vallisneri to Antonio Conti, 28 August 1727 and 10 September [1727?], in Brunelli, *op. cit.* (ref. 8), 210–12. Also, letter from Vallisneri to Riccati, 22 August 1721, in Riccati, *Carteggio* (ref. 3), 113, where

Vallisneri remarks on "la teologia fratesca" which maintains ignorance and is impervious to "il buon gusto della filosofia sperimentale e delle matematiche". For a discussion of the cynicism among intellectuals about censorship, see Rotondò, *op. cit.* (ref. 27), 1416–18. A pervasive concern among historians dealing with this period of Italian history is 'self-restraint' ('auto-censura') among intellectuals, so as to avoid the problems of censorship. How common was such restraint, and how can one document what intellectuals might have wanted to say in print? Correspondence obviously is of critical importance in these matters, as indicated above (Galluzzi, *op. cit.* (ref. 3)). This problem will be treated more explicitly below, in connection with Vallisneri's geological ideas.

37. Kurmann, *op. cit.* (ref. 1), esp. 57–61, 64–73, 141–87 for analysis of the sources, e.g., the other periodicals, used by the editors of the *Neue Zeitungen*. For the cooling of the Scheuchzer-Vallisneri friendship, *ibid.*, esp. 103–4. It is my suggestion, not Kurmann's, that this cooling had an effect on the Italian sources available to the Zurich editors. Apart from Scheuchzer, English botanist William Sherard seems to have been another friend whom Vallisneri could now and then call upon for books. See the reference to "Serard" in letter from Vallisneri to Bianchi, n.d., in Bianchi, *Carteggio* (ref. 6), 5–7.

38. See the few letters for this period in Fonds Bourguet (ref. 8), MS. 1282. For Bourguet's large number of Italian correspondents and his many trips to Italy, see Cavadini-Canonica, *op. cit.* (ref. 31), 48–50, 52. Also, F. B. Crucitti Ullrich, "Scipione Maffei e la sua corrispondenza inedita con Louis Bourguet", in Istituto Veneto di Scienze, Lettere ed Arti, *Memorie, Classe di Scienze Morali, Lettere ed Arti*, xxxiv (1969), fasc. 4.

39. Kurmann, *op. cit.* (ref. 1), 219.

40. It is difficult to assess from Generali's various works, including his footnotes to Porcia, *Notizie* (ref. 9), the number, locations, and significance of Vallisneri's foreign correspondents. In England, for example, they included Sloane and Richard Waller, but do letters to Sherard (above, ref. 37) survive? Or to Martin Lister? Vallisneri, *Lezione accademica*, in *Opere diverse* (3 vols in 1, Venice, 1715), ii, 54, referred to Lister (d. 1712) as having been "mio buon amico". The frequency with which Vallisneri sent observations to the Academia Naturae Curiosorum (above, ref. 17) suggests the possibility of a sizable German correspondence.

41. The few exceptions include both books by Rodolico, *op. cit.* (ref. 1), notably for discussion and a partial reprint of one of Vallisneri's periodical articles (below, ref. 42). Tucci, *op. cit.* (ref. 8) has greater range, including the use of manuscripts. The editor of Riccati, *Carteggio* (ref. 3) provides a useful appendix, pp. 170–4, indicating the main differences in editions of both the *Lezione* and *De' corpi marini*.

42. Letters from Cestoni to Vallisneri, 16 November 1703 and 28 December 1703, in Cestoni, *Epistolario* (ref. 6), ii, 452–4. The Latin manuscript of 1704 is no longer extant, but was later abridged, translated into Italian, and published in *Supplementi al Giornale de' letterati d'Italia*, ii (1722), 270–310, and iii (1726), 376–428. Discussion is in Rodolico, *Esplorazione* (ref. 1), 54–58, 141 n. 1, and a partial reprint, with some errors of transcription, in Rodolico, *La Toscana* (ref. 1), 315–20. See also Porcia, *Notizie* (ref. 9), 80.

43. *Galleria di Minerva*, vi (1708), 17, 151–2: reviews of Woodward, *Specimen geographiae physicae* (1704), and Scheuchzer, *Piscium querelae* (1708). Whether Vallisneri ever corresponded with Woodward is unknown, but there exists an often quoted letter from Woodward to Scheuchzer, 28 February 1724, in which Woodward asks what Vallisneri is working on. See Kurmann, *op. cit.* (ref. 1), 104–5.

44. *Giornale*, xi (1712), 199–204. The Vallisneri text is here appended to an account of a manuscript entitled "Del bagno a acqua nelle colline di Pisa", by Vibio Rustigalli, 1638 (*ibid.*, 192–9).

45. Vallisneri to Bourguet, 30 August 1721, in Fonds Bourguet (ref. 8), MS. 1282, fols 249–50. Also, Vallisneri to Bourguet, 23 September 1720, *ibid.*, fols 237–8: "la sacra scrittura a' Filosofi naturali nulla insegna." See Porcia, *Notizie* (ref. 9), 135, n. 346, and 137, n. 350. The beginnings of the Vallisneri–Bourguet friendship are obscure and may date from any of Bourguet's early visits to Italy. The first extant letter, dated 1710, seems to start *in medias res*. In his letters to Muratori, *Carteggi* (ref. 6), esp. in the years 1710–12, the theme of mixing science with faith recurs with some frequency.

46. A brief notice of the 1710 edition appeared in the *Giornale*, iv (1710), 433–4, reprinted and discussed by Vincenzo Ferrone, "Galileo, Newton e la libertas philosophandi nella prima metà del XVIII secolo in Italia", *Rivista storica italiana*, xciii (1981), 143–85, p. 163. (Ferrone also mentions, pp. 169–70, comparable Galilean themes in Muratori.) The edition of 1710 contained the letter to Christina, the *Dialogo sopra i due massimi sistemi del mondo* (1632), and some shorter pieces. According to the editor of the Riccati correspondence (ref. 3), 115, Vallisneri was citing this edition of the *Dialogo* in his letter to Riccati, 22 August 1721. But Vallisneri also was acquainted with one of the Latin editions of the *Dialogo*, which he quotes in a letter to Bourguet, December 1710, Fonds Bourguet (ref. 8), MS. 1282, fols 27–30, and in *De' corpi marini* (ref. 30), 44. The letter to Christina also circulated in many manuscript copies; see *Opere di Galileo Galilei*, ed. by Antonio Favaro (20 vols in 21, Florence, 1890–1909), v, 272–4.

47. *De' corpi marini* (ref. 30), 65, 89.

48. Vallisneri to Muratori, 30 November 1720, in Muratori, *Carteggi* (ref. 6), 224. What I have rendered "cowardly" reads: "O che coglion tedesco!" Assuming that the visitor, whose name Vallisneri mangled, was not the author, the book must have been Peter Wolfart's *Historiae naturalis Hassiae inferioris pars prima* (Cassel, 1719), which fits Vallisneri's description in all respects: beautiful copperplate illustrations of the fossils of Lower Hesse, publishing costs supported by the local ruler, prose in both Latin and German, and "una latinità che fa stomaco".

49. Vallisneri to Bourguet, 18 November 1720, Fonds Bourguet (ref. 8), MS. 1282, fols 241–2.

50. *De' corpi marini* (ref. 30), 5–11 and 15–16. The anonymous traveller was F.-M. Misson, a Protestant lawyer who emigrated to England and who is known chiefly for his *Nouveau voyage d'Italie*, first published in 1691 and then in many editions and translations. The theory of 'seeds', for example, was espoused in England by Edward Lhwyd; for this and other interpretations of fossils in the decades around 1700, see Martin Rudwick, *The meaning of fossils* (London and New York, 1972), ch. 2. For the evolution of French ideas on fossils, as these can be traced in the early volumes of the *Histoire et mémoires* of the Academy, see my article on Fontenelle, forthcoming in the *Revue d'histoire des sciences*.

51. Attacks on the French occur at the outset and at the conclusion of the first half of *De' corpi marini* (ref. 30), 5–11, 15–16, 76–77, with the intervening text devoted to Burnet and especially Woodward. Meanwhile, Vallisneri cited with approval the work of Jean Astruc (*ibid.*, 19, 43–44), Montpellier physician, and was sufficiently familiar with the Academy's *Histoire* to rely heavily upon the 1706 volume for its account of Leibniz's theory of the Earth.

52. Vallisneri to Scheuchzer, 23 September 1720 and 26 August 1721, in Kurmann, *op. cit.* (ref. 1), pp. 210–12. Whether it was actually Scheuchzer or his younger brother, also named Johann, who wanted a chair in Padua remains unclear; compare *ibid.*, 103–4, with Hans Fischer, *Johann Jakob Scheuchzer (2. August 1672–23. Juni 1733): Naturforscher und Arzt* (Zurich, 1973), 155. Kurmann's is by far the more scholarly of the two works.

53. Vallisneri seems genuinely to have wished to remain on friendly terms with Scheuchzer. Writing to Bourguet, 1 January 1722, Fonds Bourguet (ref. 8), MS. 1282 , fols 251–2, he asked if Scheuchzer had been angered by the contents of *De' corpi marini*. Later,

however, he told Bourguet (letter dated 16 September 1727, *ibid.*, fols 293–4) that, for all his admiration of Scheuchzer's "fecondissimo ingegno", he could not accept Scheuchzer's identification of the famous fossil dubbed *homo diluvii testis*. Curiously, it was Scheuchzer who heard of Vallisneri's death before Bourguet did, and he wrote to inform Bourguet (*ibid.*, fols 323–4, note appended by Bourguet to a letter from Vallisneri, dated 15 September 1729). As for Scheuchzer's point of view, Kurmann, *op. cit.* (ref. 1), 45, finds it significant that *De' corpi marini* was not reviewed in the *Neue Zeitungen* with which Scheuchzer was associated.

54. Vallisneri to Riccati, 26 July 1719, in Riccati, *Carteggio* (ref. 3), 78. Later, in a letter dated 14 August 1720 (*ibid.*, 98), Vallisneri would complain about the reactions of one *revisore* (examiner) to the manuscript of his *Istoria della generazione dell'uomo, e degli animali* (Venice, 1721). The manuscripts of both the *Istoria* and *De' corpi marini* were being examined during the same months, but the Vallisneri letters I have seen contain no reference to problems with the geological text. Among "le necessarie proteste" mentioned to Riccati, see the respectful references to miracles, the limits of human understanding, and submission to authority of the Church, in *De' corpi marini* (ref. 30), 24, 30, 76, 83–84, 89, and elsewhere.

55. For the Earth as a "machine", the uniformity of the "ordinary laws of nature", *ibid.*, 24, 31, 47, 67, 73, and elsewhere. For the separation of natural history from human sin, *ibid.*, 49: "l'effetto principale, e final del Diluvio" was to "uccidere la rubelle, e mal nata gente", and not to destroy and re-deposit "i Monti a strati sopra strati". Vallisneri suggested repeatedly that the Earth's crust had not changed significantly as a result of the Flood; if one wished to say — as both Burnet and Woodward did in their different fashions — that the Earth had changed for the worse, in order to punish mankind, Vallisneri pertinently replied (*ibid.*, 67–69) that a flood would have made the Earth more habitable, increasing the fertility of the soil. See the excellent discussion in Tucci, *op. cit.* (ref. 8), the remarks by Ferrone, *op. cit* (ref. 1), 284, and the more cautious discussion in Paolo Rossi, *The dark abyss of time*, trans. by L. G. Cochrane (Chicago and London, 1984), 75–79.

56. *De' corpi marini* (ref. 30), 106.

57. Vallisneri to Riccati, 26 July 1719, in Riccati, *Carteggio* (ref. 3), 78. Vallisneri to Bourguet, 14 August 1719 and 18 November 1720, in Fonds Bourguet (ref. 8), MS. 1282, fols 231–2, 241–2. Vallisneri to Muratori, 8 October 1721, in Muratori, *Carteggi* (ref. 6), 243. Vallisneri to Scheuchzer, letters cited above, ref. 52.

58. Vallisneri to Riccati, 7 September 1721, in Riccati *Carteggio* (ref. 3), 123. Vallisneri to Bourguet, 23 November 1717, in Fonds Bourguet (ref. 8), MS. 1282, fols 225–6, where the key sentence reads in part: "Tolta la fede, che si deve alla sagra scrittura ... chi ci assicura di questo universale Diluvio?" Also quoted in Porcia, *Notizie* (ref. 9), 135 n. 346.

59. Vallisneri did, however, use some ancient pagan literature, sometimes for topographic descriptions, but also for what the ancients supposedly "witnessed" — see *De' corpi marini* (ref. 30), 41 (Vitruvius) and especially 34 (Ovid). One of the nastier Biblical complications Vallisneri encountered as a result of his study of parasitic worms entailed controversy about whether Adam and Eve could have been so afflicted; for discussion, see Roger, *op. cit.* (ref. 1), 217–19.

60. *De' corpi marini* (ref. 30), 46. 'Obscure' was a common term in this period, however, and it generally meant a pre-literate age in *human history*, when knowledge was transmitted orally and later written down in the form of myths.

61. Antonio Conti to Vallisneri, n.d., in Conti, *Scritti filosofici*, ed. by N. Badaloni (Naples, 1972), 386–91, where the original French text of the letter is preceded by an Italian translation. The earliest date for this letter is June 1722, given Conti's allusion to the critique of Antoine de Jussieu written by Castel and published in the *Mémoires de Trévoux* of that

month. In general, see N. Badaloni, *Antonio Conti: Un abate libero pensatore tra Newton e Voltaire* (Milan, 1968) for a scholarly study of this rather odd cleric who is known to Anglophone historians of science chiefly on the basis of his encounters with Isaac Newton. Porcia's letter to Vallisneri, 24 July 1721, is in Brunelli, *op. cit.* (ref. 8), 230. For a valuable sketch of Porcia's religious attitudes, see Generali in Porcia, *Notizie* (ref. 9), 19–20.

62. Vallisneri to Bourguet, 23 November 1717 and 1 January 1725, in Fonds Bourguet (ref. 8), MS. 1282, fols 225–6, 270–1. Vallisneri to Muratori, 8 October 1721, in Muratori, *Carteggi* (ref. 6), 243; also, Tucci, *op. cit.* (ref. 8), 27. See also the remarks by Vallisneri's son in the "Prefazione" to the *Opere* (ref. 9), i, p. xx, about "il grande spazio di tempo" implied by his father's views on the sea gradually retreating from areas where sedimentary strata are now to be found.

63. Bourguet to Abauzit, 28 February 1722, in Fonds Bourguet (ref. 8), MS. 1259. This important letter was called to my attention by Kennard Bork's article (ref. 8), 61, 75, where it is cited to show that Bourguet still hoped to reply persuasively to Vallisneri's arguments. Bork and others consider Bourguet to have been a diluvial geologist, but he was neither a Burnetian nor a Woodwardian; see the elegant, delicate discussion in Jacques Roger, *Buffon: Un philosophe au Jardin du Roi* (Paris, 1989), 143. Vallisneri did not make delicate distinctions, but considered Bourguet to be too much tied to a literal interpretation of Scripture.

64. Vallisneri to Bourguet, 1 January 1722, in Fonds Bourguet (ref. 8), MS. 1282, fols 251–2: "Quando ricorriamo a' Miracoli, tutta è fornita la naturale storia."

65. See Tucci, *op. cit.* (ref. 8), 21, for one of the few surviving letters from Bourguet to Vallisneri, 12 October 1721, where Bourguet insists, in a manner entirely typical of this period, that Moses was a reliable historian. For a Galilean text in which miracles answer all questions ("tutta è fornita", quoted above, ref. 64), see Galileo, *Dialogue concerning the two chief World Systems*, trans. by Stillman Drake (Berkeley and Los Angeles, 1962), 237: "whatever begins with a Divine miracle or an angelic operation ... is not unlikely to do everything else by means of the same principle."

66. *Nova acta eruditorum*, April 1734, 167. This comment occurs in a survey of the contents of Vallisneri's *Opere*. I have not found in the *Acta* a review of the 1728 edition of *De' corpi marini*, and I have been unable to search the appropriate volume(s) for a possible review of the first edition (1721). *Mémoires de Trévoux*, December 1734, 2131; this, too, is a review of the contents of the *Opere*. What I have rendered "as perplexed as before" reads: "il laisse la difficulté dans toute sa force."

67. Review of Woodward, *Géographie physique*, trans. by Noguez (Paris, 1735), in *Mémoires de Trévoux*, February 1736, 245, 246, 253–5.

68. Vallisneri to Bourguet, 1 January 1722, in Fonds Bourguet (ref. 8), MS. 1282, fols 251–2, for allusion to "molti nemici". The editor's appendix to Riccati, *Carteggio* (ref. 3), 173–4, shows that the important changes in — actually, major additions to — *De' corpi marini* were in the biological, not the geological, parts of that book. The dispute with Andry continued unabated in this period (above, ref. 30). For all that Vallisneri at times equated Andry with the French in general or the Academy of Sciences in particular, it may be pointed out that Andry was never a member of the Academy.

69. In addition to the puzzled reactions cited above, ref. 66, Vallisneri junior included in his father's *Opere* (ref. 9), iii, 561, a letter written by Vallisneri — neither the date nor the addressee is given — which begins by tackling the allegation that Vallisneri had been too sceptical and had asked many questions but provided no answers.

70. See John Keill, *An examination of the reflections on the theory of the Earth* [by Burnet] (Oxford, 1699), notably p. 4 where Keill says that Burnet wanted him to explain "wherein

this miracle [the Flood] consisted". In reply, Keill said, "I never thought it my business to explain miracles". There is a huge literature on the English view that nature provides evidence of the existence and attributes of God; for the migration and gradual adoption of this view in Continental Europe, see Roger, *op. cit.* (ref. 1), 224–53. The anti-diluvial Frenchmen included Réaumur and Antoine de Jussieu, discussed in my forthcoming article (ref. 50). Badaloni's assertion (ref. 61), 107, that Antonio Conti adopted Jussieu's diluvial theory is based on confusion about Jussieu himself, i.e., Badaloni conflates Jussieu's marine movements with the Flood.

Baron d'Holbach's campaign for German (and Swedish) science

In 1752, when Jacob Friedrich, Freiherr von Bielfeld, produced an apologia for German intellectual attainments, he naturally composed his work in French. And he began with what hardly needed saying, namely, that few Europeans knew the German language. Nonetheless, he exclaimed, 'avec quelle avidité les ouvrages de nos chimistes sont reçus des étrangers!' This repute he attributed to German inventiveness in the discovery of, for example, phosphorus, Prussian blue, and the secrets of making fine porcelain.[1] Bielfeld might have added that Germany had also long been famous for expertise in the chemistry and technology of mining.

German publications had not gone unnoticed in Europe, since many were written in Latin. But Latin itself was, in the eighteenth century, increasingly being replaced by vernacular languages and at times was even thought to require translation into the vernacular. Publications related to mining in particular did seem to call for translation because such books could be useful to the owners and managers of mines, who might lack the traditional Latin education. Even among the élite, pertinent articles, published in French but buried in the *Mémoires* of the Berlin Academy, could more easily be overlooked than could a substantial volume in French.[2]

The combination of German repute and the relative inaccessibility of German publications began to attract the attention of great *fonctionnaires* in France shortly before 1750. In a desire to improve French industry, one such official seems to have called upon the aid of Paul-Henri Thiry, baron d'Holbach (1723-1789), a native of the Palatinate who had only recently settled in France. During the next fifteen years, from 1751 to 1765, d'Holbach would produce

1. Bielfeld, *Progrès des Allemands dans les sciences, les belles-lettres & les arts* (Amsterdam 1752), ch.1 on language and p.69-70 on chemistry.
2. The aim of reaching the non-learned reader is a recurrent theme among the authors and translators of works considered in this article. One of J. H. Pott's early papers, published in a French summary in the *Histoire* of the Berlin Academy for 1745 (Berlin 1746), p.58-61, did receive a full paragraph in the *Journal des savants*, June 1747, p.349-50 (Paris quarto edition), but Pott's work still seems to have been neglected in France until a two-volume French edition of his papers appeared in 1753.

Reprinted from *Studies on Voltaire and the Eighteenth Century* 323 (1994), by permission of the Voltaire Foundation, University of Oxford.

eight translations – often multi-volume compendia, with editorial insertions and notes – of German and Swedish works on chemistry, mineralogy, and metallurgy. In the same period, the Baron also contributed approximately 800 articles, most of them on science and technology, to Diderot's *Encyclopédie*.[3] This astonishing achievement has naturally provoked some comment by historians, although most seem a bit baffled by the time and energy d'Holbach poured into these activities. On the whole, historians have regarded d'Holbach as a transmitter and populariser of German science and technology; these activities served as prelude and adjunct to his materialist philosophy, as expounded in his *Système de la nature* (1770) and later works.[4] Two articles, concerned with the Baron's translations, have tried to place this aspect of his work in a different context from that used by biographers – namely, in what Henry Guerlac once called 'the decade of translations'.[5] More recently, Jean Ehrard, ignoring the translations and focusing on selected articles in the *Encyclopédie*, has presented d'Holbach as a critic of Buffon's 'theory of the earth'.[6]

These several approaches to the Baron's career all have validity, and yet all leave us wondering what could have prompted him to spend fifteen years engaging in labour that he himself described, referring to his translations, as 'ingrat et fastidieux'. Since all his translations appeared anonymously, as did his earliest articles in the *Encyclopédie*, one may wonder too if he had to overcome a feeling that such labour was beneath the dignity of a man of rank and wealth.[7] Whatever his misgivings and boredom, the Baron did keep going

3. Chronological lists of translations, by d'Holbach and his contemporaries, are given in the appendix below. Additional bibliographical information is in Pierre Naville, *Paul Thiry d'Holbach et la philosophie scientifique au XVIIIe siècle* (Paris 1943), p.405-20, and in Guerlac (note 5). Older literature, including Naville, underestimates the Baron's contributions to the *Encyclopédie* because his signed articles number about 400; for the attribution of unsigned articles, see John Lough, *Essays on the 'Encyclopédie' of Diderot and d'Alembert* (London 1968), ch.3.

4. Standard treatments are: Paulette Charbonnel, in d'Holbach, *Premières œuvres* (Paris 1971), p.40-50; Max Pearson Cushing, *Baron d'Holbach: a study of eighteenth century radicalism in France* (New York 1914); Naville (note 3); Virgil W. Topazio, 'D'Holbach, man of science', *Rice institute studies* 53 (1967), p.63-68; *Dictionary of scientific biography*, ed. C. C. Gillispie (New York 1970-1980), vi.468-69; W. H. Wickwar, *Baron d'Holbach* (London 1935).

5. Henry Guerlac, 'Some French antecedents of the chemical revolution', *Chymia* 5 (1959), p.73-112, esp. 98-108. A few details are added by Rhoda Rappaport, 'G.-F. Rouelle: an eighteenth-century chemist and teacher', *Chymia* 6 (1960), p.68-101, esp. 80-81. Guerlac often mentions the *Encyclopédie* as a vehicle for the ideas and information in the translations, but he does not attempt to discuss the articles.

6. Jean Ehrard, 'Diderot, l'*Encyclopédie*, et l'*Histoire et théorie de la terre*', in *Buffon 88: actes du colloque international pour le bicentenaire de la mort de Buffon* (Paris 1992), p.135-42.

7. For the phrase 'ingrat et fastidieux', see d'Holbach's preface to Antonio Neri *et al.*, *Art de la verrerie* (Paris 1752), p.viij. J.-C. Valmont de Bomare, *Minéralogie* (Paris 1762), i.xiij, found it remarkable and praiseworthy that the Baron should have engaged in such labour, rather than

in a tireless fashion that manifestly contradicts Diderot's claim that he was subject to 'passing' enthusiasms. In the next pages, I shall argue that the Baron was conducting what he considered to be an important scientific campaign, its outlines discernible only if one examines both his translations and his articles. From the translations alone, one would conclude that he simply wanted to make these works known in the expectation that they would be useful. From his many short articles in the *Encyclopédie*, there emerges what some historians have called a concern with 'mere' nomenclature or uninteresting dictionary definitions.[8] More careful reading, however, reveals a campaign with at least three major themes: (1) that nomenclature be rationalised, with a firm base in chemistry, (2) that French chemistry be reformed and broadened in subject matter and methods, and (3) that a reformed chemistry be the basis for study of the earth's history. So little is known about the evolution of d'Holbach's materialism that I think it legitimate to consider his scientific writings as having their independent rationale; that very independent coherence is the subject of this article.

Translators and translations

Henry Guerlac's 'decade of translations' was inaugurated by the chemist Jean Hellot, member of the Académie royale des sciences, at the request of J.-B. Machault d'Arnouville, controller general of finances. Concerned to encourage and rationalise French industry, Machault ordered the translation of a recent treatise on mining by Christoph Andreas Schlüter. Since Hellot knew no German, he solicited the translation from an Alsatian mining engineer named Koenig; Hellot himself then revised Koenig's prose, added information neglected by Schlüter, and contributed a lengthy preface and an introductory essay.[9] Not long after the publication in 1750 of the first volume of this translation, it seems that the new director of the Librairie, Malesherbes, made known (to d'Holbach?) 'le desir [...] de voir paroître en François les meilleurs

merely enjoying the leisure and perquisites of his rank and wealth. Historians often mention (e.g., Lough, p.111-12) that Diderot concealed the name of the new contributor to vol.ii (1751) of the *Encyclopédie*, revealing his identity in vol.iii (1753).

8. Naville, p.67, and Jacques Proust, *L'Encyclopédie* (Paris 1965), p.130. Diderot's comment on 'passing' enthusiasms occurs in a letter to Sophie Volland (?13 October 1759, *Correspondance*, ed. Georges Roth, Paris 1955-1970, ii.271).

9. Hellot's preface to Schlüter, *De la fonte des mines* (Paris 1750), esp. p.xxvij-xxix for the translation process and Hellot's extensive editing. I am unable further to identify Koenig, except to say that he was later mentioned often, with respect and gratitude, in publications and manuscripts of the mineralogist Antoine Monnet, who spent much time in Alsace. See *Dictionary of scientific biography*, ix.478-79.

Ouvrages des Allemands'. This wish, claimed the Baron, launched his own first translation and his plan, if his work was well received, to produce further translations of 'les meilleurs Ouvrages Allemands, sur l'Histoire Naturelle, la Minéralogie, la Métallurgie & la Chymie'.[10] The Baron's first venture, entitled *Art de la verrerie* (1752), presumably had some success, for he quickly proceeded to act on his announced plan.

In the next years, d'Holbach would be joined by other translators, at least some of whom may have been inspired by the Baron's example. One of these, D.-F. d'Arclais de Montamy, an old and intimate friend of the Baron, translated from the German some of the writings of J. H. Pott, a chemist whose works d'Holbach already knew and admired. In the same year that Montamy's translation appeared (1753), d'Holbach dedicated his own translation of J. G. Wallerius to Montamy.[11] The pharmacist J.-F. Demachy is not known to have been one of the Baron's friends, but when his own translation from the Latin of Johann Juncker's textbook of chemistry forced him to confront some German technical terms, Demachy consulted d'Holbach, the acknowledged expert.[12]

Another inspiration behind the translators was the teacher of both d'Holbach and Demachy, the chemist Guillaume-François Rouelle (1703-1770). Expounding in modified form the chemical doctrines of Georg Ernst Stahl, Rouelle evidently called to the attention of his pupils the existence of foreign expertise in chemistry and allied sciences. Several of the texts translated in mid-century stemmed from the Stahlian school or were deemed reconcilable with it.[13] Furthermore, one translator recounts what may not have been a unique tale. According to J.-F. de Villiers, several students asked Rouelle what works should be translated in order to improve French knowledge of metallurgy. Rouelle recommended J. A. Cramer's *Elementa artis docimasticae* (1737) and urged Villiers himself to undertake the translation.[14]

10. Neri, dedication to Malesherbes, and translator's preface, p.viij. For the role of Augustin Roux as editorial aide to d'Holbach, see Naville, p.431, n.16, or Guerlac, p.102, n.99; also remarks in appendix below, s.v. '1760 Henckel'.

11. When Montamy died in 1765, Diderot referred to him as 'le plus ancien ami du baron' (letter to Sophie Volland, 28 July 1765, *Correspondance*, v.70). See appendix s.v. '1753 Pott'. Before the publication of Montamy's translation, d'Holbach had already been recommending the works of Pott in the *Encyclopédie* (articles 'Bismuth', ii.262b-263b, 'Calcaire', ii.541b-542a). For d'Holbach's collection of books in German, see *Catalogue des livres de la bibliothèque de feu M. le baron d'Holbach* (Paris 1789; B. N., Q.8047), p.274-88; Latin and French editions of German works are listed on earlier pages of this inventory.

12. Translator's preface to Juncker, *Elémens de chymie* (Paris 1757), i.xix. For analysis of d'Holbach's circle, see Alan Charles Kors, *D'Holbach's coterie* (Princeton 1976).

13. In addition to Guerlac and Rappaport (note 5), see Rachel Laudan, *From mineralogy to geology* (Chicago, London 1987), esp. p.59-61.

14. Translator's preface in J. A. Cramer, *Elémens de docimastique* (Paris 1755), i.viij-ix. A reviewer of this translation in the *Mémoires de Trévoux*, September 1755, p.2294, remarked about

Some translators worked from Latin editions, others from German. What all shared was the conviction that French readers should become acquainted with the science and technology of Germany and (to a lesser extent) of Sweden. If d'Holbach espoused the same general aim, his writings, unlike those of his fellow-translators, also tell us in some detail how such imported knowledge could contribute to the progress of science.

Nomenclature and taxonomy

D'Holbach's choice of Wallerius for his second translation, his preface to that work, and his articles for the *Encyclopédie* all help to clarify his concern for nomenclature. In early articles with engagingly bizarre titles – 'Cachimia', 'Chalaxia', 'Chalcantham', 'Chanquo', 'Chumpi' (ii.509b; iii.18b, 19a, 139a, 403b) – d'Holbach selected terms that naturalists could find in ancient authors like Pliny and in such moderns as Paracelsus, Alonso Barba, and Boëtius de Boodt. These terms, either coined or uncritically transmitted by the authors who used them, made older books almost incomprehensible, and Wallerius provided not only a systematic taxonomy of minerals but also a concordance to this idiosyncratic vocabulary. As Louis-Jean-Marie Daubenton pointed out, in a knowledgeable tribute buried in the *Encyclopédie* ('Cornaline', iv.244b-245b), d'Holbach had performed notable service in making available to French scientists the pioneering taxonomic system of the great Swedish mineralogist. But he had also enhanced the value of Wallerius by adding French terms to the chiefly Latinate concordance. The translator himself, according to Daubenton, revealed 'une grande connoissance des minéraux', as well as 'un zele constant & éclairé pour l'avancement de la Minéralogie'.[15]

Articles seemingly on nomenclature also gave the Baron opportunities to join Diderot in raising questions about legends, popular beliefs, and unverified reports. In one of his early articles, Diderot had displayed understandable scepticism about the existence and virtues of an Abyssinian plant whose shade alone was fatal to serpents ('Assazoé', i.766b).[16] Similarly, d'Holbach doubted

Rouelle's choice of Villiers: 'C'est, en ce genre, comme si Raphaël, dans le sien, eût député Jules Romain [Giulio Romano] pour tenir sa place.'

15. Naville, p.68, refers to this article when he claims that Daubenton *asked* d'Holbach to help stabilise nomenclature (and d'Holbach responded by translating Wallerius). Nothing in Daubenton's article substantiates this claim.

16. Several of Diderot's many articles of this kind are discussed by Jean Mayer, *Diderot homme de science* (Rennes 1959), p.109. See also the remarks by Jacques Roger in *Essais et notes sur l'Encyclopédie de Diderot et d'Alembert*, ed. Andrea Calzolari and Sylvie Delassus (Milan 1979), p.244.

that a particular kind of rock fell from the clouds during hailstorms ('Brontias', ii.436b), or that 'draconites' had been suitably named in classical times as rocks found in the heads of dragons ('Draconites', v.99b-100a). Diderot had also tackled vagueness of description: the 'Acarnan' as a digestible fish (i.59a), the 'Acalipse' as either a fish or a bird (i.58a). In the latter case, he remarked: 'Voilà encore un de ces êtres dont il faut attendre la connoissance des progrès de l'histoire naturelle, & dont on n'a que le nom; comme si l'on n'avoit pas déjà que trop de noms vuides de sens dans les Sciences & les Arts, &c.' If Diderot sometimes seems amused, d'Holbach more soberly stressed the same message. 'Morochtus', for example, a name given by Pliny to a substance said to remove stains from clothing, seemed to be a kind of chalk, and naturalists had continued to use the term with varying connotations; in the Baron's judgement, this term illustrated 'la confusion qui regne' in the naming of inorganic substances, 'faute de les avoir examinées en chimiste' (x.715b). In the case of 'Onyx', names had multiplied depending on the sometimes slight differences among specimens: 'On voit par-là que les anciens lithographes ont fait tout ce qu'ils ont pû pour embrouiller les choses, en multipliant les noms sans nécessité' (xi.487b-488a). As for 'Oolite', naturalists, 'qui semblent n'avoir jamais manqué l'occasion de multiplier les dénominations, ont donné différens noms à ces sortes de pierres', depending on the size of the granules – d'Holbach's word is 'globules' – in particular specimens (xi.491b). Repeatedly d'Holbach either states or implies that chemical analysis could destroy this accumulation of verbiage by providing a firm basis for both nomenclature and taxonomy.

In forcefully expressing these views, the Baron was no innovator. Selecting chemistry as the fundamental tool of analysis placed the Baron on one side of a long dispute, but he also shared with Diderot and others – notably botanists and zoologists – the conviction that names should convey meaning, not confusion. As Charles Gillispie once remarked in this connection, 'To name is to know, not essences, nor totalities, but what we can know.'[17] One part of d'Holbach's campaign was, in short, not about 'mere' nomenclature or technical definitions, but about knowledge as expressed in names and classifications. As a contributor to an encyclopedia, he took pains to define terms found in books; as a *philosophe*, he also argued that nomenclature and taxonomy should reflect the progress of knowledge.

17. Charles C. Gillispie, *The Edge of objectivity* (Princeton 1960; reprint Princeton 1967), p.172. See also the remarks by d'Alembert, article 'Elémens des sciences', v.494b; the comment by Tarin, article 'Anatomie', quoted by Charbonnel, p.44; and the programme of analysis by J. H. Pott, *Lithogéognosie*, tr. Montamy (Paris 1753), ii.19-20.

The reform of French chemistry

In his focus on German and Swedish achievements, d'Holbach implicitly pointed out the inadequacies of French chemistry. These were of two kinds: the failure to include soils and rocks among the substances analysed, and the failure to employ extremely high temperatures as a method of analysis. French chemists were not wholly remiss, but their repertory and methods were too limited.

Arguing that chemistry provided the most crucial tool for understanding the great array of natural objects, d'Holbach broached an issue that would agitate scientists during the latter half of the century. For naturalists in the field, the obvious way to recognise what one saw was by employing such physical criteria as colour, odour, hardness, texture, and weight – as well as such minimal chemical tests as the reaction of a particular rock sample to one or another acid. Was it really necessary to transport samples to the laboratory for further tests? To this question, d'Holbach answered with an unequivocal yes. Here he agreed with Wallerius, Pott, and others – including Daubenton ('Cornaline', cited above) – that mere physical similarities of any two rocks might disguise the real differences revealed by chemistry. Or, conversely, rocks differing in appearance might be similar or identical in chemical composition. At the same time, d'Holbach conscientiously reported the various taxonomies devised for soils, rocks, and minerals by modern writers who differed among themselves on the relative importance of physical characteristics and chemical analysis. There is, he concluded, 'peu d'espérance que l'on puisse jamais concilier ces deux choses', although he made his own preference clear.[18]

Among the chemists whose writings d'Holbach and others translated into French, several gave particular attention to analyses using not only improved furnaces but especially the high temperatures generated by lenses, or so-called 'burning glasses'. French chemists had occasionally used the latter method, but chiefly for the calcination of metals; as d'Holbach pointed out, however, German experiments treated a great range of substances, and with important results. Subjecting soils and rocks to unusually high temperatures called into question the customary designation of some substances as 'apyrous', or unchangeable by heat. That category, as the Baron noted, would have to be re-defined as immunity only to ordinary degrees of heat, since nothing in nature could resist 'violent' heat.[19]

---

18. 'Minéraux', x.544b; see also 'Pierres', xii.575a, and d'Holbach's preface to Henckel, *Pyritologie* (Paris 1760), i.vj. The best short introduction to this debate as conducted in the last decades of the eighteenth century is in *Dictionary of scientific biography*, xiv.256-64.
19. 'Miroirs ardens', x.570ab; 'Pierres apyres', xii.577a; 'Pierres vitrescibles', xii.584b-585a; 'Vitrescibilité', xvii.362a; 'Vitrifiable', xvii.363b-364a. In 'Miroirs ardens', d'Holbach indicates that his article supplements an earlier one, 'Ardent (miroir)', i.623b-627b, which had been

D'Holbach was quite accurate in his critique of French chemistry. As a recent addition – and one of which the Baron wholly approved – French chemists of the 1750s had adopted a modified version of the theory associated with two Germans, Johann Joachim Becher and Georg Ernst Stahl. Embedded in this theory was a classification of three sub-species of the element 'earth', namely, the vitrifiable, the sulphureous, and the mercurial. This little taxonomic system, however useful in interpreting a range of chemical reactions, did not induce its French proponents to engage in systematic analysis of real samples of soils and rocks for their component 'earths'. Instead, experiments in the 'mineral kingdom' generally were confined to metals, bitumens, and salts – the same subjects found in Nicolas Lemery's often-reprinted *Cours de chymie* (1675). Recently, to be sure, Rouelle and some of his early disciples – notably Pierre-Joseph Macquer and Gabriel-François Venel – had been expounding and defending so-called Stahlian chemical theory, but without appreciably broadening the range of experiments available in Lemery. D'Holbach had no objections to the new theory (and he would later translate two works by Stahl), but he also urged that theory be more explicitly linked to the laboratory examination of all the inorganic materials that had already entered the repertory of chemists like Wallerius and Pott.[20]

D'Holbach's articles now and then do cite such French chemists as Homberg, the Geoffroys, and the Lemerys, but his enthusiasm was reserved for Rouelle. The many references to Rouelle, couched in terms of some intimacy ('Rouelle used to say', 'Rouelle liked to show', and similar locutions), show the Baron to have been another of the famous pupils who frequented Rouelle's courses. (Others included Diderot, Lavoisier, Rousseau, and Turgot.) For all d'Holbach's admiration, it is clear from a manuscript version of Rouelle's course, dating from 1752, that Rouelle was then innocent of the expanded

written by d'Alembert and had little to say about chemistry. For earlier French use of burning glasses for the 'solar calcination' of metals, see Henry Guerlac, *Lavoisier – the crucial year* (Ithaca 1961), p.116. Among translations, see Cramer, i.80-81; C. E. Gellert, *Chimie métallurgique*, tr. d'Holbach (Paris 1758), i.101-105; Pott, ii.6-30, esp. 24, 28-29.

20. I have elsewhere argued that the adoption and modification of Stahlian theory was the work of Rouelle, at a date that cannot quite be fixed. See Rappaport, 'Rouelle and Stahl – the phlogistic revolution in France', *Chymia* 7 (1961), p.73-102, esp. 80-81 for earths. For the limited repertory of Rouelle himself, see below, note 21. The same limitations are in Macquer, *Elémens de chymie – théorique* (Paris 1753), p.8-11, and *Elémens de chymie – pratique* (Paris 1756), topics in vol.i on the mineral kingdom. Venel used the *Encyclopédie* to defend Stahlian theory (e.g., 'Chymie', iii.408a-437b) and later began to pay attention to Hellot's edition of Schlüter and to German works available in Latin ('Docimasie', v.1a-4a). When Lemery's textbook was reprinted for the last time, its editor, another Rouelle pupil, chiefly criticised Lemery's outmoded theories, but not his still-useful experiments or their limited range: see Lemery, *Cours de chymie*, ed. Théodore Baron (Paris 1756), esp. p.11, note (p) on 'pierres'.

experimental programme of German chemistry. But Rouelle's openness to
novelty is shown in d'Holbach's preface to Wallerius (1753), where Rouelle is
mentioned as one of the two consultants – the other being the botanist Bernard
de Jussieu – about the technical aspects involved in the translation. By about
1758, Rouelle's teaching had come to include analyses of soils and rocks.[21]

The relationship between Rouelle and d'Holbach will continue to be puzzling
unless and until new sources are discovered – or the known material is re-
examined with this relationship in mind. Did Rouelle introduce the Baron to
such German works as were available in Latin, or did d'Holbach alter Rouelle's
teaching by revealing to him the riches of German-language chemistry? Both
may have happened, and answers to these questions might appear to depend
simply on ascertainable dates. Any student of either Rouelle or d'Holbach
knows that precise dates cannot be expected, but even the hope of detecting
dates begins to dissolve when one encounters in the Rouelle manuscripts a
bald assertion such as this: 'Lorsques M. Poot [*sic*] publia son ouvrage sur les
pierres et les terres, M. Rouelle travailloit depuis longtemps sur la meme
matiere, mais il a tout abandonné. On peut tirer un grand fruit des travaux de
M. Poot.' Which works of Pott: the Latin essays, the French summaries or
papers in the *Histoire et Mémoires* of the Berlin Academy, or the French
volumes translated by Montamy? Is there any way to determine if Rouelle had
actually begun, independently of Pott, those analyses for which Pott became
famous? How trustworthy are claims for Rouelle's originality?[22]

21. Muséum national d'histoire naturelle, Paris, MS. 2017: 'Procédés chimiques du cours de
monsieur Roüelle année 1752'. This is a list of experiments, without discussion. D'Holbach's
preface to J. G. Wallerius, *Minéralogie* (Paris 1753), i.viiJ. Exactly when Rouelle incorporated
soils and rocks into his teaching cannot yet be determined. Many Rouelle manuscripts bear the
same titles and dates, ultimately stemming from the notes taken by Diderot: lectures 'recueillies'
in 1754-1755, 'rédigées' in 1756, 'revues et corrigées' in 1758. But the content is not always
identical. Of two such manuscripts at the Bibliothèque nationale, one has a substantial section
of experiments on soils and rocks, but the other does not, and one expands considerably on how
soils and rocks can be classified. Compare BN, nouv. acq. fr. 4043-44, ii.278-79, with BN, ms.
fr. 12303-304, ii.240-41. By 1760, soils and rocks were assuredly part of Rouelle's repertory; see
Procédés du cours de chymie de M. Rouelle ([Paris 1760 or 1761]), erroneously catalogued at the
BN as the work of Rouelle's younger brother, Hilaire-Marin. Valuable information about the
evolution of topics treated by the two Rouelles – unfortunately, from 1760 on, not during the
1750s – is in H.-M. Rouelle, *Tableau de l'analyse chimique* (Paris 1774), p.xvj and *passim*. Rouelle
is cited or discussed in 23 of d'Holbach's articles: 'Crystal', 'Crystallisation', 'Cuivre', 'Etain',
'Fer', 'Fossile', 'Gypse', 'Lune (Hist. nat.)', 'Mercure ou vif-argent', 'Métal', 'Mine', 'Minéralis-
ation', 'Montagnes (Hist. nat.)', 'Naphte', 'Natrum', 'Or', 'Orpiment', 'Régule d'antimoine',
'Sable', 'Safre', 'Soufre', 'Tremblemens de terre', 'Vitriol'.

22. The quotation is from MS. 'Cours de chimie, ou leçons de monsieur Rouelle', Science
library, Clifton College, Bristol (England), p.959. The same statement, phrased in the first person
singular, is in BN ms. fr. 12303-304, ii.299. The casual tone may indeed be trustworthy, in view
of Rouelle's reputation for aggressively asserting his own originality: see F. M. Grimm *et al.*,

Studying the earth's past

For d'Holbach, chemistry was the key to what we now call geology and what he called 'mineralogy'. As Jean Ehrard has rightly said, the Baron considered this a virtual 'science of sciences' which had as its province not only the whole mineral kingdom, but also its past history and the history of life (fossils).[23] To modern historians, the Baron's interest in geology seems too obvious to require comment. After all, what atheist could fail to seize an opportunity to attack geological use of Noah's Flood? This d'Holbach certainly did, but his interest in geology cannot be reduced to the role of adjunct to materialism. Rather, d'Holbach the chemist saw in this science ways to explain the origins of certain features of the earth's structure. His opposition to diluvial geology, so evident in at least one of his translations, must here be supplemented by the far more informative articles he contributed to the *Encyclopédie*.

Before the 1750s, geologists commonly examined rocks and fossils *in situ*, debated the dynamics of how sediments had been deposited, and argued about how marine organisms in particular had come to be buried on land, far from modern seas and often high above sea level. By that decade, the nature of fossils themselves – were these the remains of organisms, or merely sports of nature ('lusus naturae', 'jeux de la nature')? – no longer puzzled most naturalists, so that d'Holbach evidently saw no need to do more than summarise an issue readers would be likely to encounter in books still valuable in other respects.[24] The Baron's writings show great familiarity with the literature on fossils, ranging from the most recent taxonomies emanating from Britain to the contents of Gmelin's voyage to Siberia (not translated from the German until 1767). Indeed, d'Holbach used his wealth to purchase a collection, a 'cabinet d'histoire naturelle', even if, according to Diderot, he left that collection in crates in the stables for several years.[25]

In matters of nomenclature and taxonomy, d'Holbach had little complaint

---

Correspondance littéraire, ed. Maurice Tourneux (Paris 1877-1882), ix.107. For another example of claims to priority, see note 38 below.

23. Ehrard, p.138, quoting parts of d'Holbach's 'Minéralogie', x.541b-543a.

24. 'Fossile', vii.210b; 'Jeux de la nature', viii.535ab; 'Figurées (pierres)', vi.782b-783a. The concept of *lusus* seems to have been less popular in France than elsewhere, perhaps because of the early critique by Fontenelle. For geology in the Académie royale des sciences early in the eighteenth century, see Rappaport, 'Fontenelle interprets the earth's history', *Revue d'histoire des sciences* 44 (1991), p.281-300.

25. Diderot to Sophie Volland, ?13 October 1759, *Correspondance*, ii.271. Significantly, Diderot reports that d'Holbach was joined by Montamy in organising the Baron's collection. Articles citing Gmelin are numerous, including 'Ivoire fossile', ix.63a-64a; 'Licorne fossile', ix.486ab; and 'Mammoth, os de', x.7b. The recent British taxonomists most often cited by d'Holbach are John Hill (1716?-1775) and E. Mendes da Costa (1717-1791).

about the achievements of naturalists in the handling of fossils. Just as zoologists and botanists had long been classifying living organisms in a rational manner, so too were naturalists arranging fossils into genera and species. Here the Baron even found satisfying such umbrella terms as 'ichthyolithes' (viii.482b-483a), 'nom générique donné par quelques naturalistes à toutes les pierres dans lesquelles on trouve des empreintes de poissons'.[26] Further improvement remained possible, however, as d'Holbach now and then reminded readers that Wallerius and other non-French authors had done more than his compatriots to identify and classify certain fossil forms.[27]

More important to d'Holbach than fossils was diluvialism, or the use of Noah's Flood to explain all or most of the earth's sedimentary strata. At least twice, in treating the Flood, did d'Holbach argue with the authors he was translating, and he inserted a number of anti-diluvial comments into articles for the *Encyclopédie*.[28] The article 'Déluge' itself (iv.795b-803a), however, was not by the Baron, but was largely the work of Nicolas-Antoine Boulanger. Here Boulanger provided a virtual catalogue of widely divergent interpretations, including comparisons of the Biblical text with other flood stories. D'Holbach engaged in no such exegesis, but assumed a literal interpretation of Genesis and repeatedly presented the several arguments developed by earlier naturalists to show that the Flood could not have been responsible either for sedimentary strata or for the burial of fossils. Although the subject could lend itself to polemic, the Baron's tone remained notably detached and moderate. He could even say, more than once, that

On ne peut douter de la réalité du déluge, de quelque voie que Dieu se soit servi pour opérer cette grande révolution; mais il paroît que, sans s'écarter du respect dû au témoignage des saintes Ecritures, il est permis à un naturaliste d'examiner si le déluge a été réellement cause des phénomenes dont nous parlons, sur-tout attendu que la Genèse garde un silence profond sur cet article. D'ailleurs rien n'empêche de conjecturer que la terre n'ait, indépendamment du déluge, encore souffert d'autres révolutions.[29]

26. Also 'Gryphite', vii.974b; 'Phytolites', xii.540b-541a; 'Typolites', xvi.782b-783a.
27. In addition to articles cited in note 26, see 'Glossopetres', vii.722a, where d'Holbach leaps quickly from interpreting these fossils to classifying them. For the authorship of this article, see Lough, p.126, n.1. Lough bases his attributions mainly on the literature cited (Henckel, Pott, *et al.*) and on the network of cross-references to related articles. He discusses content only in very general terms, but the consistency of content and viewpoint adds materially to evidence for the correctness of attribution.
28. Naville, p.195, says that the translation of Lehmann is the only place where d'Holbach challenged the views of his author, but see also Henckel, *Pyritologie*, i, translator's notes on p.122, 123, 131.
29. 'Fossile', vii.211a; also 'Charbon minéral', iii.190b-191a; 'Mer', x.359a; 'Montagnes (Hist. nat.)', x.674a; 'Poudingue', xiii.187ab; 'Terre, couches de la', xvi.170a; 'Typolites', xvi.782b.

The same tone was maintained when the Baron translated the diluvial treatise of J. G. Lehmann. As he explained, Lehmann's text contained much valuable information, and the author's erroneous geological theory (diluvialism) could be queried in a lengthy translator's preface.[30]

In an area of importance for the study of sedimentary strata and their origins, d'Holbach seems to have introduced French readers to a theory that had aroused much controversy in Sweden but had not yet circulated widely. (Here he even ventured to translate from the Latin, rightly assuming that Swedish academic publications were little known in France.) This theory, that the *amount* of water constituting the earth's seas had diminished in the course of time, had its origins in traditional questions about the Flood: where had all the water for the Flood come from, and where had it all gone afterwards? As a common solution to both problems, naturalists had often interpreted the Biblical 'abyss' to mean 'caverns' within the earth, from which water had emerged and into which excess water had subsequently retreated. In a mid-century alternative, Benoît de Maillet's *Telliamed* (1748) had offered a bold, controversial, non-Biblical theory of what the author called 'the diminution of the sea'. If *Telliamed* was outrageously materialistic and imaginative, the Swedes were neither. For more than fifty years before mid-century, various Swedish scientists had been gathering evidence for the changing level of the Baltic Sea during historic times and even within living memory. As cautiously interpreted by Anders Cels (Celsius) and others, the Baltic was 'diminishing' – as a logical and inevitable physical consequence, the Baltic being no landlocked body of water, all the earth's oceans must also be 'diminishing'. How to interpret the decades of evidence was provoking much debate in the Academy of Stockholm, and d'Holbach expertly presented the whole affair, reaching no conclusion, possibly because the concept of a universal ocean, steadily lowering in level, reminded the Baron too much of the Flood.[31] The Swedish theory did become widely accepted in the last decades of the century, and one may

30. J. G. Lehmann, *Traités de physique* (Paris 1759), remarks by d'Holbach, i.IX, and esp. iii.v-VJ, XXJ-XXIJ.
31. 'Mer', the first columns by d'Alembert (x.358b-359a) and the rest by d'Holbach (x.359a-361b). Given the fame of some Swedish participants (Celsius, Linnaeus), I find it inexplicable that the most important French journals devoted to book reviews (*Journal des savants*, *Mémoires de Trévoux*) should have ignored these publications. For an analysis of the debate, including the question of possible Swedish familiarity with *Telliamed*, see Tore Frängsmyr, 'Linnaeus as a geologist', in *Linnaeus: the man and his work*, ed. Frängsmyr (Berkeley, Los Angeles 1983), esp. p.125-43. D'Holbach translated the seminal paper by Celsius, in *Recueil des mémoires* (Paris 1764), i.112-29. The chief critic of Celsius, Johan Browallius, was presented in translation by Jean-Louis Alléon-Dulac, *Mélanges d'histoire naturelle* (Lyon 1763-1765), iv.94-184. I have not included this work in the appendix because its contents, insofar as these can be identified, are chiefly extracts from French periodicals.

speculate that the Baron's writings had something to do with making known work that had had little press outside Scandinavia before the 1760s.[32]

The geological issues discussed thus far have little to do with chemistry, and d'Holbach believed that linking these two sciences, as Germans were doing, would provide geology with firmly experimental, non-speculative foundations. It had long since been said, most strikingly by Fontenelle, that the chemical laboratory could legitimately be viewed as reproducing processes that had occurred in nature's past.[33] D'Holbach seized upon this concept, leaning heavily on the teachings of Rouelle and on some of his German authors. In many of his articles, he presented not quite a 'theory of the earth' but the materials and concepts from which such a theory might be erected.

In at least two articles, d'Holbach directly confronted – elsewhere he did so indirectly – the most recent and popular geological synthesis, Buffon's theory of the earth published in the first volume of the *Histoire naturelle* (1749). Buffon was charged by d'Holbach not only with being a deductive system-builder but also with ignoring the importance of chemistry. The article 'Caillou' (ii.533a-536a) begins with columns by Daubenton, Buffon's disciple and collaborator, in which Daubenton merely quotes at length Buffon's theory that the earth's core consists of igneous materials, cooled and vitrified to form glassy debris; *cailloux* are the product of the breakdown of this vitrified mass. D'Holbach began the next columns by announcing:

Nous allons ajoûter ici plusieurs observations & conjectures [...] qui se trouvent répandues dans les opuscules minéralogiques de M. Henckel, & dans le commentaire de M. Zimmermann sur ces opuscules, ouvrages Allemands, qui n'ont jamais paru en François; laissant au lecteur à décider de ce qu'elles peuvent avoir de favorable au système de M. de Buffon.

Support for Buffon proved to be non-existent, as d'Holbach went on to explain that the glassy, transparent, crystalline appearance of many *cailloux* suggested crystallisation out of a watery fluid, rather than vitrification caused by heat. Although the Baron here cited German sources, elsewhere he would emphasise Rouelle who had published classic memoirs on the crystallisation of salts.[34]

---

32. Even Linnaeus' *Oratio* (1744) on the diminution of the sea was slow to attract attention outside Sweden. For some information on editions and translations, see Linné, *L'Equilibre de la nature*, tr. Bernard Jasmin, ed. Camille Limoges (Paris 1972), p.29, n.1.

33. See Rappaport, 'Fontenelle', p.285. The popularity and quotability of Fontenelle is briefly indicated, p.299-300, with emphasis on the scientific community.

34. Also 'Glaise', vii.698b-700b, intended to supplement Daubenton's 'Argille', i.645b-647a, which deals only with the 'system' of Buffon. Although not mentioned by name in either article, Rouelle is discussed by d'Holbach in 'Crystal' and 'Crystallisation', iv.523a-524a, 529ab; he is also cited explicitly when d'Holbach treats the analogy between crystallisation and the origins of primitive formations, 'Montagnes (Hist. nat.)', x.676a. For another aqueous model to explain the formation of rocks, see 'Stalactite', xv.490a, by d'Holbach.

Having launched an intelligent critique of Buffon, d'Holbach eventually began to synthesise what he considered to be the best available evidence for the history of the earth's crust. Since he repeatedly promoted the model of aqueous crystallisation to explain the formation of many rocks, modern historians of geology would doubtless call him a 'neptunist'. But the Baron disliked and distrusted mono-causal theories. Nor should his many references to 'revolutions' make him a 'catastrophist'. As he insisted, in many of his articles, the earth's crust had been shaped by three main agents: water, fire, and air. Some transformations may have been world-wide, but many had been local; some causes had acted with violence and rapidity, but others peaceably and slowly.[35] One should begin to chart the earth's history with the formation of 'primitive' rocks, this initial phase being followed by the long history of so-called 'revolutions'. Such 'revolutions', still occurring, would require regional geological studies. In the end, geologists would have to grapple with a period of time that d'Holbach described as reaching back to prehistory, or to 'des tems dont l'histoire [histoire écrite, orale, humaine] ne nous a point conservé le souvenir'.[36]

Distinguishing 'primitive' from 'secondary' formations, d'Holbach described the former as a virtual scaffolding ('charpente') of the earth, often consisting of high mountains of unstratified and especially hard rocks. The most promising explanation of their origins seemed to be the analogy of chemical crystallisation ('Montagnes [Hist. nat.]', x.672b-673a, 676a). Characteristically, these formations contained mineral ores – an observation d'Holbach attributed to Rouelle, later 'verified' by Lehmann ('Mine', x.523a). 'Secondary' formations were typically stratified and fossiliferous, and most could be attributed to local floods or to incursions of the sea. Other post-primitive formations were volcanic, but d'Holbach, like so many of his contemporaries, considered volcanoes to be no clue to the earth's basic structure. In the language of the eighteenth century, volcanoes were 'accidents', or minor (negligible) interruptions of the regular course of nature. Study of volcanoes had, as the Baron indicated, given rise to at least one mono-causal 'system', and d'Holbach seems to have been one of the few in France familiar with the

---

35. For example, see 'Mine', x.522a; 'Révolutions de la terre', xiv.237b-238a; 'Terre, couches de la', xvi.171ab. Ehrard, p.138-40, recognises d'Holbach's statements about varieties of causes and of rates of change, but he nonetheless concludes that the Baron (in opposition to both Buffon and diluvialists) emphasised fire as well as cataclysms. That 'revolution' was not necessarily a synonym for 'cataclysm' is indicated by Rappaport, 'Borrowed words: problems of vocabulary in eighteenth-century geology', *The British journal for the history of science* 15 (1982), p.27-44, esp. p.31-38.

36. 'Ivoire fossile', ix.64a; 'Minéralogie', x.542a; 'Volcans', xvii.444b. See also d'Holbach's recommendation of the treatise by Krüger (listed below in appendix), in Lehmann, iii.XXIIJ-XXIV.

peculiar volcanic explanation of all the earth's land-forms as expounded in Anton Lazzaro Moro's *De' Crostacei e degli altri marini corpi che si truovano su'monti* (1740). Such system-building d'Holbach found deplorable. In a pithy dismissal of Moro, coupled with the authors of other mono-causal systems, d'Holbach remarked that such theorists attribute to a single cause 'des effets tout différens' (x.675a).[37]

In discussing post-primitive formations, clearly sedimentary in origin, d'Holbach attributed to Lehmann the observation that the oldest of the post-primitives consisted of what would now be called the Coal Measures (x.674b). (Parenthetically, it is worth noting that this observation may have been picked up by Rouelle, thanks to the Baron's translation of Lehmann. At some time during the 1760s, Rouelle did begin to teach his students that the Coal Measures occupied an 'intermediate' position between the primitive and the secondary formations.)[38] How to proceed further to reconstruct a history of the earth's crust entailed regional studies, as Nicolas Desmarest had already said (see below), but d'Holbach, ignoring his fellow-Encyclopedist, announced that pertinent local studies had been begun by Rouelle, for the Baron's teacher had been examining marine fossils in particular, noting that they seemed to occur in local clusters or populations ('Fossile', vii.211b). Neither Rouelle nor d'Holbach explained the significance of these observations, the Baron saying only that the pattern itself deserved attention. Rouelle's study of fossils, announced by d'Holbach, never did materialise.

In summary, one may say that d'Holbach was an alert and critical reader of geological texts. More than a populariser or a reporter of what he had read in books, he showed a willingness to evaluate his authors and even to attempt a partial synthesis of what he considered to be the best literature. That he raised questions about the Flood is not surprising, but one did not have to be an atheist to do so – even pious naturalists like Linnaeus did admit to being unable to find geological evidence for the Flood. In the last ten volumes of the

---

37. D'Holbach owned both the original Italian edition of Moro and the German translation of 1751 (*Catalogue*, p.73, 279). These publications were not reviewed in either the *Journal des savants* or the *Mémoires de Trévoux*; indeed, in the latter, July 1760, p.1695, Moro is incongruously listed among diluvialists, and the same error is in Valmont de Bomare, *Dictionnaire raisonné universel d'histoire naturelle* (Paris 1765), v.394, s.v. 'Terre'. French writers who mentioned a Moro-like theory associated it with John Ray (1627-1705), whose name d'Holbach coupled with Moro's (e.g., Buffon, *Histoire naturelle*, Paris 1749, i.522). Indications that Moro attracted some attention in Germany and England can be found in *Dictionary of scientific biography*, ix.531-34.

38. Rappaport, 'Rouelle', p.96 and n.111. One of the Rouelle manuscripts, dated 1761-1762 (but containing insertions dated 1763), makes a vague claim for Rouelle's priority, referring to 'M. Lehman, qui a tiré beaucoup de choses de M. Rouelle' (Library of the Wellcome institute, London, MSS. 3075-76, ii.37).

Encyclopédie, all of them published in 1765, there are more anonymous articles than before, but d'Holbach did not take this opportunity to become more militantly anti-religious. On the contrary, important geological articles – such as 'Mer', 'Montagnes', 'Terre' – still bore his signature (symbol), and he continued to provide careful rehearsals of publications, considerations of evidence, and moderate assessments of what conclusions one might reach on the basis of the best evidence. In other words, d'Holbach's geological writings possess a strong family resemblance to his discussions of nomenclature and chemistry: he did not hesitate to express his own preferences, but his focus stayed resolutely on scientific issues rather than on the militancy associated with his later writings.

Some conclusions

Any study of d'Holbach is severely handicapped by the paucity of information available about what one might call the 'inner' life of the Baron – his publications are numerous, his surviving correspondence virtually non-existent. We get frequent glimpses of the Baron from his contemporaries, but we know almost nothing about the evolution of those views for which he is most famous, namely, his materialism and atheism. What this study has attempted to do is to divorce d'Holbach's scientific activities, over a period of fifteen years, from his materialist militancy of the 1770s. This is not to deny that the two 'periods' of his life were connected, as his later writings do make extensive use of the science of his day (or his interpretations of that science). In the translations and articles considered here, however, there is no evidence of the Baron's later radical views on religion and society, but there is a great deal of evidence that d'Holbach shared Diderot's conviction that the sciences and technology needed modernisation on the basis of the best information available. Sharing this general point of view, d'Holbach devoted at least fifteen years to the modernisation of chemistry and allied sciences and technology.

Serious attention to d'Holbach's scientific writings inevitably raises the question of how 'influential' were the Baron's writings and translations. Difficulties are here abundant, and the next paragraphs will suggest areas needing future research.

One difficulty involves the common eighteenth-century habit of providing few if any footnotes. As a welcome exception to the rule, the Swiss pastor Elie Bertrand, in his *Dictionnaire des fossiles* (1763), adopted the antiquarian method of actually citing his sources and providing bibliographies. Almost half of the *Dictionnaire* – it was published before the last ten volumes of the *Encyclopédie* –

Baron d'Holbach's campaign for German (and Swedish) science

makes such extensive use of d'Holbach's articles and translations that one is tempted to describe Bertrand's handbook as a digest of d'Holbach.[39] At the same time, Bertrand was no 'advanced' thinker – he resisted for an unusually long time the notion that marine fossils were the remains of organisms, and he shared with Voltaire (with whom he corresponded for many years) the conviction that the earth's history exhibited not change but rather a constancy of Law and Design. Nonetheless, Bertrand contributed two articles to the *Encyclopédie*, and he seems to have submitted more items on natural history, then recovering his manuscripts when Diderot's enterprise seemed doomed by the royal prohibition of 1759.[40] For a number of reasons, then, Bertrand was a careful reader of the first seven volumes of the *Encyclopédie*, and he found the information, not the interpretations, essential for the compilation of his *Dictionnaire*.

Other scientists probably consulted the translated volumes more often than the scattered articles in the *Encyclopédie*. In chemistry and taxonomy, for example, the works of Wallerius and Pott carried great weight. Macquer's popular textbooks of the 1750s contained no discussion of soils and rocks; by 1766, however, he was using the classification and the experiments in Pott's *Lithogéognosie* (translated by d'Holbach's friend Montamy).[41] Similarly, one of the popular teachers of natural history, J.-C. Valmont de Bomare, was immune to the German current in 1758 but by 1762 had reformed his teaching to rely heavily on the nomenclature and taxonomy of Wallerius.[42] As indicated earlier, Rouelle's celebrated courses also came to include analysis of soils and rocks, although it remains conceivable that Rouelle may have embarked on this subject independently of Pott. By 1778, Pierre Bayen (another Rouelle

---

39. Bertrand, *Dictionnaire universel des fossiles propres, et des fossiles accidentels* (Avignon 1763), articles 'Bismuth', 'Bleu d'azur', 'Bley-sweiff', 'Caillou', 'Charbon minéral', 'Corne (pierre de)', and many more. Bertrand was also aware that d'Holbach was the (anonymous) translator of Wallerius and other works. The *Dictionnaire* was also published in two vols. (The Hague 1763).

40. Bertrand's articles in the *Encyclopédie* 'Joux', viii.899a-900a, and 'Keratophytes, ou Cérato-phytes', ix.119b. For vague allusions to other manuscripts intended for the *Encyclopédie*, see letters of Voltaire to Bertrand, 29 August and 4 September 1759 (D8459, D8467). It is conceivable that some of Bertrand's later contributions to the *Encyclopédie* of Yverdon (1770-1776) may originally have been intended for Diderot. See Marguerite Carozzi and Albert V. Carozzi, 'Elie Bertrand's changing theory of the earth', *Archives des sciences* (Geneva) 37 (1984), p.265-300, esp. the list of articles on p.290.

41. For Macquer's textbooks, see note 20 above. Cf. Macquer, *Dictionnaire de chymie* (Paris 1766), such articles as 'Gyps' and 'Terre' (i.567-72; ii.567-76).

42. Compare the rudimentary discussion of 'terres' in Valmont de Bomare, *Catalogue du cabinet d'histoire naturelle de M. Bomare de Valmont* (Paris 1758), p.19-25, with the same author's *Minéralogie*, i.42-45, 105-106.

pupil) could say that Pott had produced 'une révolution' among French chemists who had earlier paid little if any attention to the analysis of rocks.[43]

To assess the influence of d'Holbach – or, more generally, 'the decade of translations' – on geology is even more difficult than for chemistry. Discussion of 'primitive' and 'secondary' formations occurs during the eighteenth century in the writings of many naturalists apart from Rouelle, Lehmann, and d'Holbach; one may indeed find comparable discussions, varying in detail, in Moro, Buffon, and Arduino, and even earlier in Steno and Leibniz. The Baron himself treats this topic as a feature of Rouelle's lectures, but several authors in the decade of the 1750s seem simultaneously to have found fruitful this way of analysing the earth's crust.[44] As a general scheme, dividing the earth's crust into two (or three) large formations, representing a chronological history, became so commonplace in the last decades of the eighteenth century that to attempt to trace priorities, filiations, and influences seems almost an exercise in futility. One may say with certainty only that d'Holbach's articles, and his translation of Lehmann, made available to non-German naturalists a schematic history of the earth that other writers, not necessarily acquainted with Lehmann or d'Holbach, were finding fruitful.

To ask whether geologists in France understood d'Holbach's message that chemistry was essential to geology is to open another large field which has not yet been investigated. That chemists busily expanded their repertory and methods seems indubitable, and only one example will be presented here to suggest that the sometimes dramatic results of chemical experiments did induce geologists to pay attention to the relevance of chemistry to their own concerns.

In an early (1757) volume of the *Encyclopédie*, Nicolas Desmarest laid out a programme for field observers in matters largely geological ('Géographie physique', vii.613b-626a). Recognition of rocks here depended mainly on physical characteristics, as well as minimal tests with acids (vii.615b), but the several rock types interested Desmarest far more for their structural relationships than for their chemical composition (vii.623a). About a decade later, Desmarest's classic discussion of the basalt columns of Auvergne included a lengthy evaluation of Wallerius' taxonomy, largely based on chemistry, of volcanic rocks. In addition, Desmarest had become aware of recent experiments by Jean Darcet – another Rouelle pupil, like Desmarest himself – to subject

---

43. 'Examen de différentes pierres, publié en 1778', in Bayen, *Opuscules chimiques* (Paris 1798), ii.41; also quoted by Guerlac, 'Antecedents', p.103, n.102. Bayen goes on to explain (ii.44) his own methodological preference for solvents rather than degrees of heat.

44. See, for example, John C. Greene, *The Death of Adam* (Ames, Iowa 1959), ch.3; Laudan, p.57-58; and Gabriel Gohau, 'La "Théorie de la terre"', in *Buffon 88*, p.350.

quartz samples to very high temperatures. Darcet's various studies had been undertaken, in effect, as a continuation of the experiments of J. H. Pott. Both Darcet and Desmarest, like d'Holbach before them, had come to recognise that chemical experiments could help to determine how rocks had originated and what changes they had undergone.[45]

Without surveying the whole of French science in the 1760s and 1770s, these examples do suggest changes occurring in both chemistry and geology. Men like Darcet, Desmarest, Macquer, Rouelle, and Valmont de Bomare had been absorbing the content of translations from the German. As the repertory and methods of French chemists expanded, so too did the arsenal of analytical techniques available to geologists. Whether those 800 articles in the *Encyclopédie* played any role in these changes cannot be determined with any certainty, but the Holbachian programme explained in those articles was implicit in the texts that he and his contemporaries translated. A new body of material and ideas became available to French scientists, and the Baron's campaign bore fruit in the next decades.[46]

What of the Baron himself after his intensive scientific activity of 1751-1765? In 1761 Malesherbes asked what further work he would like to see translated, and d'Holbach's brief reply did not even include reference to those treatises he himself would translate in the next years.[47] Did Malesherbes catch the Baron in a mood of fatigue? We may never know, but we do know that d'Holbach never wholly abandoned his scientific work, as is evident from the post-1761 flood of articles for the last volumes of the *Encyclopédie*, the sporadic translations after 1765 of works by Stahl and Wallerius, and the responsibility for at least one volume of *Planches* published in 1768 to supplement the text of the *Encyclopédie*.[48] After 1765, however, d'Holbach turned increasingly to other projects, leaving us no record of what he thought about the results of his long scientific campaign.

---

45. Desmarest, 'Mémoire sur l'origine & la nature du basalte à grandes colonnes polygones', *Mémoires de l'Académie royale des sciences* 1771 (1774), p.705-75, esp. p.725 (Darcet) and 751-58 (Wallerius). The dates of presentation of this two-part memoir are given here as 3 July 1765 and 11 May 1771. See *Dictionary of scientific biography*, iii.560-61; also Guerlac, *Lavoisier*, p.79-84, for Darcet's experiments, beginning in 1768, on the combustion of that most 'apyrous' gem, the diamond.

46. Jacques Roger has suggested that the Baron's translations prepared the ground for later French acceptance of Wernerian geology (Calzolari and Delassus, p.259).

47. The exchange with Malesherbes is in Cushing, p.30-31. D'Holbach here mentions two works by Stahl, but he would later translate two quite different treatises. For the German titles of Stahl's *Traité du soufre* (1766) and *Traité des sels* (1771), see J. R. Partington, *A history of chemistry* (London 1961), ii.661, 662.

48. Lough, p.116, referring to *Planches*, vol.vi, devoted to natural history and metallurgy.

Appendix: chronological lists of translations

THE following lists were compiled using standard bibliographies (catalogues of the Bibliothèque nationale and the British Library), aided by the works of Guerlac and Naville cited above and P.-M. Conlon, *Le Siècle des Lumières: bibliographie chronologique* (Geneva 1983-). The original German and Latin titles have been omitted because many translations are actually compendia of several treatises, articles, and even selected chapters from the writings of the original authors.

The list for d'Holbach is as complete as possible for his scientific and technological works, but omits his translations (mainly from the English) of deistic and other philosophical writings; all his translations listed here are from the German, unless otherwise specified. The second list, by d'Holbach's contemporaries, cannot be assumed to be complete, and it is also limited to the period coinciding with the years of d'Holbach's own greatest activity. Information about the translators and translations has been culled from Michaud, *Biographie universelle*, from the prefaces to the translations, and from other works as indicated below.

I. Translations by d'Holbach

1752 Neri, Antonio, and Christopher Merret, Johann Kunckel. *Art de la verrerie*. Paris. Neri's original work was in Italian, Merret's translation and commentary in Latin, Kunckel's translation and commentary in German. D'Holbach claims to have compared all three (p.iij). Other texts are appended.

1753 Wallerius, Johan Gottschalk. *Minéralogie*. Paris. Appended is a translation of the same author's treatise on hydrology.

1756 Henckel, Johann Friedrich. *Introduction à la minéralogie*. Paris.

1758 Gellert, Christlieb Ehregott. *Chimie métallurgique*. Paris.

1759 Lehmann, Johann Gottlob. *Traités de physique, d'histoire naturelle, de minéralogie et de métallurgie*. Paris. As the title indicates, this is a collection. It includes prefaces by d'Holbach in vols. i and iii.

1760 Henckel, Johann Friedrich. *Pyritologie*. Paris. The second volume includes Henckel's *Flora saturnisans*, to which was added a supplement based on Rouelle's lectures and experiments on plant material, introduced (ii.159) as a perfect model of chemical analysis – a model to be emulated in dealing with minerals and animals. According to d'Holbach (i.vij), part of vol.ii was translated by a pharmacist named Charas and then revised by Augustin Roux.[49]

---

49. There was a Charas dynasty of pharmacists, founded by Moyse Charas (1619-1698), member of the Académie royale des sciences (Paris). One assumes that the Charas mentioned by d'Holbach belonged to this family.

Baron d'Holbach's campaign for German (and Swedish) science

1760 Orschall, Johann Christian. *Œuvres métallurgiques*. Paris.
1764 *Recueil des mémoires* [...] *dans les Actes de l'Académie d'Upsal et dans les Mémoires de l'Académie royale des sciences de Stockholm*. Paris. Translated from German and Latin.
1766 Stahl, Georg Ernst. *Traité du soufre*. Paris.
1771 Stahl, Georg Ernst. *Traité des sels*. Paris. Evidence that this should be added to the list of d'Holbach's translations is in Guerlac, 'Antecedents', p.108, n.115; the same evidence was subsequently noted by Partington, p.662.
1774 Wallerius, Johan Gottschalk. *L'Agriculture réduite à ses vrais principes*. Paris. Translated from the Latin, with notes added from the German edition.

II. Comparable translations by d'Holbach's contemporaries

1750 Schlüter, Christoph Andreas. *De la fonte des mines*, translated from the German by Koenig, revised and edited by Jean Hellot. Paris. Usually described as translated by Hellot, but the more complex process of preparing this volume is recounted in Hellot's preface, p.xxvij-xxix. The second volume appeared in 1753.

1752 Krüger, Johann Gottlob. *Histoire des anciennes révolutions du globe terrestre*, translated from the German by André-François Boureau-Deslandes. Paris. The translator is described in Michaud (s.v. 'Deslandes') as an aspiring philosopher and littérateur, and a member of the Berlin Academy. Nothing in this biography helps to explain his knowledge of the German language.

1753 Pott, Johann Heinrich. *Lithogéognosie*, translated from the German by Didier-François d'Arclais de Montamy. Paris. Naville (p.58) calls Montamy 'l'ami d'enfance' of d'Holbach. Implausible because of the considerable disparity of age (Montamy b.1702), but Montamy's military service may have taken him to German territory. He was aide-de-camp to Louis, duc d'Orléans, during the war of the Austrian succession (1740-1748).

1755 Cramer, Johann Andreas. *Elémens de docimastique*, translated from the Latin by J.-F. de Villiers. Paris. Villiers identifies himself (i.viij-ix) as a pupil of G.-F. Rouelle.

1757 Juncker, Johann. *Elémens de chymie*, translated from the Latin by J.-F. Demachy. Paris. Demachy, a pharmacist, had studied with G.-F. Rouelle. He indicates (i.xvij-xix) his lack of knowledge of German and how he circumvented the problem.

1759 Pott, Johann Heinrich. *Dissertations chymiques*, translated (in part) by J.-F. Demachy. Paris. Demachy apparently translated some papers from the Latin. Others had already been translated into French by Samuel Formey for publication in the *Mémoires* of the Berlin Academy, and still others were translated from the German by Formey for inclusion in this edition by Demachy (i.x-xj).[50]

1762 Marggraf, Andreas Sigismund. *Opuscules chymiques*, translated (in part) by J.-F. Demachy. Paris. Demachy's role seems to have been that of editor, as he admits (i.xvj) that the translations were produced by the author himself and by Samuel Formey. Certainly, he could not have translated those papers (11 out of 26 published here) written in German. The catalogue of the BN lists Demachy as translator, but compare Guerlac, 'Antecedents', p.104.

[1763?] *Collection de différens morceaux sur l'histoire naturelle et civile des pays du Nord*, translated from the German, Latin, and Swedish by Louis-Félix Guinement de Keralio. Paris. This rare volume was consulted at the Muséum national d'histoire naturelle, Paris, and it contains little that can be called 'natural history'. Keralio indicates (p.x-xj) that he is planning a series of translations, the next volumes to be Gmelin's account of Siberia (see below). He may have acquired his knowledge of German at the Ecole royale militaire where the subject was being taught from about mid-century.

1766 Meyer, Johann Friedrich. *Essais de chymie*, translated from the German by P.-F. Dreux. Paris. The obscure translator identifies himself as a pupil of G.-F. Rouelle and explains why he undertook this translation (i.xvii-xviii).

1767 Gmelin, Johann Georg. *Voyage en Sibérie*, translated from the German by Louis-Félix Guinement de Keralio. Paris. This is an abridged version of Gmelin, the omissions explained in Keralio's preface. One can only wonder if the many references to Gmelin by d'Holbach in the *Encyclopédie* had anything to do with Keralio's choice of text. Keralio later produced a translation of G. S. Gruner on Swiss glaciers (Paris 1770) and a single volume of excerpts from the memoirs published by the Academy of Stockholm (Paris 1772).

Intentionally omitted from this list is the multi-volume *Collection académique*, ed. Jean Berryat *et al.* (Dijon 1755-). Apart from some initial translations from the Latin of excerpts from the *Ephemerides* of the Academia naturae curiosorum, the *Dijonnais* did not tackle the publications of the Berlin Academy until vols.viii and ix (1768).

---

50. Jean-Henri-Samuel Formey came from a family of Huguenot *émigrés* and served as secretary of the Berlin Academy from 1748 until his death in 1797. See Frank A. Kafker and Serena L. Kafker, 'The Encyclopedists as individuals: a biographical dictionary of the authors of the *Encyclopédie*', *Studies on Voltaire* 257 (1988), p.140-44. For some indication of the riches of the Formey manuscripts, now becoming available, see François Moureau, 'L'*Encyclopédie* d'après les correspondants de Formey', *Recherches sur Diderot et sur l'Encyclopédie* 3 (1987), p.125-45.

XV

ESSAY REVIEW

GOVERNMENT PATRONAGE OF SCIENCE IN
EIGHTEENTH-CENTURY FRANCE

Agronomie et Agronomes en France au XVIIIe Siècle
ANDRÉ J. BOURDE
(S.E.V.P.E.N., Paris, 1967, 3 vols.)

The title of André Bourde's monograph, although an accurate statement of the major focus of the study, does little to suggest its rich content. 'Agronomy' means not only scientific farming, but, in eighteenth-century France, the efforts made to *introduce* scientific farming; reform of any kind under the Old Regime, however, usually raised a host of social, economic, political, and intellectual problems. And Bourde has successfully treated all these problems in a synthesis so masterful that a reviewer who singles out any portion runs the risk of distorting the whole. Specialized literature on what would seem at first glance to be an incredible variety of topics has been brought together to describe a reform movement whose coherence and magnitude has never before been fully explored.

In contrast to the more famous Physiocrats, Bourde's agronomists were less doctrinaire and more empirical, and their proposals appealed strongly to royal ministers interested in bringing about economic improvement without endangering the foundations of government and society. Just how moderate some of these reforms were, how much desired by royal officials, and yet how difficult to put into practice are matters which emerge in the course of the monograph. Bourde emphasizes the role of the agronomists in areas central to the reform of agriculture: crop rotation, enclosures, the cultivation of wastes, new crops, new implements and methods, fertilizers, livestock. He also treats the dissemination of information, the establishment of schools for the teaching of agricultural techniques, and the progress of veterinary medicine. Throughout there are enlightening comments on the local variations in laws and customs, the piecemeal attempts of provincial intendants and others to alter the "routine" of the peasantry, the attitudes of peasants and of special interest groups (such as the Parlements and Provincial Estates) to particular reforms, and the work of individual landowners whose initiative provided some impulse towards the spread of new ideas and methods. Certain intellectual currents—the prestige of science, admiration of the pastoral literature of ancient Rome, the mania for everything 'exotic'—are analyzed for their relevance to both the increasing popularity of agricultural pursuits and the impact of the agronomists in particular. In all volumes, but especially in the section on Bertin's ministry, there is considerable discussion of the role of the central government which was receptive to new ideas, interested in encouraging experiment, earnest in its desire to do virtually anything to promote the well-being of the realm; and there are numerous examples of the problems faced in attempting to put reforms into effect without infringing on the rights, substantially altering the laws and customs, or endangering the welfare of individuals or groups.

Most of the monograph deals with the period 1750–1789. Beginning, however, with the classic *Théâtre d'agriculture* (1600) by Olivier de Serres, Bourde analyzes the ideas and the policies pursued in the ages of Sully and Colbert, the two most important periods in French agriculture before 1750. The discussion of Serres shows clearly the relative lack of ability of his seventeenth-century imitators and the slow rate at which new and reliable information was added to that available in 1600. Among the more significant sources of such new information were the gardening manuals and botanical works of LaQuintinie and others. At the same time, government policies seem not to have evolved in any particular direction; instead—always excepting Sully and Colbert—successive ministries displayed a haphazard, occasional, and relatively minor interest in agricultural affairs.[1] Agronomy as a movement, therefore, dates from the 1750s, with the work of Duhamel du Monceau (and others) and the advent of Bertin as Controller General of Finances.

As Bourde observes, 1789 does not constitute a turning-point in the history of agricultural reform. To have carried his subject beyond that date, however, would have meant considerable duplication of good studies already available, particularly those by Octave Festy.[2] There is no doubt that Bourde could add fresh insights into the later period, but he probably chose wisely in restricting himself to occasional comments on those problems which were studied without interruption and without major changes of institutions during the first year or two of the Revolution.

It is now evident that Bourde's earlier volume, *The influence of England on the French agronomes, 1750–1789* (Cambridge, 1953), was, in effect, a preliminary sketch of a subject which enters prominently into *Agronomie et agronomes*. In the newer work, more space is given to an analysis of Duhamel du Monceau's interpretation of the writings of Jethro Tull and to both Anglophilia and Anglophobia in France.[3]   The whole subject of foreign influences is also broadened to include brief but suggestive discussions of the agronomists' knowledge of the agriculture of the Low Countries, China, and elsewhere, with full recognition of the differences between appreciation of genuine innovations in agriculture and propagandistic admiration of the supposed moral virtue and enlightenment of the Chinese.

One of the most valuable features of these volumes is the extensive bibliography (iii, 1657–1730) which is a virtually exhaustive catalogue of major and minor original sources and of subsequent scholarly works. Bourde's text and footnotes include analyses of the content of treatises, manuals, and dictionaries, attempts to assess their popularity, and discussions of the differences in successive editions of the more important works.   He also analyses the many agricultural periodicals—and the agricultural content of publications like the *Journal des savants*—which often flourished for only a few years and are now not readily available. This thorough combing of the printed literature is supplemented by selective use of documents in the Archives Nationales and a few other repositories.

Since the reviewer must blame as well as praise, a word ought to be said about two important defects, one remediable and the second less so, in these otherwise extraordinary volumes.   The lack of an index will come as no surprise to readers of French monographs; the analytical table of contents, however logically arranged, is a completely inadequate guide to the complexities of Bourde's subject.   The second problem has no easy solution, since it involves the perennial question of the extent to which reformers made any impact on the French countryside.   Bourde is, of course, well aware of the difficulties here, and he seems torn between his desire to defend the effectiveness of the reform movement and his knowledge of the restrictive influences of illiteracy, tradition and custom, and the laws.   Indeed his analysis of the obstacles to reform is so penetrating that to ask just what *was* achieved is all the more pressing.   He might here have tried to measure his findings against the picture presented by Arthur Young.   Granted that it is difficult to deal with the peasantry, more information is available about their social betters, and one might ask, for example, if Young was correct in his repeated assertions that agricultural improvement was generally being neglected by the proprietors of great estates who preferred life in Paris or Versailles and left the "real work" to the *métayers*.   Many of Bourde's agronomists and improving landowners were, in fact, aristocrats.   Bourde, however, concludes that agronomy was essentially a "bourgeois" movement, thereby ignoring his own evidence and missing a challenging opportunity to assess the most influential and the most valuable treatise on the state of French agriculture in 1789.[4]

Agronomie et agronomes, like other major works of scholarship, treats some subjects exhaustively, while at the same time it opens up new areas for research and provides material and ideas for the re-examination of more familiar subjects. Bourde's perceptive allusions to international affairs, commerce, and philosophy, his comments on local laws and customs, his references to the complexities of the governmental bureaucracy, and his characterizations of reformers and royal officials constitute a perpetual invitation to his readers to use this monograph as a starting point for the further study of all these topics, with agriculture playing a subordinate rather than a dominant role.

For historians of science, this monograph sheds invaluable light on the reputation of science during the eighteenth century. In an agricultural context, 'science' sometimes meant botany or chemistry, but more often it meant "to do things scientifically", to observe, to experiment, to draw up Baconian axioms, and to test these axioms in practice. Holding such views put the agronomists solidly alongside the scientists of the period—indeed, many scientists were agronomists—and strengthens the picture we are acquiring of the prevalence of empiricism during the Enlightenment. That the sciences were held in great esteem has long been acknowledged by historians of the eighteenth century, and various aspects of this subject have been explored in scholarly studies of the image of Isaac Newton, the mania for 'cabinets' of natural history specimens, the popularity of public and private lectures and demonstrations, the proliferation of scientific dictionaries and manuals, and the scientific terms, images, and ideas in the writings of the philosophes. From this literature there has emerged a complex pattern: science was, for some, a fashionable hobby, or a model for the study of man and society, or a symbol invoked to support a particular philosophy, or the orderly search for knowledge about the natural world.[5]

Perhaps the greatest remaining gap in our knowledge of this subject is the question of governmental attitudes towards science. Although Bourde offers innumerable examples of scientific talent being recruited by Bertin and his colleagues—and other historians have contributed to this list—government patronage of science still awaits investigation. This topic is, to be sure, a special instance of the activities of a paternalistic government which was constantly awarding pensions, grants, and financial privileges; but to separate science from this broader context is justifiable for an age which attributed great virtues to the scientific enterprise. Considering a royal minister like Turgot separately from his philosophical friends who did not occupy high administrative positions is also justifiable, since a royal minister possessed opportunities to put his beliefs into practice in a way that the private citizen often did not. It may be less permissible to follow Bourde in limiting such a study to the period 1750–1789, but the earlier date, if not the later one, does seem to me to correspond approximately to certain shifts in both the government and the scientific community. Not only was the sheer number of scientists increasing rapidly after about mid-century, but the evidence now available also suggests that government officials, more and more interested in reform, were turning for aid to the scientists far more regularly than was the practice during the late seventeenth and early eighteenth centuries.

There are many avenues of approach to the study of government patronage of science, and each has been tried on a limited scale. Thus, for example, monographs devoted to the development of certain industries, such as mining and textiles, contain valuable information about the application of science to technological reform. The establishment and subsequent history of such royal institutions as the École des Mines and the Académie des Sciences have also been studied with care, as have certain scientific enterprises of the size and importance of the expeditions to observe the transits of Venus. In addition, there are good biographies of some of the key figures, such as John Holker.[6] This brief survey could be lengthened, but it will be more revealing to give some indication of the formidable projects still to be undertaken. One of these is surely a serious study of Daniel Trudaine and his son. Despite their importance as royal ministers and patrons of science, the Trudaines emerge from relative obscurity only in works devoted to other subjects, such as the history of the École des Ponts-et-chaussées, the textile industry, and the beginnings of systematic reform in agriculture. The variety of activities in which they engaged rightly suggests that their influence was pervasive, and these activites ought to be brought together and analyzed in a way that would illuminate some of the darker corners of governmental policies.[7] A second case in point is that of André Thouin, who enters into scholarly literature as little more than the man who took his orders from Buffon. For some sixty years, however, and most of them after Buffon's death, Thouin exercised considerable authority at the Jardin du Roi. He also corresponded with naturalists and directors of botanical gardens in many countries, distributed seeds and seedlings to virtually anyone who requested them, trained and supplied gardeners to aristocratic amateurs, French colonial administrators, and scientific expeditions, worked to preserve the gardens and collections of exotic plants at Versailles and elsewhere during the Revolution, and was a fairly prominent member of scientific and agricultural societies.[8]

Rather than prolong this list of desiderata, I should like to offer some general comments about the available literature and about what I believe ought to be done. We now have a large and growing number of studies which focus upon the changes in particular areas of the French economy; such works tend to emphasize the economy, and governmental efforts to direct or "encourage" it, with the role of science and scientists usually subordinated and sometimes ignored. We also have studies of certain royal officials and their reform policies; here again, while the use of scientific expertise is sometimes mentioned, such considerations are always subordinated to matters of economic and political concern. For the scientists, we have a few biographies and a small literature dealing with instances of scientists serving on special committees consulted by royal ministers.[9] However admirable these studies, they tend only to leave us with the general impression that royal officials were willing to use scientific talent and that they sought expert advice whenever a suitable occasion arose. These studies also suggest that such advice was sought for purposes almost wholly practical: to judge the work of inventors, to propose remedies for specific crises (especially in agriculture), or to devise new (or improve old) techniques for the benefit of industry or agriculture.

Among the questions I should like to see investigated are: To what extent and for what purposes did government officials recruit scientific talent? How did a scientist go about seeking and obtaining royal patronage? How did officials distinguish between the men of ability and the less talented, between the practicable scheme and the technically less feasible one? (How often, in fact, did they try to make such distinctions, and how often did talent weigh less than more personal considerations?) Can one discern particular programs and policies, or did officials generally respond to the needs, projects, and suggestions of the moment? How much confidence did ministers have in scientific research? How often did they support "pure" sciences in contrast to obviously useful ones like agronomy? What effects did patronage have on the work of the scientists?

Unfortunately, I cannot answer these questions, but I can supply a few tentative comments based upon those examples which have come to my attention. All too often, in fact, my comments will have to take the form of additional questions. In the following discussion, I shall be using 'patronage' broadly to mean not only cash subsidies but also recognition involving the publication of a book by the Imprimerie royale or the award of a royal brevet. At the same time, I shall restrict myself to the patronage of scientists, excluding the large numbers of inventors of mechanical contrivances or "secret processes" for which patents or subsidies were sought.[10]

What did science mean to government officials? Presumably, men like the Trudaines, Bertin, Turgot, and Malesherbes shared with their contemporaries the belief that the proper application of science would produce progress of a desirable, durable, and reasonable kind. There are, however, peculiar problems involved in the study of the attitudes and policies of royal ministers. Men of state were not given to the writing of tracts explaining their philosophies, and such tracts, if they did exist, would probably furnish statements of ideals rather than of practical possibilities.

However committed to support of science or to reform in general, a royal minister had to be even more concerned with the inadequacies of his budget. He also had to consider social and economic problems which the private citizen-reformer might ignore; as Bertin said, referring to the proposals of the agronomists: "It is between the protection that is due to the suffering part of humanity and the greater benefit to agriculture as a whole, that it is difficult to decide."[11] Making such decisions proved to be difficult even in times of acute crisis, and the machinery of royal absolutism worked slowly, funds were generally limited, and discussion did not always result in action. Furthermore, when a royal minister did decide to seek expert advice, he faced the problem of selecting the most appropriate expert.

The final problem mentioned above often had an easy solution which dated back at least as far as the era of Colbert and Louvois: one consulted that established body of experts, the Royal Academy of Sciences. Many variations of this practice are to be found in the eighteenth century. Daniel Trudaine seems to have devised an original method, first helping to train and then employing as consultants the mining experts Guillot-Duhamel, Jars, and Monnet. Committees of experts were often convened by royal officials—or by the Academy of Sciences or some other institution acting at government request—to study particular problems; in fact, for a guide to some of the special committees established in the last decades of the Old Regime, one can hardly do better than consult the index to volume six of the *Oeuvres de Lavoisier*. There is every indication that the use of scientific expertise increased markedly as the century progressed, and one striking example of this shift in attitude on the part of royal officials is the contrast in the way they handled the agricultural crises of 1709 and 1784: the earlier crisis was alleviated by grants of funds and supplies of grain, while that of 1784 resulted in the creation of a committee of scientists and administrators to handle the emergency and to plan for the future.[12]

The consultation of experts is not as simple as it may sound. Looking back, we find scientific "eminence" easier to determine than was the case in the eighteenth century. One indication of eminence was membership in the Academy of Sciences, but that test was not infallible since some scientists of ability were not academicians and some academicians had acquired their positions more as a result of lobbying than as a reward for talent. Furthermore, prominent scientists and agronomists were themselves in disagreement over matters ranging from the validity of the new chemistry to the value of marl as fertilizer. If the experts disagreed, how was a bureaucrat to decide among them? When Bertin consulted his chief mining expert, Antoine Monnet, about the advisability of establishing a school of mines, Monnet replied that this was unnecessary; yet Bertin continued to work towards such a foundation. Why did Bertin reject the advice of his own protegé? And when the École des Mines was founded, after Bertin's retirement, its director was B. G. Sage, a chemist who had the respect of virtually none of his fellow scientists. The whole career of Sage is an example of influence outweighing ability, but how that influence was exerted remains obscure.

Problems of major concern were not the only ones to reach the government bureaux. There was also a steady stream of lesser requests for aid, all of them requiring consideration, and none of them important enough to require the consultation of a Lavoisier. An unusual document published years ago by Maurice Tourneux provides us with a catalogue of such applications submitted to the Maison du Roi in a short period of time. Although very revealing, the document does have serious limitations. First, information about the applicants is accompanied by recommendations from the pen of an anonymous *premier commis*, but the minister's final decisions are not always recorded. In addition, this is a limited sample because the ministers of the Maison du Roi apparently considered certain kinds of patronage beyond their jurisdiction. (Thus, one project was referred to the Bureau du Commerce, a local plague was to be handled by the local authorities, and a cure for rabies was simply not "de notre ressort".) Finally, there is no indication of how the budget for that period—and the date of the document is uncertain—might have affected the recommendations and decisions.[13]

The Tourneux document testifies to the remarkable perspicacity of the *commis*, and it also gives some information about the principles which guided his judgment. Although aware of the new, curious, and possibly useful discoveries being made in natural history, he also seems to have been aware of the general lack of enthusiasm in scientific circles for the work of naturalist Buc'hoz and geologist Giraud–Soulavie; similarly, he could differentiate between the medical quackery of one man and another physician who performed experiments "en homme éclairé". On the whole his preference was for "ouvrages utiles" rather than literary works, however morally instructive the latter might be; but he also distinguished between the consistently useful activities of a Parmentier and the lesser ones of an inventor who deserved a subsidy but certainly not a pension. And in commenting on Pierre-Charles LeMonnier's application for aid in astronomical and navigational research, he remarked: "Ce sont des sciences exactes qu'il est important d'encourager".

The alert *commis* was evidently basing his judgments upon advice of some kind, perhaps the gossip in literary and scientific salons, or the reading of journals and book reviews, or the informed opinion of a personal friend. Of these three possibilities, the latter is often the easiest to examine, the most elusive, and the most important.

At present, we have only fragmentary information about the existence of some influential personal connections, and we know even less about the effectiveness of such influence. The careers of Buffon and Louis–Guillaume LeMonnier provide good illustrations of these problems. About the latter, we can say little because of a lack of published documents, but as "Premier médecin ordinaire du Roi" his influence at Court was apparently great, and he played an important role in obtaining royal patronage for more than one "voyageur-naturaliste".[14] For Buffon the documents are abundant, and historians have often remarked upon the special esteem in which he was held by the ministers of the Maison du Roi and upon the political talents he displayed in his campaign to expand the Jardin du Roi. To build up the collections of the Jardin, he obtained brevets which entitled many naturalists to call themselves "Correspondants du Jardin du Roi", and one of the protegés helped in this way was the young Lamarck. Buffon's influence with the Imprimerie royale (part of the Maison du Roi) was considerable, but just how he exerted an influence which was in effect tyrannical remains mysterious. Virtually the only works on natural history published by the Imprimerie royale in the period 1749–1789 were editions of Buffon's *Histoire naturelle* and Lamarck's *Flore française*.[15] Malesherbes disagreed radically with some of Buffon's scientific ideas, and certainly the influence of Malesherbes—before, during, and after the years in which he was Directeur de la Librairie—must have been far from negligible.[16] And yet, as far as we know, Buffon's judgment seems to have prevailed.

Buffon and LeMonnier—and Condorcet during the period of Turgot's ministry—are unusual examples of scientists in close personal contact with some of the highest officials of the Old Regime. Probably more typical is the case of Laplace who found it necessary to marshal the support of government officials, influential friends, scientific colleagues, and even the widow of the man to whose post he aspired. Lobbying of this kind was not only customary, but was also essential because of the keen competition for the available positions, pensions, and grants. Furthermore, some scientific projects, by their very nature, required the co-operation of more than one government department; or, at least, to seek the support of more than one department increased a scientist's chances of acquiring patronage. Thus, for example, the travelling naturalist interested in foreign trees was wise to address himself to the Ministry of the Navy (for his own transportation and to plead the benefits to shipbuilding), Ponts-et-chaussées (concerned with trees to be planted along royal roads), the Directeur des Bâtiments (tree nurseries and royal parks and gardens), the Controller General of Finances, or any combination of these four. The sheer complexity of this process suggests that the Old Regime needs its Sir Lewis Namier. And, as two historians have remarked, the process also suggests that astute lobbying could be a more important factor in success than was the possession of scientific talent.[17]

The influence of any one or more individuals upon one or more potential patrons was often modified by a number of unforeseeable circumstances. At the Imprimerie royale, the budget had to be borne in mind, and the publication of a scientific treatise might involve expensive illustrations, tables, diagrams, and symbols. Crystallographer Romé de l'Isle had both talent and influence and was known to be in need of money, but he was repeatedly denied government assistance; Buc'hoz, on the other hand, was less talented and equally needy—whether he had powerful protectors is unknown—and was granted a subsidy apparently because he had already obtained a royal *arrêt* to protect him from importunate creditors.[18]

Lest the officials of the Old Regime be thought to have been entirely lacking in principles, it should be pointed out that considerations of national policy exerted an important influence upon their decisions; apart from wanting to encourage progress of various kinds, royal ministers also had some specific aims in mind, and these aims were largely responsible for the encouragement they gave to useful or potentially useful projects. One prime consideration was the strengthening of the French economy so that ultimately France might become as self-sufficient as possible and a major exporting nation as well. French losses in the War of the Austrian Succession and then in the Seven Years' War had served to emphasize in a concrete way such problems as: the advisability of reorganizing and building up both army and navy, the need to expand overseas trade, and the desirability of recovering the international power and prestige rapidly being lost to England. All these interests would be served by improvements in the domestic economy and by such expedients as sending vessels on missions of geographical and commercial exploration. Concerns like these do much to explain government interest in the research of Parmentier to improve food supplies for the military, the conservation and expansion of sources of industrial fuel, the work of Buffon and others on trees suitable for shipbuilding, the organization of the Régie des poudres, the importation of English industrial secrets and the mining techniques of Sweden and the German States, the attempts to acclimatize in France or in French colonies the exotic crops of the Far East or the Levant or South America, the research of agronomists to improve crop yields or breeds of sheep, the efforts of naturalists and scientific societies to learn more about the natural resources of France, and dozens of other projects of major and minor importance.

What evidence is there that government patronage was also extended to "pure" research, or to research which might eventually yield practical dividends but which, for the moment, might be considered a luxury? In a regime perpetually near bankruptcy, could the government afford to subsidize "pure" research? These questions would seem to have an obvious answer when one recalls the elaborate astronomical and geodetic expeditions undertaken at government expense by members of the Academy of Sciences in the seventeenth and eighteenth centuries. Such ventures were primarily scientific and too expensive to pay for themselves in practical benefits, although there was always the possibility of improvements in navigation and even in military strategy. (In the latter connection, the Cassini survey of France was supported by Louis XV for military reasons.) In a sense, however, these expeditions did "pay" in terms of national prestige and glory; on the one hand, they contributed immediately to progress in the most advanced of the exact sciences, while, on the other, they revealed to the world the enlightened views of Louis XIV and the readiness of his successor to maintain the traditions of French scientific hegemony. When Tourneux's *premier commis* emphasized the importance of supporting the exact sciences, he was acknowledging their prestige in a manner quite typical of his age.

However one explains the rationale behind government sponsorship of the expeditions of Clairaut, LaCondamine, Pingré, and others, patronage in this area seems to me to be more easily understandable than does government interest in the activities of the "voyageurs-naturalistes". Although very popular in the eighteenth century, natural history had none of the prestige of astronomy, and its attraction—for royal ministers and naturalists alike—lay principally in its "curious" subject matter and potential usefulness. Naturalists gathered during their voyages plants of botanical interest or ornamental value, or plants which might serve some agricultural, commercial, or industrial purpose. Since most such "exotic" productions had to be acclimatized in France, there was great risk of the voyages failing in their aims. Keeping the plants alive during the return to France posed difficult problems, and, even when this was accomplished, the plants might then succeed only in hothouses and thus be unsuited to cultivation on a wide scale. Furthermore, when a naturalist went overseas, there was always the possibility that a naval disaster would mean the loss of his collections and perhaps his life. For these and other reasons the readiness of government officials to sponsor such travels seems to me remarkable. Certainly, the Treasury did not always invest heavily in these expeditions, since a single naturalist could easily embark on a French vessel already bound on some other mission. But there were at least three expeditions which can serve to dramatize the problems outlined above: Dombey's voyage to South America, Michaux's to North America, and Lapérouse's circumnavigation of the globe. All three, but especially the latter two, were undertaken in periods of considerable fiscal distress; but despite the risks of failure, these voyages were elaborately financed and planned.

The search for plants of botanical or ornamental, rather than useful, importance was one aim of some of the earliest naturalists—Tournefort, for example—whose missions were undertaken "par ordre du Roi". By the 1760s, however, interest in useful plants had intensified as a result of concern with agricultural reform and the balance of trade. It was thought that soil unsuited to grain might well be used for rice or pineapples or nutmeg trees; not only would agriculture prosper, but France would then find it unnecessary to import these commodities. At the same time, and especially after the first voyage of Captain Cook, scientific and geographical discoveries were also becoming increasingly important to government sponsors concerned that France should not lag behind England in prestige and enlightenment. Competition with England apparently influenced Bougainville in 1763, and his example is said to have inspired the Abbé Terray and other officials to arrange the expedition of Yves Kerguelen in 1771.[19]

The various reasons for sending naturalists abroad were, of course, not necessarily in conflict, for every traveller could be instructed and expected to make scientific discoveries *and* to bring back useful plants. For the government patron, therefore, the real problem was whether the results of a voyage would justify the expense involved. Given a choice, Bertin clearly preferred investing in projects which would yield practical benefits as quickly as possible. In organizing provincial societies of agriculture after 1760, he instructed the intendants to select members well versed in the practice of agriculture and to avoid recruiting academicians. Historians have suggested that Bertin had little faith in pure research—or, more bluntly, in the empty verbiage of academies—or that he simply did not want to duplicate the agricultural work already going on in some of the provincial scientific societies. Both interpretations seem to me to have elements of accuracy, and I would suggest that Bertin was perfectly willing to concede merits to scientific research but that he was unwilling to wait the requisite number of years for theory to become applicable to practice.[20]

The mission of Ko and Yang illustrates a slightly different side of Bertin's character. He wanted the two Chinese to study the various industries of France so that they might compare and report on the industrial secrets of China. One historian has described this episode as the beginnings of "la collaboration technique franco-chinoise", and the phrase is an accurate statement of Bertin's principal motive.[21] Once in China, however, Ko and Yang found that they were not permitted to learn the secrets desired by Bertin; but the project was not abandoned, and for some twenty years Bertin collected from these two and other correspondents any available information about China as well as handsome specimens of *chinoiserie*. His initial motives may have been largely practical, but it is also clear that, like his contemporaries, Bertin was fascinated by the exotic East.

Turgot's interest in Ko and Yang took a significantly different form. In 1766 he drew up a questionnaire for the travellers which reveals that he wanted to collect systematically information of various kinds: scientific, technological, and cultural.[22] Two years later Turgot asked one

Desurgy if similar questions could be applied to "L'univers entier", and Desurgy replied with a project for a Bureau of Correspondence which would gather information from all corners of the world. There would be no need to send voyagers, Desurgy argued, because the Bureau would use the services of local inhabitants and of Frenchmen living abroad, and such information would probably be more accurate than that collected by casual visitors. In only two or three years, excellent results could be expected for all branches of knowledge. Turgot's reaction to this proposal is unknown, although there is reason to think that it aroused his interest for a time.[23] Certainly, the idea that information of all kinds should be collected systematically was common enough in the eighteenth century. Royal ministers had long made periodic attempts to gather statistical data about the whole kingdom, and many writers were advocating the systematic collection of mineral specimens, plants, and agricultural and other implements, all to be housed in museums open to the public. There is some evidence that one such project for provincial natural history museums was presented to Turgot, during his term as Controller General, by Chazerat, then intendant of Auvergne; but Turgot's reaction remains unknown.[24]

It is hardly surprising that Turgot should have been the recipient of proposals such as Desurgy's and Chazerat's. From his early days at the Sorbonne, he had earned a reputation as an enlightened philosopher and a scientific amateur of some ability. As intendant of Limousin, he actively supported the agricultural society in Limoges, founded a short-lived school of veterinary medicine, tried to convert his peasants to some of the ideas of the agronomists, and conducted a lively scientific correspondence with Condorcet. As Controller General, his activities included the organization of the Régie des poudres (placed under Lavoisier's direction), the creation of a Chair of Hydrodynamics in Paris, and the sponsorship of more than one natural history expedition. During much of his public career, he regularly sought the advice of scientists and especially of his good friend Condorcet.[25]

Many of Turgot's activities and ideas find their parallels in the careers of Bertin and Daniel Trudaine, and Turgot often differed from them in degree rather than in kind. He tended to be more interested in science for its own sake, more concerned with the systematic collection of information, more a philosopher than Bertin or Trudaine. Although fully convinced of the benefits to be derived from the sciences, Turgot's vision was broader in that he was willing to invest in research which might bring eventual, rather than immediate, practical results.

These complex motives characterise Turgot's policies towards "voyageurs-naturalistes". During his twenty-one months as Controller General of Finances, he sent two naturalists overseas and contemplated sending a third. St Edmond, en route to the Indies, was lost at sea, and the voyage of Gouan to Persia apparently never progressed beyond an idea expressed in correspondence. But Joseph Dombey's journey to South America has received careful study and can be discussed here with some necessary simplification in detail.[26] In many ways, the case of Dombey differs little from that of earlier voyagers: he was attached to a large expedition, given a salary, and instructed to bring back plants both useful and botanically interesting. An unusual feature of the mission is Turgot's personal concern with its planning, manifested by his consultations of Condorcet, Thouin, and A.-L. de Jussieu. In a manner foreshadowing the more systematic preparations for the Lapérouse expedition, Turgot was apparently trying to be sure that Dombey would consider all the possibilities appropriate to the regions he was to explore. This voyage was unusual, too, in that it seems to have been the first elaborate, lengthy, government-sponsored one in which natural history was not subordinated to astronomy or some other concern. (Even the larger expedition to which Dombey was attached was a natural history expedition, financed by Spain.) At least, I believe that detailed comparison would show earlier voyages to have been more modest, the naturalist sometimes put to work as an employee of the Compagnie des Indes (Adanson and Sonnerat) or sent along as a kind of appendage to another mission (Joseph de Jussieu and Commerson). The voyages of Dombey and St Edmond are all the more remarkable when one recalls that Turgot incessantly counselled reductions in government expenditure, but he clearly had his own list of budgetary priorities, and science was high on that list. As botanist Gouan remarked more than once, when Turgot fell from power, the sciences lost one of their most powerful protectors.[27]

Attributing to Turgot's influence a dramatic shift in government policy would, of course, be rash; it is curious and impressive, nonetheless, to find that overseas expeditions seem to have increased in scale and frequency during the last years of the Old Regime. This is the period in which the Michaux and Lapérouse voyages began, while other, less dramatic, expeditions continued to take place; at the same time, there was even a rather odd attempt to arrange for the systematic exchange of scientific information between France and the United States.

Writing after the French had invested heavily in the American Revolutionary War, Malesherbes remarked that, although it would be desirable to have the government send a naturalist to explore North America, it might be more realistic "d'envoyer le Voyageur non pas aux frais du Roy mais à ceux d'une Société d'amateurs comme font Souvent les Anglois".[28] Soon after the conclusion of the war, however, royal officials began to plan the expedition of André Michaux. Government interest in North American flora and fauna, in fact, dated back at least to the 1770s, when information was being sought from French consuls and from the governors of the thirteen colonies. In addition to the traditional desire for trees as ornaments and for planting along the *routes royales*, France was facing a growing need for forest trees for both fuel and shipbuilding; furthermore, Malesherbes and others had been experimenting with varieties of American trees which would flourish in marshy land, and he was urging this method of employing waste lands (rather than draining them and then perhaps finding the soil poor). Since the war had cut France off from England, its normal supplier of North American plants, to establish or strengthen existing contacts without an English intermediary was a desirable course of action.

Although all these factors were present in the Michaux voyage, the scale of the expedition and the purely scientific element in it should not be overlooked. How important scientific research was to Michaux's government patrons cannot yet be assessed, but the voluminous correspondence preserved at the *Archives Nationales* and the time, energy, and money expended on both sides of the Atlantic certainly suggest that this mission was of great significance to its sponsors. Michaux, in fact, was directed to maintain secrecy about his activities so that he would not be distracted by demands for seeds or seedlings from influential persons who had acquired a taste for foreign plants.[29]

The Lapérouse expedition, which sailed in the same year as Michaux, 1785, has attracted much attention because the mysterious disappearance of the two vessels remained an international *cause célèbre* for several decades. More significant here, however, is the fact that the mission was almost entirely scientific in its aims. Scientists and learned societies were directed to draw up instructions for the voyagers, and two ships were supplied with an enormous reference library and any quantity of scientific equipment.[30] Even more than in the case of Michaux, precautions were to be taken to maintain secrecy about the discoveries of the expedition, and the motives here remain obscure despite the hints given by contemporaries. To prevent useful discoveries from being divulged indiscriminately was, in any event, standard procedure at the time, and it is possible that Lapérouse's sponsors wanted the opportunity first to scrutinize the scientific results of the expedition for their potential practical value.

Before concluding, I should like to consider one more scientific enterprise of the 1780s, the creation of the Académie des Sciences et Beaux-arts des Etats-unis. This short-lived institution, established in Richmond, Virginia, was planned by one Quesnay de Beaurepaire, grandson of François Quesnay. Little is known about the long campaign involved in the founding of the Academy, but Quesnay did arouse the interest of prominent Frenchmen and Americans and succeeded in obtaining the endorsement of the Paris Academy of Sciences. (Whether endorsement also meant a grant of funds is not clear.) The purpose of the institution was both to collect scientific information and specimens for transmission to France and to be an educational foundation somewhat similar, in the scientific part of its activities, to the Jardin du Roi. One professor, physician and naturalist Jean (or John) Rouelle, was dispatched to Richmond in 1788, but this seems to have been one of the few concrete steps taken to implement the project.[31] So much obscurity surrounds this Academy that it might better have been omitted from this discussion; but the fact that so ambitious and systematic a project coincided in time with the equally ambitious and systematic explorations of Michaux and Lapérouse is intriguing and perhaps of some significance.

Until more evidence is available, it would be premature to insist that the French government of the 1780s was increasingly concerned with the patronage of science for its own sake. Indeed, the evidence we do have suggests that motives continued to be mixed and that the claims of both pure and applied research were receiving royal attention. At the same time, the case of the "voyageurs-naturalistes" seems to me to provide an unusual opportunity, a stage in miniature, for the examination of government policies. To undertake such an examination would require, first, the compilation of detailed information about the planning and purpose of each voyage and especially about the extent to which government funds were involved. Lacking, too, is reliable information about whether notable practical results followed the earlier voyages and could thus be expected of the later ones. In botany, of course, the voyages contributed to the classification schemes of Linnaeus, A.-L. de Jussieu, and others; but what can be said about the cultivation in France of rice and pineapples, nutmeg and cloves, and the swamp trees of Virginia?[32] If we find—and I suspect that this will prove to be the case—that the tangible results were not great enough to justify the personal and financial risks, and that, nonetheless, the government invested more heavily in these expeditions as the years went on, then we shall be left with a problem as intriguing as it is complex. How much patronage was granted out of continuing concern for national prestige? How important was a possibly indiscriminate taste for everything "exotic"? Did the encyclopedic desire to collect information from "L'univers entier" influence ministers less philosophical than Turgot? Or is it fruitless to search for general policies pursued by royal officials? Will historians have to be satisfied when they have "Namierized" the Old Regime by studying the networks of personal connections and the process of lobbying? Or will historians perhaps discover that the scientists, rather than the ministers, played an important role as the unconscious framers of policy when, in the course of their lobbying, they attempted to justify their own research by formulating and appealing to what they believed to be the interests of France as a whole?

REFERENCES

1. Cf. Bourde, *Deux registres* (*H. 1520–H. 1521*) *du contrôle général des finances aux Archives nationales (1730–1736): Contribution à l'étude du ministère d'Orry* (Gap, 1965).

2. Festy, *L'agriculture pendant la révolution française: Les conditions de production et de récolte des céréales* (Paris, 1947), and other works by the same author.

3. Also, Bourde, "L'agriculture à l'anglaise en Normandie au XVIIIᵉ siècle," *Annales de Normandie*, viii (1958) 215–233.

4. *Agronomie*, iii, 1563. Bourde does make use of Young's *Travels*, but generally for specific details or to point out Young's occasional errors.

5. A recent contribution to this literature is the important article by Henry Guerlac, "Where the statue stood: Divergent loyalties to Newton in the eighteenth century", in *Aspects of the eighteenth century*, ed. Earl R. Wasserman (Baltimore, 1965). Also, *Enseignement et diffusion des sciences en France au XVIIIᵉ siècle*, ed. René Taton, in the series "Histoire de la pensée" (Paris, 1964). Cf. Duhamel du Monceau, in *Agronomie*, i, 251.

6. See Henry Guerlac, "Some French antecedents of the chemical revolution", *Chymia*, v (1959) 73–112, and Harold T. Parker, "French administrators and French scientists during the Old Régime and the early years of the Revolution", in *Ideas in history*, ed. R. Herr and H. T. Parker (Durham, N. C., 1965). Roger Hahn's *The anatomy of a scientific institution: The Paris Academy of Sciences, 1666–1803* is to be published by the University of Michigan Press. See also C. C. Gillispie, "The natural history of industry", *Isis*, xlviii (1957) 398–407.

7. Above, n. 6; G. de la Fournière, "Les Comités d'agriculture de 1760 et de 1784", *Bulletin du Comité des Travaux historiques et scientifiques: Section des sciences économiques et sociales, 1909* (Paris, 1910), 94–121; and S. Delorme, "Une famille de grands Commis de l'Etat, amis des Sciences, au XVIIIᵉ siècle: Les Trudaine", *Revue d'histoire des sciences*, iii (1950) 101–109.

8. Thouin's papers are at the Muséum d'Histoire Naturelle. In addition to the academic eulogies pronounced in 1825 by Georges Cuvier and A.–F. Silvestre, see G. Bertemes, "Correspondance de Linné Père et Fils avec André Thouin", *Bulletin de la Société d'histoire naturelle des Ardennes*, xxx (1935), and P.–A. Cap et al., *Le Muséum d'histoire naturelle* (Paris, 1854).

9. Probably more has been done on this subject in connection with Lavoisier than with any other scientist; see W. A. Smeaton, "New light on Lavoisier: The research of the last ten years", *History of science*, ii (1963) 51–69.

10. Valuable details about the inventors are to be found in the works cited above, n. 6, and in the various monographs by Shelby T. McCloy.

11. Quoted by Marc Bloch, "La lutte pour l'individualisme agraire dans la France du XVIIIᵉ siècle", *Annales d'histoire économique et sociale*, ii (1930) 524; translation by C. B. A. Behrens, *The Ancien Régime* (London, 1967), 175.

12. S. T. McCloy, *Government assistance in eighteenth-century France* (Durham, N.C., 1946), ch. 1, and Henri Pigeonneau and Alfred de Foville, *L'administration de l'agriculture au Contrôle général des finances (1785–1787): Procès-verbaux et rapports* (Paris, 1882).

13. Tourneux, "Un projet d'encouragement aux lettres et aux sciences sous Louis XVI", *Revue d'histoire littéraire de la France*, viii (1901) 281–311. I am indebted to Roger Hahn for calling this article to my attention.

14. Cap, *op. cit.*, 32–34; A. Lasègue, *Musée botanique de M. Benjamin Delessert* (Paris, 1845), 53–60; and Auguste Chevalier, *La vie et l'oeuvre de René Desfontaines* (Paris, 1939), 191–195. Some of LeMonnier's correspondence is at the Muséum d'Histoire Naturelle. See Y. Laissus, "Lettres inédites de René Desfontaines à Louis–Guillaume LeMonnier", *Comptes rendus du 91ᵉ Congrès national des Sociétés savantes, Rennes, 1966* (Paris, 1967), 153–169.

15. A catalogue, arranged chronologically, is to be found in Auguste Bernard, *Histoire de l'Imprimerie royale du Louvre* (Paris, 1867), 123–263. Pierre Sonnerat, for example, presumably had to publish his own *Voyage aux Indes orientales et à la Chine, fait par ordre du Roi, depuis 1774 jusqu'en 1781* (2 vols., Paris: L'auteur, 1782).

16. P. Grosclaude, *Malesherbes, témoin et interprète de son temps* (Paris, 1961). Malesherbes, *Observations . . . sur l'histoire naturelle générale et particulière de Buffon et Daubenton* (2 vols., Paris, 1798); published posthumously, this work was begun soon after 1749 and is an attack on Buffon's anti-Linnaean views.

17. Denis I. Duveen and Roger Hahn, "Laplace's succession to Bézout's post of *Examinateur des Elèves de l'Artillerie*: A case history in the 'lobbying' for scientific appointments in France during the period preceding the French Revolution", *Isis*, xlviii (1957) 416–427. This article begins with a valuable discussion of some of the ways in which a scientist could earn a livelihood.

18. Tourneux, *op. cit.*

19. Gilbert Chinard, ed., *Le voyage de Lapérouse sur les côtes de l'Alaska et de la Californie (1786)* (Baltimore, 1937), p. x. Chinard goes on to say that government interest in exploration declined again after Kerguelen's voyage and revived after 1783; I must disagree, as will appear below.

20. Emile Justin, *Les sociétés royales d'agriculture au XVIII⁰ siècle (1757–1793)* (St.-Lo, 1935). Similar views were held by Bertin in connection with the geological survey he commissioned; see R. Rappaport, "Guettard, Lavoisier, and Monnet: Conflicting views of the nature of geology", in *The history of geology before Darwin*, ed. Cecil J. Schneer, in press.

21. J. Dehergne, *Les deux chinois de Bertin: L'enquête industrielle de 1764 et les débuts de la collaboration technique franco-chinoise*, unpublished thesis (Paris: École Pratique des Hautes Études, 1965). I am indebted to M. René Taton for calling this study to my attention and allowing me to consult it.

22. *Oeuvres de Turgot et documents le concernant*, ed. G. Schelle (5 vols., Paris, 1913–1923), ii, 523–533.

23. Archives nationales, Chartrier Tocqueville, 154 AP II, dossier 149, pcs 37, 37¹. I am grateful to M. le Comte de Tocqueville for permitting me to consult these papers.

24. P.-J.-B. LeGrand d'Aussy, *Voyage d'Auvergne* (Paris, 1788), 25–26.

25. Douglas Dakin, *Turgot and the Ancien Régime in France* (London, 1939); Roger Hahn, "The chair of hydrodynamics in Paris, 1775–1791: A creation of Turgot", *Actes du dixième Congrès international d'Histoire des Sciences* (Paris, 1964), ii, 751–754; and Charles Henry, ed., *Correspondance inédite de Condorcet et de Turgot, 1770–1779* (Paris, 1883). Also, Henry Guerlac, *Lavoisier—The crucial year: The background and origin of his first experiments on combustion in 1772* (Ithaca, N.Y., 1961), ch. 5.

26. E.-T. Hamy, *Joseph Dombey* (Paris, 1905); Bertemes, *op. cit.*; and *Oeuvres de Turgot*, iv, 88. Also, Muséum d'Histoire Naturelle, MS 1987, pc 523.

27. Muséum d'Histoire Naturelle, MS 1992, pcs 425, 426. In these letters, Gouan couples the names of Turgot and Malesherbes.

28. Chartrier Tocqueville, dossier 157, pc 67. A valuable discussion of Malesherbes as naturalist and agronomist is in Grosclaude, *op. cit.*

29. A study of the two Michaux, begun by the late Léon Rey, is in process of completion by Yves Laissus; Gilbert Chinard gives a brief description of its content and valuable information on earlier Franco-American relations in his "André and François-André Michaux and their predecessors," *Proceedings of the American Philosophical Society*, ci (1957) 344–361. Also, Julia Post Mitchell, *St. Jean de Crèvecoeur* (New York, 1916), 101–106.

30. *Voyage de La Pérouse autour du monde*, ed. M. L. A. Milet–Mureau (4 vols. and atlas, Paris, an V [1797]). The list of equipment published in vol. i should be compared with that at the Muséum d'Histoire Naturelle, MS 1928, i, no. 1; the savants generally were given all they asked for and more.

31. H. B. Adams, "L'Académie des Etats-unis de l'Amérique", *The academy*, ii (1887) 403–412. On Rouelle, see Henry Sigerist, "The rise and fall of the American spa", *Ciba symposia*, viii (1946) 313–326. The Academy was not meant to limit itself to matters scientific, as its title indicates, but the Committee of Correspondence in Paris was dominated by scientists and the first (and only) professorial appointment was to a scientific chair.

32. Cf. Bourde, *Agronomie*, esp. ii, 597–741.

THE LIBERTIES OF THE PARIS ACADEMY OF SCIENCES, 1716–1785

THE Duc de Choiseul, consulted in 1768 about a forthcoming election in the Academy of Sciences, offered his advice to the Academy's secretary and then declared: "Furthermore you understand that this observation has no bearing whatever on the other candidates or on the votes in the Academy which must retain all their freedom and integrity."[1] The idea that the Academy possessed freedom of choice was expressed from time to time throughout the century by royal ministers whose duties included "protection" of the several royal academies. Thus, in 1726, the abbé Bignon replied to a solicitation for membership in the Academy of Sciences that he would do all he could to preserve freedom of choice in elections. Another such solicitation in 1784 was transmitted to the Academy's president with the ministerial reminder that it was "essential not to constrict" freedom of choice.[2] These sentiments were shared by many academicians, for reasons like those of chemist P.-J. Macquer: "I have suffered on certain occasions from seeing conniving mediocrity win out over talent unsupported by patronage. . . . I have often feared . . . that if the liberty of elections were hampered, this dangerous mediocrity would prevail and ultimately bring about a shameful ruin of the Academy."[3] A retrospective assertion of academic liberties was eventually to be used in defense of the institution on the eve of its abolition, when Lavoisier declared that "even under the old regime, . . . a kind of respect . . . guaranteed the sanctuary of the sciences against the invasion of despotism."[4]

1. Choiseul to Fouchy, 21 May 1768, in Archives of the Academy of Sciences (hereafter AAS), Paris, dossier Bézout.
2. Bibliothèque nationale (BN), MS fr 22234, fol. 28. Archives nationales (AN), O¹*495, fol. 186. Also, BN, MS fr 22231, fols. 82, 104; 22234, fol. 25; and Roger Hahn, *The Anatomy of a Scientific Institution: The Paris Academy of Sciences, 1666–1803* (Berkeley, 1971), p. 81.
3. Hahn, *Anatomy*, p. 82.
4. *Oeuvres de Lavoisier*, 6 vols. (Paris, 1862–93), IV, 618.

However earnest their statements, ministers and savants well knew that academic elections were often engineered. Bignon, for example, more than once informed interested parties that he was "working at" a particular election and hoped in time to sway the minds of the voters. Similarly, Grandjean de Fouchy, during his years as Perpetual Secretary, more than once offered advice to colleagues and ministers about how posts in the Academy might be juggled in order to achieve desired elections or promotions. And more than one academician wrote directly to the Minister of the Maison du roi in efforts to obtain a prior commitment for an election or a promotion.[5] In short, while both sides claimed that the Academy had and ought to have certain liberties, both commonly took part in the manipulation of elections.

Conflicting statements about the Academy as an institution—its liberties in principle, the intrigues in practice—have their parallel in descriptions of the tenor of academic life and the behavior of academicians. Early in the century, Bernard de Fontenelle had repeatedly eulogized the objective, detached, unworldly savant, whose very profession raised him above pettiness.[6] More often, however, academicians analyzed their colleagues in less flattering terms. Not only are judgments fallible, declared C.-G. de Malesherbes, but voting academicians can also be swayed by "authority" and even by "nepotism." Professional bias can sometimes blind us to merit, Antoine-Laurent Lavoisier admitted, while Joseph Dombey more bluntly observed that men pursuing similar careers are naturally jealous of one another. Rivalries, jealousies, and competitive place-seeking all had a stimulating effect upon the ego and prompted J.-L. Lagrange to remark that pretensions seemed always inversely proportional to merit.[7] And yet, Lavoisier assures us, bias is not pervasive because to be a scientist requires "the habitual use of reason."[8]

Like these academicians, modern historians have reached no consensus about the nature of academic life or the way in which the institution functioned. There has been, for example, some tendency to minimize in-

5. H. Omont, ed., *Lettres de J.-N. Delisle au comte de Maurepas et à l'abbé Bignon* (Paris, 1919), p. 33. Anonymous account of the activities of Bignon and others, n.d. [1725], in AAS, dossier P. LeMonnier. Jean Torlais, *Réaumur* (Paris, [1936]), p. 212. Fouchy, MS "Etat Present de la classe anatomique," AAS, dossier 1758. Draft from Fouchy to [St. Florentin?], n.d., AAS, dossier 27 June 1770. Letters to LeRoy, Arcy, and Ayen in AN, O^{1*}412, fols. 323–4, 356; 414, fol. 1107.

6. *Oeuvres de Monsieur de Fontenelle*, 5 vols. (Paris, 1825), esp. I, 83, 96–97, 173–175; II, 100, 250. Hahn, *Anatomy*, pp. 42, 56.

7. Malesherbes to d'Alembert, 1769, in C. Henry, ed., "Correspondance inédite de d'Alembert," *Bullettino di bibliografia e di storia delle scienze matematiche e fisiche*, 18 (1885), 569. Lavoisier MS, AAS, dossier 28 July 1784. E. T. Hamy, *Joseph Dombey* (Paris, 1905), pp. 86–87. *Oeuvres de Lagrange*, ed. J.-A. Serret, G. Darboux, L. Lalanne, 14 vols. (Paris, 1867–92), XIII, 206.

8. Above, n. 7. He was addressing his colleagues.

trigue and to focus instead on merit as the most important criterion in academic elections. At the same time, while one historian claims that merit was also a criterion for promotion within the ranks, another describes promotion as the automatic concomitant of seniority.[9] As for the relationship between crown and Academy, one historian tells us that royal ministers often set aside the Academy's judgment and sometimes appointed a new member not nominated by the voters. Elsewhere we are told that royal influence was confined largely to the choice of members in certain ranks and categories: *honoraires, associés libres,* or *surnuméraires.*[10] On the whole, historians agree that the Academy did not possess complete freedom of choice, but few have recognized that such liberty was thought to exist and was deemed essential by ministers and savants.[11]

The purpose of this study is to examine elections and promotions in the Academy in order to determine what degree of internal freedom that body possessed. Consideration will be given to electoral practices and the extent to which the voters followed rules prescribed for the company in 1699 and 1716. Ministerial influence, before and after balloting, will also be analyzed: how often was such influence exercised, for what reasons, and with what response from the academicians? Given the frequent claims to liberty, and the equally frequent admission that voting was manipulated, some effort will be made to determine what ministers and savants had in mind when they invoked academic liberties.

Certain limitations are imposed on this study by the nature of the subject and the available materials. Discussion will be confined to the "working ranks" in the Academy: assistants, associates, and pensioners (*adjoints, associés, pensionnaires*). Since these were the only ranks required to be productive, the quality of this part of the membership determined the scientific quality of the institution; indeed, assertions of liberty were generally made in behalf of the working ranks. In addition, examples will be drawn almost exclusively from the period 1716–1785, a period in which these ranks and the rules governing procedures underwent no change.

Evidence for this inquiry is both plentiful and severely limited. The Archives of the Academy provide lists of nominees for every post, but they

9. Jean Torlais, *L'abbé Nollet, 1700–1770* (Paris, [1954]), pp. 52–53. Hahn, *Anatomy*, p. 80.

10. Hahn, *Anatomy*, pp. 80–81, 98, 98n. S. L. Chapin, "Les associés libres de l'Académie royale des Sciences: Un projet inédit pour la modification de leurs statuts (1788)," *Revue d'histoire des sciences,* 18 (1965), 7–8. E. Maindron, *L'ancienne Académie des sciences: Les académiciens, 1666–1793* (Paris, 1895), pp. 4–5.

11. Those who discuss the matter usually refer not to ministers but to specific crises which provoked an assertion of liberties by academicians. Cf. K. M. Baker, "Les débuts de Condorcet au secrétariat de l'Académie royale des sciences (1773–1776)," *Revue d'histoire des sciences,* 20 (1967), 229–280; W. C. Ahlers, "Un chimiste du XVIIIe siècle. Pierre-Joseph Macquer (1718–1784): Aspects de sa vie et de son oeuvre" (Thesis, Ecole Pratique des Hautes Études, 1969); Hahn, *Anatomy,* pp. 81–82.

rarely contain the numerical results of balloting, summaries of debates on candidates, or explanations of decisions by the voters and the king. Private correspondence, although abundant, too rarely fills in such gaps. Academicians and ministers alike were, after all, usually in or near Paris, and delicate matters such as academic politics were better reserved for conversation.[12] Gaps in the evidence mean that certain questions are historically unprofitable. Even in well documented cases, it remains impossible to identify voting blocs in the Academy; at best, one can discover the views of only a handful of the thirty or more voters in a given election.[13] As a result, there is at present no way to answer such questions as: What influence was exerted by the *honoraires,* the pensioners, the savants in particular disciplines? How often did scientific issues sway the voters? What was the role of personal friendships or enmities, of philosophical outlook, of patronage? Partial answers to these questions are sometimes available and are used here when evidence permits; but the major focus in these pages is on patterns and policies rather than on factions, and on the Academy's relations with the crown rather than on the intricate networks of influence and protection.

The *règlement* of 1699 spelled out for the first time the structure of the Academy, rules governing the members, and procedures for elections and promotions; subsequent modifications were then incorporated into letters-patent in 1716, after which date no change was to affect the working ranks until 1785.[14] Six classes were established (geometry, astronomy, mechanics, anatomy, botany, chemistry), each to consist of seven members, the seven distributed in three ranks: two assistants, two associates, and three pensioners. All, but especially the pensioners, were to present periodic proof of research, and all had to reside in Paris, a precaution meant to ensure regular attendance at meetings and frequent communication among members. Before an election, the class concerned was to draw up a list of several names; the list was then offered to all the pensioners and honorary members whose vote determined the two or three names to be submitted to the king. Tallying of votes was done in secret, and the Academy and king alike were merely presented with names listed preferentially. Final choice among the nominees was made by the king, or, rather, in his name. It is worth noting that each list of candidates for promotion had to include the name of at least

12. I am also under no illusion that I have exhausted all available materials. My reliance on the Archives of the Academy has been facilitated and made pleasurable by Mme Pierre Gauja and her staff, to whom I can but inadequately express my gratitude.

13. Cf. Baker, "Débuts de Condorcet," pp. 235–236.

14. Léon Aucoc, *Institut de France. Lois, statuts et règlements concernant les anciennes académies et l'Institut, de 1635 à 1889* (Paris, 1889), pp. lxxxiv–xcv. A discussion of the application of the rules through about 1753 is in Jean Hellot, MS "Collection de ... Reglemens et deliberations par ordre de Matieres," AAS, dossier Règlements. There are two such MSS, one in Hellot's hand and the other a fair copy; I have followed the pagination of the latter.

one nonmember; the Academy was thereby enjoined to consider outsiders who might be more talented than the members.

After more than thirty years without defined rules and procedures, the membership was understandably pleased to be guaranteed a certain permanence and freedom from royal or ministerial caprice. The architect of the *renouvellement* of 1699, Jean-Paul Bignon, was appointed President of the Academy for that year, and the otherwise unemotional pages of the *procès-verbaux* record that this news was received "with great joy."[15] To be sure, the period 1699–1716 saw a certain amount of internal adjustment taking place. A few members were expelled for prolonged absence from meetings, while others, who held posts requiring absence from Paris, were made to choose between membership in the Academy and the posts in question. Furthermore, when the letters-patent of 1716 abolished the rank of *élève* and created six posts at the assistant level, those *élèves* in excess of six were permitted to stay on as "supernumerary" assistants; during the next few years, the normal process of attrition reduced the working ranks to the statutory forty-two.[16]

With a constitution in hand, the Academy in 1716 presumably had only to follow the detailed rules, while the royal protector had only to watch that this was done. Since the following pages deal with the relationship between Academy and crown, it is appropriate now to introduce the several men who served as intermediaries between the king and the savants. For the years 1716–1749 these men were Jean-Paul Bignon (d. 1743) and Jean-Frédéric Phélypeaux, Comte de Maurepas. Contemporaries and historians agree that both were vitally concerned with progress in the sciences. In addition, both were kept well informed about academic life, Bignon by his friends Fontenelle and R.-A. F. de Réaumur, Maurepas by Bignon and perhaps by such protégés as Pierre Bouguer, Buffon, and H.-L. Duhamel du Monceau. Since Bignon and Maurepas consulted each other constantly, it would be difficult to determine which of the two bore the greater responsibility in matters affecting the Academy.[17] After the fall of

15. AAS, MS "Procès-verbaux des séances" (hereafter PV), 4 February 1699. The *règlement* was read to the Academy that day. See Fontenelle, *Oeuvres*, I, 63, 75.
16. The supernumeraries of 1716 were Bomie, Bragelongne, G. Delisle, Deslandes, and Winslow. Cf. *Index biographique des membres et correspondants de l'Académie des sciences* (Paris, 1954 and other editions). Also, Aucoc, *Institut*, pp. xciii–xciv, and Hellot MS, pp. 14–16, 34–35.
17. Neither man has been studied adequately. For Bignon, see J.-J. Dortous de Mairan, *Eloges des académiciens* (Paris, 1747), pp. 288–313; Torlais, *Réaumur*, pp. 204–215; Fontenelle, *Oeuvres*, I, 62; AAS, dossier Bignon. For Maurepas, M. Filion, *Maurepas ministre de Louis XV (1715–1749)* (Montreal, [1967]); A. Picciola, "L'activité littéraire du comte de Maurepas," *Dix-huitième siècle*, 3 (1971), 265–296; *Oeuvres de Condorcet*, ed. A. C. O'Connor and M. F. Arago, 12 vols. (Paris, 1847–49), II, 167, 466–498. For his relations with Bignon, see esp. BN, MS fr 22231, fol. 138; 22234. fol. 1. The later reputation of Maurepas, when recalled to power in 1774, differs from that of his earlier ministry.

230

Maurepas in 1749, the ministers charged with academic affairs were the Comte d'Argenson (1749–1757), the Comte de St. Florentin, later Duc de La Vrillière (1757–1775), Malesherbes (1775–1776), Amelot de Chaillou (1776–1783), and the Baron de Breteuil (1783–1787). Intellectually, only Argenson and Malesherbes cannot be called nonentities, but neither played as important a role in the Academy's life as did their less impressive ministerial colleagues. If little of relevance is known about St. Florentin, Amelot, and Breteuil, a contemporary assessment suggests that about St. Florentin, at least, there is little to be known: "He is accessible, he answers letters punctiliously, he listens goodnaturedly, he is neither hard nor unfair by nature, . . . [but] one could not be more limited than he is."[18]

On what basis did the Academy draw up its recommendations for the election of new members? How often was the Academy's preference ignored by the ministry, and why? What evidence is there of governmental pressure applied before an election took place? How did the Academy respond to ministerial intervention?

In the seven decades after 1716, the Academy held elections to fill approximately one hundred vacancies at the lowest rank. The *règlements* specified that candidates had to make themselves known to the membership by "some work of their composition," reported on favorably by an ad hoc committee of members.[19] According to Bignon, this meant that candidates of proven merit were to be preferred to those of future promise.[20] That proof of merit was desired by the membership, too, is suggested by J.-N. Delisle in discussing with Bignon the candidacy of Delisle's own assistant at the Observatoire du Luxembourg. Louis Godin, he explained, was a competent astronomer but had not yet demonstrated his ability to do independent research; the then vacant post in the Academy ought not to be filled until a more tried candidate could be found. The post remained vacant for several years.[21]

Rules and policies notwithstanding, the minutes of meetings until about mid-century seldom record the presentation of memoirs by candidates or reports on them by the membership. The existence of such lacunae, however, may mean only that records are incomplete or that candidates offered few or many memoirs depending upon how well known they were to the members.[22] In later decades, the presentation of memoirs seems to have

---

18. *Journal de l'abbé de Véri,* ed. Jehan de Witt, 2 vols. (Paris, 1928–30), I, 112. P. Grosclaude, *Malesherbes, témoin et interprète de son temps* (Paris, [1961]). Fouchy, "Eloge de M. le Comte d'Argenson," *Histoire de l'Académie royale des sciences,* 1764 (1767), 187–197. A. Doria, *Le comte de St.-Florentin: Son peintre et son graveur* (Paris, 1933), pp. 15–19.

19. Aucoc, *Institut,* p. xciv. Numbers of vacancies and certain other large figures are approximate because of some gaps and anomalies in the records.

20. BN, MS fr 22234, fol. 1.

21. Archives of the Observatory of Paris (hereafter AOP), B.1.2, fol. 42. The vacancy was unfilled from late 1719 until July 1725; Delisle's recommendation dated from 1722.

22. Abbé Nollet, in MS "Réponses aux observations de Mr. l'abbé Nollet," fol. [5v], AAS, dossier d'Alembert.

occurred more regularly, but still not invariably. Even when candidates did follow the rules, and the records are at their fullest, the historian can still be left in doubt about the influence of scientific merit upon the balloting. Selected examples will illustrate this problem.

In 1725, Pierre Maloet and F.-J. Hunauld were the declared candidates for a vacancy in the *class of anatomy,* and both offered memoirs to be reported on by academicians. The scheduled report on Hunauld's work either was not prepared or has been lost, while that on Maloet was equivocal: he seemed to have abilities, but had tackled subjects too complex for his powers. Maloet was the Academy's first choice for the vacancy. Within three weeks, Hunauld was again a candidate, this time for a vacancy in the *class of chemistry.* Bignon believed that Hunauld had proven his worth, and Hunauld's eulogist later asserted that he had been given a post in chemistry, rather than anatomy, because the Academy wished to elect a man of such talent even if he had to be placed in an inappropriate class.[23] In short, why was Maloet elected? We do not know, but are faced with the possibility that his talents played a less than decisive role.

At a later date, botanists Jean Descemet and Jean-Baptiste Lamarck stood for election after both had presented several memoirs and Lamarck the manuscript of his *Flore française.* All the surviving reports are eulogistic, recommending publication under the Academy's auspices. Why was Descemet first on the list of nominees for the vacancy? Again, we do not know, but there is some likelihood that his earlier appearances on such lists had given him a claim to "seniority" among the eligible candidates.[24]

The election of Lavoisier in 1768 provides another example of the difficulties inherent in discovering the criteria used by voters. In retrospect it is obvious that Lavoisier had greater talent than Gabriel Jars, but this could not then have been apparent to many members of the Academy. While his abilities were appreciated by Macquer and J.-E. Guettard, and his memoirs praised by the *rapporteurs,* Lavoisier was but an interesting novice when compared with Jars who had behind him several years as a metallurgist and government consultant. If proven merit was to be preferred, Jars probably had the stronger credentials in 1768, but Lavoisier was the Academy's first choice.[25]

Preelection campaigning within the Academy was doubtless a normal fact of life, and such evidence as we have again suggests both that criteria other than merit were important and that the motives of the voters are often undetectable. In 1752, for example, chemist Jean Hellot admitted that he favored the candidacy of C.-F. Geoffroy because he had admired Geoffroy's

23. PV, 30 June, 28 July, 11 & 29 August 1725. BN, MS fr 22234, fol. 1. Mairan, "Eloge de M. Hunauld," *Hist. Acad.,* 1742 (1745), 207. Hunauld in 1728 shifted to the class of anatomy.
24. PV, 6 & 24 February, 24 & 28 April, 5 May 1779. The seniority of nonmembers is discussed below.
25. PV, 11, 14, & 18 May, 1 June 1768.

232

father and would thus do all he could to help the son.[26] At a later date, the aid of Jean d'Alembert was recruited in the election of Nicolas Desmarest. What kind of persuasion d'Alembert employed, however, is unknown. Personally unable to judge the quality of Desmarest's geological work, he was convinced that Desmarest was enlightened and that his research would create distress in the corridors of the Sorbonne. To advance the career of a fellow philosophe always loomed large in d'Alembert's mind, but there is no reason to suppose that such arguments would have carried weight in the Academy.[27]

While voters might be influenced by matters other than scientific competence, it is worth noting that the hierarchical social structure of the Old Regime had at most a negligible effect upon voting. Or, more accurately, votes might be swayed by the knowledge that a candidate had influential patrons, but the rank or status of the candidate himself was relatively unimportant. Early in the century there was, to be sure, some discussion of whether men of rank, possessing scientific ability, could with dignity occupy a post of assistant or associate.[28] There is no evidence, however, that such concerns persisted among academicians. The special category of *honoraires*, created in 1699, was available for titled savants who cared about these distinctions. Although honorary members were usually men of rank or status (*gens en place*) and, at best, scientific amateurs, the existence of this category may well account for the apparent absence of social pressure in elections to the working ranks. At the other end of the social scale, humble birth did not distress the voters, as long as candidates were of "honorable estate." One aspirant, J.-B. LeRoy, thus felt obliged to reassure his future colleagues that he would utterly renounce his career as a clockmaker.[29]

Ministerial interference, before and after balloting, supplies another ingredient in an already murky picture. Perhaps the simplest—certainly the most easily identified—form of intervention was the ministry's occasional decision not to abide by the expressed wishes of the voters. During the Bignon-Maurepas period, in six of about fifty elections the Academy's first

26. BN, MS fr 12305, fol. 374. Argenson, too, is here said to have been swayed by "nepotism." This was the case later referred to by Malesherbes (above, n. 7).
27. Henry, "Correspondance inédite de d'Alembert," pp. 529, 570. *Voltaire's Correspondence*, ed. T. Besterman, 107 vols. (Geneva, 1953–65), LV, 141 (no. 11129). The limited influence of the philosophes in the Academy is treated by A. Birembaut, "L'Académie royale des sciences en 1780 vue par l'astronome suédois Lexell (1740–1784)," *Revue d'histoire des sciences*, 10 (1957), 149–150.
28. Delisle-Louville letters of 1718 and 1724–25, AOP, B.1.1, fols. 81–88; B.1.2, fols. 134, 138. Cf. C.-P. Duclos, "Honoraire," in Diderot and d'Alembert, *Encyclopédie* (Paris, 1751–1780), VIII, 291–292.
29. LeRoy to [Mairan?], 17 August 1751, AAS, dossier LeRoy; quoted in J. Bertrand, *L'Académie des sciences et les académiciens de 1666 à 1793* (Paris, 1869), pp. 74–75. Also, Macquer, in Hahn, *Anatomy*, p. 82, and A. Doyon and L. Liaigre, *Jacques Vaucanson, mécanicien de génie* (Paris, 1966), pp. 219–220, 440.

choice was not confirmed by the king; after 1749 this was to occur only once. Although Bignon and Maurepas wanted to maintain a high level of proven talent in the Academy, their rejections of the voters' wishes did not stem from any independent evaluation of merit. Instead, some degree of merit and a variety of pressures influenced these decisions. In 1725, for example, the younger brother of astronomer J.-N. Delisle, Delisle de la Croyère, was the Academy's second choice and was confirmed by the king. He and Godin, the Academy's first nominee, had equivalent credentials, both having served as assistants to the elder Delisle. But the Delisles were on the eve of a scientific mission to Russia, and the elder brother had convinced Maurepas that regular communication with the Academy would be best ensured if both voyagers were members.[30] Far different was the situation in 1729 when the younger Saurin, son of a member, was twice the Academy's first nominee and was twice denied confirmation. Ministerial approval was first given to Pierre Mahieu who, it was said, had no desire to enter the Academy and was astonished to find himself in its ranks. The second effort of Saurin *fils* resulted in the vacancy remaining unfilled for two years, at which time A.-C. Clairaut, the Academy's second choice in 1729, was confirmed. This odd series of events suggests that antagonism toward one or both of the Saurins was the motive behind ministerial actions.[31] The confirmation of the abbé Nollet in 1739 and of Lamarck in 1779 probably were cases of influence in high places. Nollet, unlike his competitor Mignot de Montigny, was well known to such influential academicians as Réaumur and C.-F. de Cisternay Dufay, and his teaching of experimental physics was already earning the admiration of the court. Lamarck's principal protector was Buffon whose recommendations carried great weight with the Minister of the Maison du roi.[32]

After mid-century the case of Lamarck was unique, and the evidence suggests that successive ministries were confirming both the first and second nominees when votes were close or when Bignon or Maurepas might have chosen only the second candidate. The election of chemists H.-T. Baron and

30. AOP, B.1.2, fols. 153, 158–160. Delisle actually convinced Cardinal Fleury and others to persuade Maurepas, who had the final voice in such matters. Also, J. Marchand, "Le départ en mission de l'astronome J.-N. Delisle pour la Russie (1721–1726)," *Revue d'histoire diplomatique,* October–December 1929.
31. PV, 3 & 7 September 1729. Fouchy's later account (AAS, dossier 28 July 1784) contains some inaccuracies, but stresses the offensive behavior of Saurin *père.* Maurepas had been willing to confirm Clairaut as early as 1728, if the Academy so desired (BN, MS fr 22231, fol. 82). For Mahieu, see the unpublished study by E. Bonnardet in AAS, dossier Mahieu. Saurin *fils* (Bernard-Joseph) became a lawyer, poet, and dramatist.
32. Torlais, *Nollet,* pp. 26–31, and *Réaumur,* pp. 81–83. Also, AN, O¹*384, fol. 46. For Lamarck, see Hahn, *Anatomy,* p. 81; E. T. Hamy, *Les débuts de Lamarck* (Paris, [1908]), p. 27; Buffon correspondence of 1777, in AN, O¹*488, passim; and Joseph Laissus, "La succession de LeMonnier au Jardin du roi," *Comptes rendus du 91ᵉ congrès national des sociétés savantes,* Rennes, 1966 (Paris, 1967), Section des sciences, t. I, Histoire des sciences, pp. 146–147.

234

C.-F. Geoffroy in 1752, astronomers Jean-Sylvain Bailly and E.-S. Jeaurat in 1763, and Lavoisier and Jars in 1768 saw very close results in the balloting; all six became members, three as "supernumeraries." A fourth example of this kind dates from 1778 when both C.-M. Cornette and P.-E. Fontanieu received ministeral confirmation. At that time, however, the Academy chose to question the appointment of Fontanieu, and its reasons for doing so are discussed below in the context of the events of the 1770's. Here it is sufficient to note that ministerial confirmation of two names, when votes were close, was said by one academician to be the normal expectation in 1768.[33]

Governmental pressure and persuasion before an election were probably of regular occurrence, although in this area the evidence is fragmentary. In 1723, for example, Bignon and Réaumur seem to have been responsible for the election of Dufay; indeed, before he had attended a single meeting of the Academy, efforts were being made to promote him. But why Dufay was so esteemed at that stage in his career is not known; nor is it known how the voters were persuaded in his behalf.[34] When Buffon was elected in 1733, one of his friends remarked: "I have known for a long time that a post in the Academy of Sciences was being arranged for him."[35] Unlike Dufay, Buffon had presented two well-received memoirs to the Academy and had two nonentities for his competitors. At the same time, he was engaged in research of interest to Maurepas. These various factors all doubtless played a part in his election, but the nature and importance of ministerial pressure in his behalf remain obscure.[36]

If preelection pressures too often elude our grasp, two instances, both dating from 1770, show that subtle pressures were at times replaced by virtual orders to elect certain candidates. Two vacancies occurred among the astronomers in less than one month, and the Academy nominated, to fill the first post, Charles Messier and J.-D. Cassini (Cassini IV). In the letter confirming the election of Messier, St. Florentin announced that the king "would have simultaneously confirmed Cassini [for the second post, but] ... His Majesty had not wanted to disrupt regular procedures or

33. BN, MS n.a.f. 21015, fols. 198–199. The context here is a particular promotion, not an election.

34. "Lettres de Dufay à Réaumur," *Correspondance historique et archéologique*, V (1898), 306–309. Two of the four letters published here are now in AAS, dossier Dufay; all lack the year, but clearly concern his election and possible promotion. The latter did not take place in 1723, but the post remained vacant for a year, after which Dufay obtained it. For assessments of his abilities as scientist and courtier, see Fontenelle, *Oeuvres*, II, 386–7, 392, 396–7; BN, MS n.a.f. 21015, fols. 6–9; P. Brunet, "L'oeuvre scientifique de Charles François Du Fay (1698–1739)," *Petrus Nonius*, 3 (1940), 77–95.

35. Quoted in H. Monod-Cassidy, *Un voyageur-philosophe au XVIIIᵉ siècle: L'abbé Jean-Bernard Le Blanc* (Cambridge, Mass., 1941), p. 478n.

36. Lesley Hanks, *Buffon avant l'"Histoire naturelle"* (Paris, 1966), pp. 39–42. Despite Buffon's claim, Fouchy was not hastily transferred to another class to make room for him in the Academy; cf. BN, MS fr 22229, fols. 219–220.

undermine in the slightest way the rules of the Academy, which [should] itself elect Cassini." A week later Cassini received an overwhelming vote.[37] Equally forceful were efforts in behalf of chemist Balthazar-Georges Sage, who is said to have attracted the interest of Louis XV himself. Sage's abilities and reputation were called to the attention of the Academy by one of the minister's underlings, who repeatedly urged the Academy to hasten its procedures and elect Sage before disbanding for its annual vacation. Although no unseemly or unlawful haste followed, Sage was indeed elected.[38]

The foregoing discussion of elections suggests that more evidence would merely disclose more intrigue. Until the 1770's, however, pressure applied by the ministry seems to have excited little anxiety among academicians, even when, as in the case of Sage, the candidate was thought to be mediocre.[39] Similarly, intrigue within the Academy itself was to arouse misgivings only at an equally late date.

The one electoral problem to attract attention as early as mid-century involved the judgment of merit. To seek men of merit had always been an acknowledged principle, but whether the principle was consistently applied in practice was a question called to the Academy's attention in 1759 by the Chevalier d'Arcy, by d'Alembert in 1769, by J.-C. Borda in 1770, by d'Alembert, Arcy, and Montigny in 1778, and by Borda and Condorcet in 1784. Although the proposals by these academicians treated several issues, all explicitly or implicitly raised questions about how and by whom merit was to be judged. So many talented men were cultivating the sciences, d'Alembert argued, that the Academy could afford to be more selective in its choices. The Academy, he added, had never demanded sufficient proof of talent, and it therefore stood in need of strengthened rules and new voting procedures.[40] Voting reforms were also advocated by other critics, who pointed out that the preferential ballots then in use often gave results detrimental to the most talented candidates. As Condorcet put it, the existing system was so defective that the Academy "winds up electing not the most worthy candidate, but the man whom the majority thinks not unworthy."[41]

37. PV, 14 July 1770. There were precedents for the kind of royal appointment avoided here: the Comte de Brancas in 1758, J.-F.-C. Morandi in 1759.

38. Letters of Mesnard de Chousy to Fouchy, AAS, dossier Sage. Also, P. Dorveaux, "Apothicaires membres de l'Académie royale des sciences. XI. Balthazar-Georges Sage," *Revue d'histoire de la pharmacie*, 23 (1935), 156–159.

39. A balanced assessment is in Henry Guerlac, "Sage," in *Dictionary of Scientific Biography*, ed. C. C. Gillispie (New York, 1975), XII, 63–69.

40. D'Alembert's two memoirs on this subject are in AAS, dossier d'Alembert and dossier Règlements; the latter is published in C. Henry, ed., *Oeuvres et correspondances inédites de d'Alembert* (Geneva: Slatkine Reprints, 1967), pp. 35–50. See Hahn, *Anatomy*, p. 132.

41. Condorcet's observations on a proposal by Borda, in *Hist. Acad.*, 1781 (1784), 31. See Duncan Black, *The Theory of Committees and Elections* (Cambridge, 1958), pp. 156–159, 178–180, 184, and Baker, "Débuts de Condorcet," pp. 254–255. For the ballots then in use, see R. Hahn, "Quelques nouveaux documents sur Jean-Sylvain Bailly," *Revue d'histoire des sciences*, 8 (1955), 343–344.

236

At the same time, more than one critic suggested that personal or professional prejudices and the disenfranchisement of associates had also had adverse effects in elections. Should associates vote in elections? Should the associates in the same class as the candidates be allowed to vote? Would prejudices within a class result in the names of talented candidates never being presented to the eligible voters?[42]

Although all these questions were debated during meetings of the Academy, not always could the members bring to a vote proposed reforms of the statutes. In any event, most members remained unconvinced that certain changes were necessary or desirable. The broadened franchise in particular was no more acceptable by 1785 than it had been in 1759 or 1769. A revised form of preferential voting was eventually adopted, however, and an effort was made in 1784 to reduce the possibly biased influence of each class in assessing the talent of nonmembers in the same scientific discipline.[43]

Such debates suggest not only that members were concerned with merit, but that merit, however poorly judged, had in fact entered into consideration throughout the century. And yet it is equally clear that other factors also played a decisive role in the minds of voters and ministers. Indeed, it is rare to find any discussion of merit in connection with particular candidates; instead, one is confronted by the common but empty phrase that a given candidate is "worthy" (*un homme de mérite*). For the most part, one may assume that candidates were sufficiently familiar to the scientific community to obviate the necessity for discussion, or that discussion was simply not committed to writing. Nonetheless, one wonders how the *honoraires* and other voters were persuaded of the abilities of candidates in sciences outside their own competence. In the absence of more concrete evidence, the best argument for supposing that merit was a vital criterion in elections is the fact that France's most eminent scientists were all members of the Academy. Admittedly, a good many mediocrities can also be found in the ranks, but this was no doubt inevitable while the available pool of scientific talent remained small; and the number of mediocrities was to decrease in time. By 1784, Lavoisier could rightly claim that "no illustrious savant had died without being a member of the Academy," even if some had not achieved membership as early in life as their abilities warranted.[44]

42. In addition to works cited above, see Bertrand, *L'Academie*, pp. 76–77; Hahn, *Anatomy*, pp. 130–132; and memoirs by Condorcet, Lavoisier, and Fouchy, in AAS, dossier 28 July 1784.

43. PV, 28 July 1784; Black, *Theory of Committees*, p. 180; Hahn, *Anatomy*, pp. 98–101.

44. Lavoisier MS, fol. 3v, in AAS, dossier 28 July 1784. The one exception which comes to mind is crystallographer Romé de l'Isle who was never elected. His reputation is discussed in *Journal inédit du duc de Croÿ 1781–1784*, ed. Grouchy and Cottin, 4 vols. (Paris, 1906–1907), III, 19; *Correspondance inédite de Condorcet et de Turgot 1770–1779*, ed. C. Henry (Paris, 1883), pp. 98, 117–118; Hahn, *Anatomy*, p. 181; and John G. Burke, *Origins of the Science of Crystals* (Berkeley, 1966), pp. 62–63, 80.

Vacancies in the two upper ranks of the Academy require a somewhat different treatment from those at the lowest level, since at no time in the period 1716–1785 did the Academy select an outsider to move directly into the upper ranks. On three occasions, outsiders were indeed confirmed as associates, but all had been the second choices of the voters. On the single occasion when an outsider was thought too distinguished for the lowest rank, the Academy proposed that he be named a "retired" (*vétéran*) associate. Furthermore, only once did the Academy wish to promote an assistant directly to the rank of pensioner.[45] In short, virtually all vacancies in the upper ranks were filled by members from the ranks immediately below.

While one historian has claimed that promotions were based on merit, another maintains that seniority was the normal criterion; still others have wondered why extraordinary merit did not bring rapid promotion.[46] In fact, of some 140 promotions in seven decades, more than three-quarters did occur in order of seniority; the extent to which this was the result of conscious policy remains to be determined. Most exceptions to the seniority pattern date from the Bignon-Maurepas era, when there is evidence that merit may have played some role in promotions. Seniority was also of importance in those early decades, but the adoption of a seniority system seems to date from about mid-century.

When seniority is calculated on the basis of a member's date of election, the records of the Bignon-Maurepas era reveal a considerable number of what came to be called *passedroits*: the passing over, or ignoring the "rights," of a member in order to promote someone with less seniority. A rough count puts the number of *passedroits* at twenty-five, or about 40 percent of all promotions before 1749; and the figure swells if one adds those outsiders named to the associate rank and the single assistant promoted to pensioner. After 1749 there were only six such cases, or about 8 percent. The great majority of *passedroits*, both before and after 1749, stemmed not from ministerial decisions but from the actions of the voters themselves; that is to say, the "irregularly" promoted academician was generally the Academy's first choice for the post.[47] That the number of cases diminished after mid-century may reflect a consistently higher level of talent in the Academy and an increased likelihood that members deserved promotion. At the same time, evidence to be presented below does reveal a shift of

45. The four outsiders named associates were Dortous de Mairan, Bouguer, Bertin, and Jean Darcet (PV, 17 December 1718, 1 September 1731, 9 May 1744, 31 March and 24 April 1784). *Vétéran* status was used only for members (not aspirants) who for some reason could no longer take active part in academic affairs. The unusual promotion was that of Jean Hellot (1739).
46. Above, n. 9; Bertrand, *L'Académie*, pp. 82–83; T. L. Hankins, *Jean d'Alembert: Science and the Enlightenment* (Oxford, 1970), p. 137.
47. These percentages are maxima. I have examined about half of the promotion lists in PV, especially those which other evidence suggested might be irregular; it is thus likely that the unexamined half contains few additional irregularities.

policy among the voters who grew ever more reluctant to assess the merits of their colleagues; instead they moved in the direction of automatic promotion on the basis of seniority. By 1769, d'Alembert could say that for a mediocre member to remain unpromoted was a mere "metaphysical" possibility. And Condorcet later remarked that to deny promotion to the senior associate in his class meant "a kind of censure" of that member rather than a genuine election.[48]

Circumstances surrounding the earlier *passedroits* suggest that the Academy was enforcing two policies: members must be productive, and they must reside in Paris. The two least active members, the luckless Mahieu and the abbé Terrasson, were repeatedly passed over in favor of such men as P.-L. M. de Maupertuis, C.-M. de LaCondamine, P.-C. LeMonnier, and d'Alembert. Chemist G.-F. Boulduc and physicians J.-M.-F. Lassone and J.-B. Senac had to spend much time at Versailles, while the Delisle brothers seemed to be extending indefinitely their mission to Russia; all were passed over in promotions, three of them repeatedly.[49] Although other *passedroits* are not as readily explained, there is no evidence that the membership felt any reluctance to ignore the claims of seniority. After 1749 the six cases of this kind fall into no discernible pattern, but it is worth noting that at least three of the six provoked some debate about the damage done to seniority rights.[50]

The idea that seniority constituted a claim to promotion was certainly present from a very early date, witness Fontenelle's comment in 1705 that to be an *élève* was tantamount to inheritance or survivorship (*une espèce de survivance*).[51] Fontenelle, however, was pointing not to length of service but to sheer membership, and this preference for members over outsiders can be considered a form of seniority. The same sentiment was made more explicit when, in 1725, an anonymous observer summarized Maupertuis' eligibility for promotion: "as a member, [he] deserves to be preferred to an outsider."[52] Once elected, members seemingly bore an indelible mark of merit.

Although seniority defined as length of service can be found in some early documents, emphasis on this criterion becomes apparent only after mid-

48. *Oeuvres...de d'Alembert*, p. 42. Condorcet MS, fol. [1r], in AAS, dossier 18 December 1784. Also, Jeaurat to Academy, 17 November 1779, AAS, dossier Jeaurat.

49. Cf. E. Bonnardet, "Jean-Baptiste Terrasson," *Bulletin de l'Oratoire de France*, 36 (October 1939), 328–337, and *Enseignement et diffusion des sciences en France au XVIIIᵉ siècle*, ed. R. Taton (Paris, [1964]), p. 278. For the posts held by Boulduc et al., see the entries in *Index biographique*. For the Delisles, below, n. 66.

50. The five promoted members were d'Alembert (1756), Charles Bossut and Condorcet (1770), Condorcet (1773), Bossut (1779); the sixth, Darcet (1784), was named an associate without ever having been an assistant. The debates of 1756 and 1779 will be treated below; documents of 1779 imply that similar issues arose in connection with Bossut's promotion in 1770. Bossut (MS in AAS, dossier 1 December 1779) says *passedroits* always aroused lively protest; no such evidence has been found before 1756.

51. Fontenelle, *Oeuvres*, I, 127. Also, AOP, B.1.1, fol. 85.

52. AAS, dossier P. LeMonnier.

century. One manifestation of this change dates from 1752 when chemist G.-F. Rouelle made a serious bid to move from the lowest rank directly to the highest. The Academy often did include on lists of nominees for the rank of pensioner the name of one assistant, but reactions to Rouelle's candidacy suggest that the practice was only a matter of form. Rouelle based his campaign on his own merits, and such pretensions induced Jean Hellot— himself the sole example of a promotion of this kind—to advise Macquer, also an assistant, to enter into the competition. Significantly, however, Hellot counseled Macquer not to solicit promotion, but to ask for second place on the list of nominees; such modest behavior would offset the outrageous conduct of Rouelle. Macquer did follow this advice, but neither he nor Rouelle was nominated and the senior associate was promoted.[53]

Since both Rouelle and the second associate, P.-J. Malouin, engaged in this campaign, there was clearly some remaining hope that considerations other than seniority could influence promotions; but such hope, as clearly, was fading. What may well have been the last gasp in behalf of merit dates from 1756, when d'Alembert, the second associate in his class, requested promotion. Perhaps because he foresaw difficulties, d'Alembert in fact proposed that he either be moved to a new *class of physics,* elevated to "retired" pensioner, or made a supernumerary pensioner in his own class. When the Academy and ministry granted the latter request, it was on condition that he receive no pension until his seniors were "placed." There seems to have been no protest from Montigny who had just suffered a *passedroit,* but other academicians are said to have objected to what they deemed a violation of custom. The next year the ministry was to promote Montigny on grounds of his seniority.[54]

Increasing concern about seniority is especially evident in 1758 when, not for the first time, two assistants elected to membership on the very same date contended for a single vacancy at the associate level. Earlier cases of this kind had apparently passed unnoticed, while the Lalande-LeGentil affair of 1758 provoked soul-searching among some of the voters. Neither astronomer could claim to be senior and, furthermore, the votes had been tied when both were elected in 1753. Instead of attempting to evaluate the work of his fellow astronomers, Grandjean de Fouchy resorted to counting up the number of memoirs published by each and trying to decide if all the memoirs had really dealt with astronomy rather than mathematics or

53. Letters of Hellot and others, BN, MS fr 12305, fols. 193, 360, 374, 376–7; MS fr 12306, fol. 212; MS fr 9134, fol. 80; MS fr 23226, fol. 7. An unusually early (1719) reference to seniority rights is in AN, O¹*368, fol. 103.

54. PV, 24 & 31 March 1756; 7 & 14 December 1757. Condorcet, *Oeuvres,* II, 595; III, 74–75. Information and misinformation are in Hankins, *Jean d'Alembert,* p. 73; R. Grimsley, *Jean d'Alembert (1717–1783)* (Oxford, 1963), pp. 158–9; J. Mayer, "D'Alembert et l'Académie des sciences," in *Literature and Science,* Proceedings of the Sixth Triennial Congress, Oxford, 1954 (Oxford, 1955), p. 202.

physics; and he had to conclude, after all, that both had done what was required of them as members. He then went on to suggest a series of rearrangements which would allow both to be promoted. When the Academy nominated only J.-J. Lalande, Fouchy wrote to the Cardinal de Luynes, president for that year, to say that His Eminence's absence from the meeting had brought "malheur" upon the Academy because some members thought it important that Guillaume LeGentil be promoted simultaneously.[55] Whether additional maneuvering followed is unknown. Lalande was promoted.

The subsequent history of the idea of seniority adds little of novelty, although it does much to reveal the temper of academic life and the way in which considerations of seniority hardened within the Academy. To be sure, recognition that there was a seniority system dawned slowly upon some members who insisted that it was still possible to "strive" for promotion.[56] On the whole, however, promotion had become so automatic that even a flagrant case of charlatanism and academic bad taste, which earlier would have provoked "displeasure" and penalties, only aroused discussion but no punishment of the offender who was promoted when his turn came.[57] Nonetheless, there did arise on one occasion a genuine contest for promotion. This was so unusual by 1779 that the several competitors went so far as to present their credentials on the floor of the Academy. This unprecedented display sheds a harsh light on academic politics.

The vacancy of 1779 occurred among the pensioners of the *class of geometry,* and the declared candidates were all associates in the three "mathematical science" divisions: Bézout, Bossut, Jeaurat, and LeGentil.[58] Since Bossut was the only one of the four in geometry, and shifting from class to class had been prohibited since 1768, theoretically there should have been no contest at all. For obscure reasons, however, the Academy decided that Bézout, too, was a legitimate candidate for the post. Bézout, Jeaurat, and LeGentil all claimed seniority if they could be allowed a change of class.

55. Fouchy to [Luynes], n.d., and MS "Etat present de la classe astronomique," in AAS, dossier 1758.

56. Duhamel du Monceau, in Hahn, *Anatomy,* p. 131, and Malesherbes, in *Bullettino di bibliografia,* 18 (1885), 569; both Duhamel and Malesherbes were replying to proposals for reform.

57. See Dorveaux, "Sage," pp. 216–223, for details of Sage's behavior during and after 1778. Lexell, in Birembaut, "L'Académie vue par Lexell," pp. 156, 163–164, offers some explanation for the lack of any punitive action. For an early case of penalized misconduct, see Condorcet, *Oeuvres,* II, 85–87.

58. Relevant documents are in AAS, dossiers Jeaurat, LeGentil, and 1 December 1779 (MS by Bézout which gives some indication of Bézout's point of view). The order of events can be found in PV, 24 November through 7 December 1779, but the arguments in the dossiers were not transcribed into the minutes. Cf. Lexell, in Birembaut, "L'Académie vue par Lexell," pp. 151–152.

Jeaurat went on to hint at his own merit by giving the voters a list of his publications. And Bossut explicitly mentioned merit, only to dismiss the issue. The *règlements*, he remarked, permit the Academy to select freely, to name even an outsider to the upper ranks; but this option had been provided in the expectation that it would rarely be used. If the Academy departed from custom and promoted Bézout, he would no doubt fill the post competently, but competence was a "personal consideration" which should not concern the voters; indeed. to introduce such matters was to open the door to rivalry, bias, and arbitrariness. Seniority, he added, was by far the safer criterion, and he, Bossut, was the senior associate in geometry. The Academy did its duty and nominated Bossut, proposing at the same time that Bézout be made a supernumerary pensioner in his own class of mechanics. Both proposals were confirmed.

When the weighing of merit is said to arouse the worst passions, we have come a long way from the spirit of the *règlements* and from Fontenelle's image of the objective scientific intellect. In 1785, however, Lavoisier showed himself to be well aware of the concerns of his colleagues when he made every effort, in his proposed reform of the Academy's structure, to see that no member would feel his seniority rights to have been prejudiced. That he succeeded to a remarkable degree helps to explain why the reforms were approved.[59]

After mid-century, knowledge of the Academy's seniority system spread among both nonscientists and aspirants to membership. In three instances, unsuccessful candidates for election were even to claim that they had "rights" to a seat in the Academy because they had been candidates for more years than their competitors. A fourth, Jeaurat, was elected after a campaign in which he described himself as "antedating all my competitors as much by my publications as by my solicitations for membership." According to Grandjean de Fouchy, such arguments did carry weight with some members.[60] Among nonscientists d'Alembert played a major role in publicizing the seniority system. As a supernumerary pensioner, he expected to receive automatically the pension vacated by the death of Clairaut in 1765. When St. Florentin proved to be dilatory, d'Alembert informed his friends and the public at large that seniority alone, apart from his abilities,

---

59. Lavoisier, *Oeuvres*, IV, 574–577, 581–582. Even in 1791, when revolutionary orthodoxy required that talent be rewarded, seniority and other factors continued to weigh more heavily in promotions; cf. AAS, dossier A.-L. de Jussieu, MS dated 7 December 1791 and signed by Jussieu, Jeaurat, and Portal.

60. AOP, B.1.8, fol. 174. Also, Fouchy MS, AAS, dossier 28 July 1784; PV, 11 March 1778 (Demachy); AAS, dossier Descemet, and PV, 1, 8, & 12 February 1783; R. Davy, *L'apothicaire Antoine Baumé (1728–1804)* (Cahors, 1955), p. 64, and P. Dorveaux, "Apothicaires membres de l'Académie Royale des Sciences. XII. Antoine Baumé," *Revue d'histoire de la pharmacie*, 24 (1936), 347.

242

ought to have secured him the pension. Unaware of bureaucratic complexities, d'Alembert's friends were ready to believe his claim that he was a victim of persecution.[61]

As the seniority system hardened, occasional voices of criticism were to be heard in behalf of the idea of promotion as a reward for talent. Not surprisingly, those who espoused this cause belonged to or sympathized with the small contingent of philosophes in the Academy. Bossut himself, during the debate of 1779, was careful to explain that the seniority system was safest given the Academy's structure, but that a reformed structure would allow the use of more suitable criteria for advancement. Despite earlier emphasis on his own seniority rights, d'Alembert, too, questioned the system when in 1769 he proposed reforms in the Academy's organization. Indeed, both d'Alembert and a leader of the opposition at the time, the abbé Nollet, admitted that merit was the better criterion for promotion. But when d'Alembert, perhaps to mollify his audience, declared that most members were talented and would not, in a reformed Academy, be permanent occupants of a low rank, he confirmed his opponents in their belief that the system was not in need of reform.[62]

Thus far, discussion of promotions has centered around the policies of the Academy rather than the ministry, for the evidence suggests that in this realm the Academy enjoyed considerable autonomy. Although ministerial pressure may have been exerted in this area from time to time, such interference was apparently rare. If we look at the careers of court favorites, men for whom pressure ought to have been applied, it is striking that they did not achieve more rapid promotion than their colleagues; on the contrary, those who spent much time at Versailles sometimes had their promotions delayed or denied. Furthermore, the ministry seldom adopted the expedient of confirming the Academy's second choice in promotions; it is worth noting, however, that most such examples date from the Bignon-Maurepas era.[63]

The reasons behind relative ministerial indifference remain obscure. In 1773, St. Florentin was to explain to an ambitious academician that mem-

61. See any edition of d'Alembert's correspondence of 1765 with Voltaire, Catherine II, and Frederick II. Also, Lagrange, *Oeuvres*, XIII, 38–40, 42, 46, 48. D'Alembert to *Journal encyclopédique*, 15 August, 1 October, 15 December 1765, and in L. Bachaumont, *Mémoires secrets*, 36 vols. in 18 (London, 1784–1789), II, 249, 280. Biographers usually adopt d'Alembert's view of the affair. One reason for St. Florentin's delay was the fact that d'Alembert's claims were being disputed by Arcy and Vaucanson, and the Academy had to debate the issues. PV, 18 May, 3, 7, & 14 August, 16 & 23 November 1765. Also, AN, O¹*407, fols. 274, 320–321; BN, MS n.a.f. 9544, fols. 2–3, and n.a.f. 21015, fols. 164–165, 167–168. Diderot was one of the few outsiders aware of some of the complexities; cf. Mayer, in *Literature and Science*, p. 204.

62. D'Alembert and Nollet MSS cited above, nn. 22, 40.

63. Promotions of Couplet *fils* (1717), Louville (1719), J.-L. Petit (1725), LaCondamine (1739), and Cassini III (1745). For Jacques Vaucanson, see below, nn. 64, 73. Court favorites will be treated below in the context of the residence requirement.

bership in the Academy was a sufficient distinction for any savant, and it may be that the several ministers believed rank in the Academy to be less important than the fact of membership. More positive evidence reveals that when the Academy began to emphasize the seniority system, the ministry endorsed and upheld this policy. In fact, St. Florentin twice felt obliged to remind the Academy that senior associates had claims to promotion.[64] It seems reasonable to conclude that, by adopting the Academy's policy, the ministry in effect limited its own opportunities to intervene in promotions.

Before attempting to generalize about the Academy's evolution, two more aspects of electoral practice must be discussed: the residence requirement and the appointment of supernumeraries. Both can be used to reveal the extent to which the members and the ministry thought it necessary to enforce the *règlements,* and both have some bearing on the problem of ministerial interference in the Academy's internal affairs.

As noted above, members of the working ranks were supposed to reside in Paris, to attend meetings regularly, and to give evidence of continuing research. Prolonged absence was permissible for such activities as scientific missions, provided that prior consent had been obtained and the absence was of some fixed duration. By 1716 several members had been "excluded" for failure to live up to one or another of these provisions. After 1716, however, exclusions were virtually unheard of, and normal procedure became to retire or "veteranize" absentees, a practice which allowed them to retain some of the privileges of membership. Alternatively, an absentee might be passed over in promotions; increasing reluctance to inflict *passedroits* meant that this expedient was rarely used.[65]

The policies just outlined were, in fact, applied very erratically to all types of absentees. Courtiers L.-C. Bourdelin, Malouin, and L.-G. LeMonnier became pensioners, while G.-F. Boulduc, C.-C. Angiviller, Lassone, Senac, and Joseph Lieutaud all retired before reaching that rank. Among the voyagers, LeGentil was veteranized after an absence of ten years, the Delisles after sixteen, and Joseph de Jussieu after fifteen; at the same time, candidates might be elected to the assistant level while they were away on long missions or when it was known that their departure was imminent.[66] The

64. Cf. promotions of Montigny (PV, 14 December 1757) and Vaucanson (PV, 16 & 23 November 1765, 20 May 1768). Also, letters of St. Florentin in PV, 29 July 1772, and in AN, O¹*415, fols. 939–940. St. Florentin became Duc de La Vrillière in 1770; his earlier title is retained here to avoid confusion.

65. Hellot MS, pp. 14–16. A rare exception to the residence rule at the very outset was Denis Dodart in 1699, but this was not to set a precedent and did not help Joseph Sauveur's case (PV, 28 February 1699).

66. For absence on mission, see the stipulations in the Delisle brevet, 22 June 1725, AAS, dossier J.-N. Delisle, and Marchand, "Le départ en mission de Delisle." In 1733 a post of pensioner was being "held" for Delisle should he return from Russia; when he delayed, he was veteranized (BN, MS n.a.f. 9186, fols. 10–11). Also, Fouchy MS, AAS, dossier 1758, and A. Chevalier, *La vie et l'oeuvre de René Desfontaines* (Paris, 1939), pp. 32–33, 36. Dombey was

anatomist Bertin was retired shortly after an apparent nervous collapse, while J.-B. Winslow remained on the active roster for some time after his health had failed. Other absentees, among them Buffon, Alexis Fontaine, and d'Alembert, seem never to have been threatened with retirement.[67] In the majority of cases, it is likely that the retirement of absent members was hastened or slowed depending on the availability of other attractive candidates whom the Academy wished to elect or promote. This is evident in 1758–59, following the death of Antoine de Jussieu and the retirement of his brother Joseph; after a dozen years of unchanging composition, the class of botany could be reshuffled to acquire four new young members. Similarly, the retirement of LeGentil in 1770 enabled the astronomers to acquire Cassini IV.[68]

The ministry was quite as erratic as the Academy in enforcing the rules. During the Bignon-Maurepas era, there was some concern about having too many astronomers away on missions at the same time, and the Academy was twice instructed to bring the active roster to full strength by electing supernumeraries. But no objection was raised to the election of Joseph de Jussieu in 1742, although he had already been in South America for seven years. That no one seems to have queried the absence of Buffon and d'Alembert is hardly surprising. Presumably d'Alembert was too eminent, too much in the public eye, and too likely to voice his complaints. Buffon was not only too eminent, but he was a ministerial favorite; the problems created by his performing his duties as treasurer from his estate in Burgundy were quietly resolved in 1772 by the appointment of Mathieu Tillet to serve as his adjunct. Despite such examples, ministerial laxity remains inexplicable, and Condorcet may have been a wise diagnostician in saying that as long as academicians produced *bons ouvrages,* "a milder and wiser administration [than that of Louis XIV] seems to have relaxed the strictness of the rules."[69]

The appointment of supernumeraries has been variously described as a vehicle for royal interference and as a sign of stress in the Academy's structure. Disregarding the unique circumstances of 1716, the first supernumerary was named in 1731 and was to be followed by twenty-three more, the last of them in 1784. Among the earliest such appointments, two fall into a class of their own, the Academy having been told to elect additional mem-

advised to seek election while on mission (Laissus, "La succession de LeMonnier," p. 152). Jussieu was retired on condition that he remain eligible for reinstatement on his return, and LeGentil was to be reinstated in 1772.

67. Gabriel Bory proposed in 1788 that the Academy adopt guidelines in allowing members to become *vétérans* (Chapin, "Les associés libres," p. 13), and his proposal resembles rules already in effect in the Academy of Inscriptions.

68. For the astronomers, see Fouchy to [St. Florentin?], n.d., AAS, dossier 27 June 1770. Also Fouchy, MS "Etat present de la classe anatomique," AAS, dossier 1758.

69. Condorcet, *Oeuvres,* II, 87.

bers because too many astronomers were away on missions.[70] A second group of three was appointed because of electoral problems; in one instance, some confusion had attended the counting of ballots, and in the others votes had been tied.[71] Seven received their appointments by request of the Academy,[72] and seven more were the result of ministerial decisions to confirm the Academy's first and second choices to fill the available vacancies.[73] Of the five thus far left unaccounted for, one was Jean Darcet whom the Academy thought too distinguished for election to the lowest rank; the minister thereupon decided that supernumerary associate would be suitable. Two, Tillet and Condorcet, were made supernumerary pensioners when they became permanent officers of the Academy. Finally, L.-G. LeMonnier was promoted at ministerial insistence, and Toussaint Bordenave was a royal appointee not nominated by the Academy. In short, only in the last two cases can the ministry be described as having exercised unusual interference in the normal workings of the Academy; and, as we shall see, these two appointments were to arouse opposition.

While all ministers were willing to create supernumeraries, Maurepas recognized at an early date that the practice ought to be restrained lest the careful structure of 1716 be distorted.[74] In 1778—under circumstances discussed below—Amelot was to promise that he, at least, would no longer initiate such appointments. But the anxiety expressed by Maurepas was not to recur to ministers or members until 1784, after five supernumeraries were created within twelve months. The limited number of seats in the Academy had always meant that rapidity of promotion depended on the mortality or retirement rate among senior members, and some academicians of the 1780's had grown impatient to reach pensioner rank.[75] Earlier in the century, one could hope to rise rapidly by shifting from one class to another, but this practice had been halted by royal order in 1768.[76] Thereafter, the

70. PV, 7 April 1731, 25 June 1735. Fouchy and Cassini III were elected.
71. PV, 28 March 1733, 29 July 1752, 19 & 29 January 1763. Clairaut, C.-F. Geoffroy, and Jeaurat were named.
72. In chronological order, d'Alembert (1756), Arcy (PV, 27 June 1770; AAS, dossier 27 June 1770; Bossut MS, AAS, dossier 1 December 1779; AN, O¹*412, fols. 325, 356), Bézout (1779), LeGentil (PV, 20 February and 4 March 1782), M.-J. Brisson (PV, 4 December 1782), René Desfontaines (AAS, dossier 1783, plumitif des séances, 26 February), Bailly (PV, 11 December 1784).
73. Hellot (1739), Montigny (1757), Lavoisier (1768), Deparcieux (1768), Angiviller (PV, 29 August 1772), Fontanieu (1778), Périer (PV, 19 March 1783). Angiviller and Fontanieu will be treated below. Deparcieux was the Academy's first choice in 1768, but Vaucanson, the second nominee, had seniority and had lost an earlier bid for promotion in 1765.
74. BN, MS fr 22229, fols. 219–220.
75. Most requests for promotion came from the associates; a typical explanation is that by Brisson, in PV, 4 December 1782.
76. The shifting of classes was not as often the road to promotion as was then assumed. There had been thirty-two shifts in the period 1719–1768; only eighteen meant a change in rank. Twelve of the eighteen passed over members with greater seniority, but the latter were almost invariably men bypassed more than once. For continuing debate on the possibility of

only way to rapid advancement became to request supernumerary status. Such requests might of course be denied, but denials were rare and a sufficient number was granted after 1780 to make plain to the interested parties that the practice was becoming an abuse. The Baron de Breteuil, perhaps weary of special pleading, suggested in 1784 that future requests of this kind should not be forwarded to him unless approved by two-thirds of the voters. A more complex solution was then offered by Condorcet who proposed that an associate be forbidden to ask for the rank of supernumerary pensioner unless he had been an associate for twenty years. In the end, intricacies were abandoned, and a committee of academicians asked the minister to forbid the Academy to ask for supernumeraries. This prohibition became part of the new *règlements* adopted in 1785.[77]

Surveying the history of the Academy's electoral practices, a few clear patterns emerge. In the election of new members, the rules required—and everyone agreed in principle—that candidates demonstrate their abilities. In fact, however, a variety of pressures often did influence the voters. That such pressures tended to be of a personal kind—that is, rivalries, loyalties, the use of influence—is hardly surprising when it is recalled that the few surviving accounts stress that personalities rather than scientific issues dominated meetings of the Academy. Further research would probably reveal that philosophical issues and professional differences also played their part in the outcome of elections. Although there is evidence that merit was not ignored, the relative importance of this criterion cannot be determined with any precision. At the same time, in filling the two upper ranks, the Academy did at first show some willingness to weigh merit or at least the proper performance of academic duties; after mid-century, however, members were emphasizing seniority and avoiding debate on the abilities of their colleagues. Bossut was doubtless correct in thinking that the seniority system reduced strife within the Academy, but the system also effectively stifled the spirit of the *règlements*.

The Academy's relationship to its protector also underwent some change during these decades. Broadly speaking, it is likely that ministerial interference was more frequent than we know, and it is apparent that such interference was more characteristic of the Bignon-Maurepas era than afterwards. Denying the Academy's first choice was an expedient used rarely throughout the century, but most instances antedate 1749. Preelection campaigning probably occurred regularly under every ministry, but the surviv-

shifting classes after 1768, see AN, O^{1*}412, fols. 325, 356; Bossut MS, AAS, dossier 1 December 1779; PV, 7 December 1779.

77. Breteuil in PV, 24 April 1784, and quoted in Bertrand, *L'Académie*, pp. 71–72. Condorcet's proposal included other conditions, one of them a two-thirds vote (AAS, dossier 18 December 1784), A Breteuil proposal of late 1784 is in Lavoisier, *Oeuvres*, IV, 592–3. The minister's attitude was provoked by the effort of Charles to become a supernumerary assistant; for the refusal of requests by Charles and Antoine Baumé, see AN, O^{1*}495, fols. 43–44; 489, fol. 91. The committee's solution is in Lavoisier, *Oeuvres*, IV, 570.

ing evidence suggests that here, too, the practice was more common before 1749.

To judge from the content and the tone of early documents, the Academy enjoyed a close working relationship with Maurepas and especially with Bignon; put another way, ministerial surveillance was constant and was seemingly accepted without resentment by the savants. The later decades, by contrast, might be characterized as a period of relative ministerial compliance. Not only did the ministry adopt the seniority system developed by the Academy, but the number of times that the Academy's first and second nominees were both confirmed may indicate reluctance to dispute the Academy's decisions or to exercise independent judgment. Similarly, while Amelot promised to create no more supernumeraries, such requests originating in the Academy generally received approval. Indeed, this decline in active surveillance was to suggest to the aging Grandjean de Fouchy that the Academy had long since entered into an agreement with the ministry by which the voters' expressed preferences would always be granted; Fouchy's memory was unreliable, but his statement seems an accurate reflection of the atmosphere in which the Academy functioned.[78] Compared with Bignon and Maurepas, only Malesherbes among the later ministers seems to have combined an informed judgment in scientific matters with an intimate knowledge of academic affairs. If Argenson was a cultivated patron of letters, St. Florentin, Amelot, and probably Breteuil were uninspired courtiers. It is thus understandable that the initiative should have passed to an Academy willing to regulate itself.

When ministerial interference became rarer, and often cruder, such interference was to provoke crises and protests. The first disturbance of this kind dates from 1758 and was an isolated affair, while the decade of the 1770's was marked by a series of conflicts between the Academy and the ministers of the Maison du roi. Were the ministers acting as "despotically" as critics alleged? On what grounds did the Academy defend itself? Did these conflicts alter the relationship between Academy and crown?

The crisis of 1758 began when St. Florentin urged that L.-G. LeMonnier be promoted to supernumerary pensioner. Aware of LeMonnier's absenteeism, St. Florentin privately admitted that the Academy's rules were thus "not favorable" to him, while Fouchy, after rehearsing possible precedents, declared to a colleague: "It would perhaps be unfair always to enforce the rules rigidly. But it would be a mockery of the law to go so far as to think that it does not apply to an academician who for six years has sent [the Academy] no work, especially when his competitor is as faithful and hardworking as M. Guettard."[79] Fouchy's nice distinction between bending and breaking rules was not ignored by St. Florentin, who promised the

78. AAS, dossier 28 July 1784.
79. Fouchy to [Morand?], n.d., AAS, dossier 1758. St. Florentin to LeMonnier, 2 August 1758, AN, O^1*400, fols. 454–455. Also, PV, 22 & 29 July, 5 August 1758. A draft of the

Academy that the LeMonnier promotion would set no precedent. If LeMonnier and St. Florentin got their way, it can hardly be a coincidence that the next year was marked by the retirement, on grounds of absenteeism, of two courtier-academicians.[80] Eventually, however, all parties returned to the usual erratic enforcement of residence rules, and more than a decade of calm was to follow before the Academy faced another objectionable example of ministerial intervention.

Conflicting precedents, complicated by other issues of principle and personality, marked the crises of the 1770's. The first of these is well known since it involved the career of Condorcet. Fouchy proposed to St. Florentin in 1773 that Condorcet be made his assistant, to succeed fully to the post of secretary when Fouchy should retire or die. St. Florentin approved and directed the Academy to discuss the suitability of this appointment. Votes were divided on the issue, with a vocal minority, supporters of the candidacy of astronomer J.-S. Bailly, protesting that for the Academy to approve an appointment which already had royal consent was a travesty of an election. Condorcet did obtain the post, but in 1776, upon Fouchy's retirement, asked that a new vote be taken in order to test freely the Academy's wishes. Although Bailly and his friends were present at the new election, no rival candidate was proposed and Condorcet's appointment was unanimously endorsed.[81]

Three features of this conflict are especially pertinent here. In the first place, the Academy itself was divided by personal and philosophical differences, with the majority supporting Condorcet and St. Florentin. Secondly, the minority invoked what were deemed to be the relevant precedents, and discovered that Fouchy and his predecessor, Dortous de Mairan, had each been the sole nominee presented for the crown's approval; they pointed out that Condorcet, by contrast, was the crown's nominee presented for the Academy's approval. St. Florentin, however, realized that a new precedent had been established in 1772 by the appointment of Mathieu Tillet as assistant to the treasurer. Tillet's appointment had been suggested to the minister by Buffon, and the Academy had then given its approval without murmur.[82] Academicians were thus not on the safest of grounds in citing precedent, a matter which may help to explain the third important

---

Academy's protest, probably by Morand, is in AAS, dossier 29 July 1758. LeMonnier and Guettard had equal seniority.

80. For the retirement of Lassone, St. Florentin to Fouchy, 3 February 1759, AAS, dossier Lassone; for Lieutaud, Condorcet, *Oeuvres*, II, 401, 404, and AAS, dossier Joseph Lieutaud.

81. Baker, "Débuts de Condorcet." The Condorcet letter sought by Baker (p. 254n) exists in a draft in BN, MS n.a.f. 5151, fol. 92. For Macquer's role see Ahlers, "Un chimiste," pp. 61–63.

82. PV, 12 & 19 December 1772; AAS, dossier 1772. The precedent of Tillet, invoked by St. Florentin, is not mentioned by Baker in his otherwise admirable analysis. For Tillet's reputation, see Lexell, in Birembaut, "L'Académie vue par Lexell," pp. 161–162.

feature of this affair: the fact that, apparently for the first time, members asserted their right to freedom of choice in elections. This claim, however, was so much the position of a frustrated minority that it should not be taken to represent the principles of most members in 1773. Indeed, it was Buffon, responsible for Tillet's appointment, who led the protest in behalf of liberty, while, ironically, the nucleus of philosophes in the Academy was using that kind of political pressure which, in principle, it deplored.[83]

Once launched, the claim to freedom of choice in elections had a brief but dramatic career. When surgeon Toussaint Bordenave, who had not been nominated for an existing vacancy, was appointed supernumerary assistant in 1774, the membership united in its protest.

> The Academy has seen with sorrow the appointment of Bordenave to a post of assistant in Anatomy without his having been nominated by the Company. By unanimous decision, the Academy asks [St. Florentin], in his capacity as academician, president [of the Academy], and Minister, to make known to His Majesty that this appointment is wholly unprecedented in the Academy's history and entirely contrary to the rules which the Monarch himself has given the Company and which have always been observed; and that, besides, such an appointment henceforth makes voting pointless and, in interfering with the liberty of elections, heralds the destruction of the Academy.[84]

This time the precedents were unambiguous, and Bordenave was shifted to "veteran" status two weeks later. It is to be noted, however, that another court favorite, Angiviller, had recently (1772) been made a supernumerary associate without ever having been an assistant; but Angiviller had been the Academy's second nominee for the post, and the appointment aroused no comment. Worth noting, too, is the fact that Angiviller's nomination had been engineered, in the expectation that he would be confirmed. Despite the Academy's assertion of its liberties, one may reasonably suspect that had pressures succeeded in placing Bordenave's name on the list of nominees, there might have been no crisis in 1774.[85]

The third rebellion of the 1770's once more involved the position of L.-G. LeMonnier. A supernumerary pensioner, he could normally have expected to move to regular pensioner status upon the death of Bernard de Jussieu in 1777. The absenteeism of LeMonnier, the terms of the Academy's protest, the reply of Amelot—all were virtually the same as in 1758. Again the ministry was temporarily victorious, although later in the year another court

83. That Condorcet was not himself entirely devoted to the Academy's rights and liberties is made clear by Baker, "Débuts de Condorcet," esp. pp. 253-254, 259. See also the remarks on Arcy in Condorcet, Oeuvres, II, 387, and quoted in Hahn, Anatomy, p. 131, also p. 134.

84. Fouchy MS, n.d., AAS, dossier Bordenave. Also, PV, 16, 19, & 26 March 1774, and BN, MS fr 9134, fol. 99. For Macquer's role see Ahlers, "Un chimiste," pp. 65-69. Condorcet, Oeuvres, II, 542, later misrepresented the issues in his eulogy of Bordenave.

85. LeRoy to X, n.d. [1772], AAS, dossier LeRoy. Macquer (B.N., MS fr 9134, fols. 100-101, 133-134) implies that there was so much pressure before the election that the voters feared to nominate Bordenave lest the same happen in future.

favorite and absentee was to retire from active membership and LeMonnier was to do the same in 1779. The brief flirtation with academic liberties in 1773 and 1774 had given way to the Academy's more customary citation of rules and precedents.[86]

The last flurry of the decade took place in 1778 when Cornette and Fontanieu were nominated to a vacancy in the *class of chemistry* and both were confirmed, Fontanieu as supernumerary. Although others had been named in similar fashion, the Academy decided to protest against the appointment of a supernumerary. Fontanieu was veteranized, and the minister promised to refrain from such appointments in future.[87]

Reviewing the several crises of the decade, it should now be apparent that only the Bordenave affair is a clear case of unprecedented ministerial interference. And it is the one case in which the Academy was unanimous in claiming that its liberties were in jeopardy. Uniquely true of the Bordenave affair, in addition, is the rapidity with which the minister retreated. By contrast, the debates about Fontanieu continued for many months, while the appointment of Condorcet and the promotion of LeMonnier can be described as temporary ministerial victories followed by solutions acceptable to the Academy. Indeed, a glance at the years after 1778 reveals that relations between the ministry and the Academy were little affected by the crises just discussed. The year 1779 saw the confirmation of Lamarck, the Academy's second choice, and this was accepted with the normal lack of protest. Six supernumeraries were confirmed, five of them by request of the Academy.[88] And no one seems to hae mentioned the residence rule when botanist René Desfontaines was elected, although he was known to be planning a long foreign expedition. In short, the Academy had reverted to normalcy, and the ministry, except in the case of Lamarck, gave evidence of return to compliance with the Academy's wishes.

In view of the Academy's behavior before 1773 and after 1778, one must inevitably ask why there was so concentrated a period of rebellion and what meaning is to be attached to the idea of academic liberties. The fact that these crises arose in the 1770's suggests a parallel with public sentiment about "liberty" and "despotism" in the years after the Maupeou coup of 1771. Indeed, one observer specifically applied these political terms to the

---

86. For the retirement of Angiviller, PV, 6 & 10 December 1777; AAS, dossier Angiviller; BN, MS fr 12305, fol. 15. For LeMonnier's efforts not to retire in 1777, and the sequel: PV, 15 November 1777, 19 December 1778; AAS, dossier 1777, entries for 29 November in the *plumitifs* by Condorcet and Lavoisier; AAS, dossier L.-G. LeMonnier, letter to his brother, 28 November 1777; BN, MS n.a.f. 5152, fols. 289–290.

87. PV, 14 & 18 March 1778; 20, 27, & 30 January 1779.

88. See the lists given above, nn. 72, 73. After 1778 only J.-C. Périer may not have been a case of academic request, but no evidence on this subject has been found. His inclusion in n. 73 (rather than n. 72) was an educated guess.

Bordenave affair.[89] Such language, however, seems not to have entered into academic vocabularies—a circumstance true even in the Académie française, dominated by the philosophes, during bitterly contested elections in 1772.[90] Unlike the literary institution, the Academy of Sciences included few known philosophes and many opponents of its own leading philosophe, d'Alembert. Furthermore, had there been an important heightening of political awareness among the savants after 1771, such sentiments ought to have left some trace in connection with the appointments of Tillet and Angiviller the next year. No such trace has been found. However provocative the political parallel, there is no evidence that these large issues affected thinking about the Academy's internal affairs during this decade. In fact, a disenchanted visitor to the meeting rooms in the Louvre was to remark in 1773: "It is a pity ... that personal rivalries are more important there than matters of science, and that people hardly listen to one another, so much is each concerned with his own aims, with his own business, and with attending meetings only out of fashion or self-interest."[91] Narrowly personal concerns may indeed help to explain the acceptance of the Tillet and Angiviller appointments (both men were well liked), just as they help to explain some of the antagonism toward Condorcet.

If academicians were often self-centered, they also were able on occasion to enlarge their vision in defense of their institution. Except in 1774, the specific issue in each crisis was, in fact, quite insignificant: the residence of pensioners, royal appointment of the secretary, the creation of a supernumerary. When seen against a background of past ministerial behavior and academic willingness to bend the rules and to submit to pressures, the selection of such matters as grounds for protest makes little sense. These crises do become more comprehensible if viewed in the light of growing ministerial neglect after 1749. The academicians of the 1770's were protesting less against specific ministerial actions than against the fact that the ministry was once more becoming very active and exerting pressures to which the members were no longer accustomed. What is more, the relatively delicate maneuvers of Bignon had been replaced by the bluntness of St. Florentin.

When texts and contexts are examined, it is clear that the Academy did not demand that complete independence which *la liberté des suffrages* im-

89. Bachaumont, *Mémoires secrets,* XXVII, 209–210. Details are garbled in this account. A greater variety in public sentiment after the coup than is usually presented by historians is indicated by D. Hudson, "In Defense of Reform: French Government Propaganda during the Maupeou Crisis," *French Historical Studies,* 8 (1973), 51–76.
90. An excellent account of the royal veto of the elections of Jacques Delille and J.-B.-A. Suard, and the sequel, is in Lucien Brunel, *Les philosophes et l'Académie française au dix-huitième siècle* (Paris, 1884), esp. pp. 244–260.
91. *Journal de Croÿ,* III, 54–55.

plies. The only occasion on which "liberty" was emphasized by more than a few members was in connection with Bordenave, and the word was then intended to mean both freedom from undue pressure and the necessity to abide by the rules and precedents. The latter emphasis is more characteristic of other crises, when academicians were asserting their right to be the interpreters of the *règlements* and of precedents. Within these guidelines, whose existence guaranteed a degree of academic autonomy, members were well aware that pressure would still be exerted by their colleagues, by aspirants to membership, and by the ministry. Indeed, when three academicians—two of them with reputations for radicalism—tried to find remedies for the exercise of "authority" by ministers, they could only suggest that the Academy should be careful to nominate men of talent so that no minister would be tempted to use his powers or judgment.[92] Clearly, then, the Academy's objections were not to pressure, but to excessive pressure; and the only effective recourse was to insist on the rules and precedents. Clearly, too, the binding force of the *règlements* was meant to apply more to the ministry than to the Academy itself; the best illustration of this attitude is to be found in the issue of supernumeraries, with the academicians requesting such appointments and the ministry denied the initiative in these matters.

The royal bureaucracy probably deserves some of the credit or blame for inadvertently teaching the academicians about their rights. It will be recalled that *la liberté des suffrages* was being defended by Bignon as early as 1726. At times, to be sure, such pronouncements were merely polite refusals to exert the influence the ministers knew they possessed. On the other hand, if scientific competence were to be judged, Bignon and Maurepas knew the academicians to possess more expertise than they, and they therefore realized that a certain freedom of judgment was essential in the Academy. Above all, ministers knew it to be their duty to enforce the provisions of the *règlements,* and they were careful to inform the Academy whenever particular irregularities were intended not to constitute precedents. In a sense, then, ministerial notions of academic liberties were remarkably similar to those expressed by the academicians. To exert pressure was considered normal, while undue pressure or departure from precedent was a matter requiring special explanation. These lessons may well have been reinforced in the

92. D'Alembert, Arcy, and Montigny, 1 April 1778, in Bertrand, *L'Académie,* pp. 76–77. This document is unique in also attacking the problem of electioneering within the Academy itself. The original text cannot now be located, but PV, 1 April 1778, reveals that the debate centered not on elections to the working ranks but to the categories of *associé libre* and *associé étranger.* When the Revolution was some months old, Cassini IV suggested another solution to the problem of authority: the Academy should present only one nominee, to be confirmed or rejected by the king. This was the system in use in the Académie française where rejections had occasionally aroused both academic protest and public agitation. See R. Hahn, "L'Académie royales des Sciences et la réforme de ses statuts en 1789," *Revue d'histoire des sciences,* 18 (1965), 15–28.

1750's by the reading to the Academy of Jean Hellot's history of the application of the *règlements* from 1699 through about 1753. Certainly, the first LeMonnier affair, in 1758, revealed that some savants had acquired a good working knowledge of rules and precedents.

By the 1780's, both the Academy and the ministry might be diagnosed as victims of arteriosclerosis. The Academy's repeated refusals to countenance a variety of reform proposals were usually couched in terms of predictions that to change the rules would impair the quality of the institution.[93] On a lower level, members did not hesitate to remind each other that reform might endanger seniority rights. Simultaneously, in using precedents to defend its growing autonomy, the Academy appealed to the letter of the law and the inviolability of custom. Self-interest, the weight of custom, a narrow legalism, and a sense of corporate and individual rights pervaded the membership during the decade before the Revolution. It may seem ironic that some earlier concern for merit had become a minority cause, and a radical one, even while the intellectual climate of the Old Regime was moving in the opposite direction. The Academy, however, had gone through a process not unknown in other facets of life under the Old Regime: a young institution, capable of some flexibility and some spirit of innovation, had become an established corporation. That the ministry, too, was suffering from the same complaint is evident from St. Florentin's remark to an importunate academician: "one must introduce nothing new."[94] No minister—and certainly no one of the caliber of St. Florentin—could follow the example of Bignon in 1699 and arrive with a new set of rules in hand. Academicians instead had to be wooed into accepting certain changes, and in 1785 Breteuil and Lavoisier appealed to the members' self-interest and emphasized that to reject a new *règlement* would be to spurn a token of royal generosity toward a valued corporation.

93. For a reasoned defense of certain electoral practices, see Lavoisier, AAS, dossier 28 July 1784.
94. St. Florentin to Arcy, 20 May 1770, AN, O¹*412, fol. 356.

INDEX

T - #0473 - 101024 - C0 - 224/150/20 - PB - 9781138382619 - Gloss Lamination